Guidebook to the
Extracellular Matrix and Adhesion Proteins

A computer system will be available from October 1993 to accompany

Guidebook to the cytoskeletal and motor proteins

and

Guidebook to the extracellular matrix and adhesion proteins

Due to the rapid pace of biological research, the editors and publishers of this book believe it is important that its readers are kept informed of recent developments on these proteins. For this purpose, we have established a computer database that can be accessed throughout the world-wide computer Internet system. This computer database will not include the full entries shown in this book and its companion volume; instead the authors have been asked to add, periodically, any new information on their protein that has been obtained or published since they wrote their original entry. Authors will be asked to deposit updates in this database from October 1993.

The protein update system can be accessed through the Internet by the information server program called Gopher, developed by the University of Minnesota Computer and Information Services Department. If you do not have the Gopher client program installed on your computer, contact a computer system specialist at your institution for more information. You can also receive Gopher software for a variety of computer platforms via anonymous FTP from **boombox.micro.umn.edu** in the **/pub/gopher** directory. Questions or comments about Gopher can also be sent by e-mail to **gopher@boombox.micro.umn.edu**.

Instructions for accessing the protein update system:

1. Using the gopher program, contact the Gopher server account located at **itsa.ucsf.edu** (port 70). On some systems, type **gopher.itsa.ucsf.edu**.
2. Choose **Reseacher Tools**.
3. Choose **Protein Data Base** (for the *Guidebook* series).
4. Follow the instructions.

This protein update directory will also be made available by anonymous FTP from **itsa.ucsf.edu** (use your own e-mail address as the password). The file will be located in the directory **/pub/protein**.

Guidebook to the Extracellular Matrix and Adhesion Proteins

Edited by
Thomas Kreis
University of Geneva, Geneva, Switzerland
and
Ronald Vale
University of California, San Francisco, USA

OXFORD NEW YORK TOKYO
OXFORD UNIVERSITY PRESS

Oxford University Press, Walton Streeet, Oxford OX2 6DP

Oxford New York Toronto
Delhi Bombay Calcutta Madras Karachi
Kuala Lumpur Singapore Hong Kong Tokyo
Nairobi Dar es Salaam Cape Town
Melbourne Auckland Madrid

and associated companies in
Berlin Ibadan

Oxford is a trade mark of Oxford University Press

Published in the United States
by Oxford University Press Inc., New York

First published 1993
Reprinted 1994

A catalogue record for this book is available from the British Library

Library of Congress Cataloging-in-Publication Data
Guidebook to the extracellular matrix and adhesion proteins/ edited by Thomas Kreis and Ronald Vale.
1. Extracellular matrix proteins. 2. Cell adhesion molecules.
I. Kreis, Thomas. II. Vale, Ronald.
QP552. E95G85 1993 574.8'.7—dc20 93–7053

ISBN 0 19 859933 1 (Pbk)

Printed and bound in Great Britain by
Butler & Tanner Ltd, Frome and London

Contents

PART 1 EXTRACELLULAR MATRIX PROTEINS

PART 2 CELL ADHESION AND CELL-CELL CONTACT PROTEINS

Contributors

Steven M. Albelda, The Wistar Institute, Philadelphia, Pennsylvania, USA
W. Scott Argraves, Biochemistry Laboratory, American Red Cross, Rockville, MD, USA
John Attia, Center for Research in Neuroscience, Montreal General Hospital Research Institute, 1650 Cedar Ave., Montreal, Quebec, Canada H3G 1A4
Merton Bernfield, Joint Program in Neonatology, The Children's Hospital, Harvard Medical School, Boston, MA, USA
Albertus D. Beyers, MRC Cellular Immunology Unit, University of Oxford, UK
Joy Burchell, Epithelial Cell Biology Laboratory, Imperial Cancer Research Fund, London, UK
Shukti Chakravarti, The Eye and Ear Institute of Pittsburgh and the Department of Ophthalmology, University of Pittsburgh, Pittsburgh, PA 15213, USA
Ruth Chiquet-Ehrismann, Friederich Miescher Institute, CH-4002 Basel, Switzerland
Guido David, Center for Human Genetics, University of Leuven, Campus Gasthuisberg O&N, Herestraat 49, B-3000 Leuven, Belgium
David T. Denhardt, Department of Biological Sciences, Rutgers University, New Brunswick, USA
Kurt J. Doege, Shriners Hospital, Department of Biochemistry and Molecular Biology, Oregon Health Sciences University, Portland, Oregon, USA
Kurt Drickamer, Department of Biochemistry and Molecular Biophysics, Columbia University, New York, NY 10032, USA
Robert Dunn, Center for Research in Neuroscience, Montreal General Hospital Research Institute, 1650 Cedar Ave., Montreal, Quebec, Canada H3G 1A4
Eva Engvall, La Jolla Cancer Research Foundation, La Jolla, CA 92037, USA
R. Alan B. Ezekowitz, Division of Hematology/ Oncology, Children's Hospital and Dana Farber Cancer Institute, Dept. of Pediatrics, Harvard Medical School, Boston, MA, USA
Larry W. Fisher, Bone Research Branch, NIDR, NIH, Bethesda, MD, USA
A. de Fougerolles, Center for Blood Research, Harvard Medical School, Boston, MA, USA
Werner W. Franke, Institute of Cell and Tumour Biology, German Cancer Research Center, D-6900 Heidelberg, Germany
Andrew J. W. Furley, Center for Neurobiology and Behavior, HHMI at Columbia University, New York, USA
David R. Garrod, Cancer Research Campaign, Epithelial Morphogenesis Research Group, Department of Cell and Stractural Biology, University of Manchester, M13 9PT, UK
Paul F. Goetinck, Cutaneous Biology Research Center, Massachusetts General Hospital East, Building 149, 13th Street, Charlestown, MA 02129, USA
Corey S. Goodman, Howard Hughes Medical Institute, Department of Molecular and Cell Biology, University of California, Berkeley, CA, USA
John R. Hassell, The Eye and Ear Institute of Pittsburgh and the Department of Ophthalmology, University of Pittsburgh, Pittsburgh, PA 15213, USA
Martin E. Hemler, Dana Farber Cancer Institute, Boston, MA, USA
Michael T. Hinkes, Joint Program in Neonatology, The Children's Hospital, Harvard Medical School, Boston, MA, USA
Susan Hockfield, Section of Neurobiology, Yale University School of Medicine, New Haven, CT, USA
Michael Hortsch, Department of Anatomy and Cell Biology, University of Michigan, Ann Arbor, Michigan, USA
Richard Hynes, Howard Hughes Medical Institute and Center for Cancer Research, Department of Biology, Massachusetts Institute of Technology, Cambridge, MA, USA
Thomas M. Jessell, Center for Neurobiology and Behavior, HHMI at Columbia University, New York, USA
Fernando Jiménez, Centro de Biología Molecular, CSIC-UAM, Madrid, Spain
Tom St. John, Programs in Cell Adhesion and Structural Cell Biology, ICOS Corp., 22021 - 20th Avenue SE, Bothell, WA 98021, USA
Judith P. Johnson, Institute for Immunology, Goethestraße 31, D-8000 Munich 2, Germany
Jürgen Kartenbeck, Institute of Cell and Tumour Biology, German Cancer Research Center, D-6900 Heidelberg, Germany
P. J. Koch, Institute of Cell and Tumour Biology, German Cancer Research Center, D-6900 Heidelberg, Germany
Arthur D. Lander, Department of Brain and Cognitive Sciences, Department of Biology, Massachusetts Institute of Technology, Cambridge, MA 02139, USA
Gordon W. Laurie, Department of Anatomy and Cell Biology, University of Virginia, Charlottesville, VA 22908, USA
Jack Lawler, Vascular Research Division, Department of Pathology, Brigham and Women's Hospital and Harvard Medical School, Boston, MA, USA
Dan R. Littman, Dept. of Microbiology and Immunology and the Howard Hughes Medical Institute, UCSF, San Francisco, CA, USA
U.J. McMahan, Department of Neurobiology, Stanford University School of Medicine, Stanford, CA 94305, USA
Robert P. Mecham, Departments of Cell Biology and Medicine, Washington University, St Louis, MO, USA
Arthur M. Mercurio, Laboratory of Cancer Biology, Deaconess Hospital, Harvard Medical School, Boston, MA, USA

Jussi Merenmies, Department of Medical Chemistry and the Institute of Biotechnology, University of Helsinki, Helsinki, Finland
Deane F. Mosher, Departments of Medicine and Physiological Chemistry, University of Wisconsin, 1300 University Avenue, Madison, W1 53706, USA
Diana G. Myles, Department of Physiology, University of Connecticut, Health Center, Farmington, CT 06030, USA
Peter J. Newman, Blood Research Institute, The Blood Center of Southeastern Wisconsin, Milwaukee, WI, USA
Yoshifumi Ninomiya, Department of Molecular Biology and Biochemistry, Okayama University Medical School, Okayama, Japan
Björn Öbrink, Department of Medical Cell Biology, Medical Nobel Institute, Karolinska Institute, Stockholm, Sweden
Åke Oldberg, Department of Physiological Chemistry, University of Lund, P.O. Box 94, S-22100 Lund, Sweden
Bjorn Reino Olsen, Department of Anatomy and Cellular Biology, Harvard Medical School, Boston, MA, USA
Laurelee Osborn, Biogen Inc., Cambridge, MA, USA
David L. Paul, Dept. of Anatomy and Cellular Biology, Harvard Medical School, Boston, MA, USA
Michel Piovant, Laboratoire de Génétique et Biologie Cellulaire, CNRS, Marseille, France
Paul Primakoff, Department of Physiology, University of Connecticut, Health Center, Farmington, CT 06030, USA
Charles W. Prince, Department of Nutrition Sciences, University of Alabama, Birmingham, Alabama, USA
Jody A. Rada, The Eye and Ear Institute of Pittsburgh and the Department of Ophthalmology, University of Pittsburgh, Pittsburgh, PA 15213, USA
Fritz G. Rathjen, Zentrum für Molekulare Neurobiologie, Hamburg, Germany
Heikki Rauvala, Department of Medical Chemistry and the Institute of Biotechnology, University of Helsinki, Finland
Louis F. Reichardt, Department of Physiology and Howard Hughes Medical Institute, University of California, School of Medicine, San Francisco, CA 94143-0724, USA
John Roder, Samuel Lunenfeld Research Institute, Mount Sinai Hospital, 600 University Ave., Toronto, Ontario, Canada M5G 1X5
Steven D. Rosen, Department of Anatomy and Program in Immunology, University of California, San Francisco, CA 94143-0452, USA
Steve Rosenman, Programs in Cell Adhesion and Structural Cell Biology, ICOS Corp., 22021 -20th Avenue SE, Bothell, WA 98021, USA
Zaverio M. Ruggeri, Roon Research Center for Arteriosclerosis and Thrombosis, Department of Molecular and Experimental Medicine, Committee on Vascular Biology, Research Institute of Scripps Clinic, La Jolla, California, USA
Urs Rutishauser, Case Western Reserve University, School of Medicine, Cleveland, Ohio, USA
Melitta Schachner, Department of Neurobiology, Swiss Federal Institute of Technology, Hönggeberg, 8093 Zurich, Switzerland
Barry D. Shur, Department of Biochemistry and Molecular Biology, The University of Texas, M. D. Anderson Cancer Center, Houston, TX, USA
Jaro Sodek, MRC Group in Periodontal Physiology, University of Toronto, Toronto, Ontario, Canada
T. A. Springer, Center for Blood Research, Harvard Medical School, Boston, MA, USA
Ross W. Stephens, Department of Virology, University of Helsinki, SF-00290, Helsinki, Finland
Masatoshi Takeichi, Department of Biolphysics, Kyoto University, Kyoto, Japan
Palmer Taylor, Department of Pharmacology 0636, University of California, San Diego, La Jolla, CA 92093, USA
Joyce Taylor-Papadimitriou, Epithelial Cell Biology Laboratory, Imperial Cancer Research Fund, London, UK
Matthew L. Thomas, Department of Pathology, Washington University, School of Medicine, St. Louis, MO 63110, USA
Rupert Timpl, Max-Planck-Institut für Biochemie, Martinsried, Germany
Bryan P. Toole, Department of Anatomy and Cellular Biology, Tufts University, Boston, MA 02111, USA
Mike Tropak, Samuel Lunenfeld Research Institute, Mount Sinai Hospital, 600 University Ave., Toronto, Ontario, Canada M5G 1X5
Antti Vaheri, Department of Virology, University of Helsinki, 00290 , Finland
Jerry Ware, Roon Research Center for Arteriosclerosis and Thrombosis, Department of Molecular and Experimental Medicine, Committee on Vascular Biology, Research Institute of Scripps Clinic, La Jolla, California, USA
Alan F. Williams, MRC Cellular Immunology Unit, University of Oxford, UK
Hee-Jong Woo, Laboratory of Cancer Biology, Deaconess Hospital, Harvard Medical School, Boston, MA, USA
Dieter R. Zimmermann, Institute of Pathology, University of Zürich, Schmelzbergstrasse 12, 8091 Zürich, Switzerland

Preface

The biology of the 1980s and 1990s may well come to be remembered as the era of discovery of new proteins. Recent advances in molecular biology, genetics, and protein purification have conspired to accelerate the rate at which cellular proteins and their amino acid sequences are being identified. Such efforts, combined with information from genome sequencing projects, will ultimately lead to the identification of the entire repertoire of proteins that govern the workings of the cell.

Now that the floodgate of discovery of new proteins has been opened wide, the amount of new information on cellular proteins is exceeding the capacity of assimilation of most scientists. At the same time, it has become imperative for research workers to expand their knowledge base, since interactions between previously unconnected sets of proteins are being uncovered at a rapid pace.

These considerations motivated us to compile the 'Guidebook to the Cytoskeletal and Motor Proteins' and the 'Guidebook to the Extracellular Matrix and Adhesion Proteins' which should serve both seasoned scientists and students alike. Each class of proteins is prefaced by a general introduction that describes their overall functions and some of the interesting questions that challenge workers in the field. The biological and structural attributes of about 200 individual proteins, or groups of closely related proteins, are concisely described by investigators who participated in their discovery or characterization. Information about purification methods, assays of activity, and reagents available to study the proteins (such as antibodies and cDNA clones) are provided, together with a list of key review and research articles. We have included only the well characterized 'structural' proteins in these books; primarily those which have been purified, sequenced, and characterized with specific antibodies. Regulatory proteins that modulate the cytoskeleton or cell adhesion have, by and large, not been included in this edition. We also acknowledge that our coverage of structural proteins is regrettably incomplete, and we apologize to investigators whose protein of interest is not found in these volumes. We are happy to receive suggestions for new entries or other improvements that can be incorporated into the next edition.

We are indebted to the more than 240 authors of the entries without whose collaboration and contributions this project would not have been possible. We would also like to thank Drs D. Cleveland, B. Geiger, D. Louvard, M. Osborn, T. Pollard, L. Reichardt, J.P. Thiery, J. Tooze, and K. Weber for their advice and helpful assistance. We are very grateful to L. Hymowitz and C. Kjaer for excellent secretarial services.

November, 1992

Thomas E. Kreis (Geneva)
Ronald D. Vale (San Francisco)

1

Extracellular matrix proteins

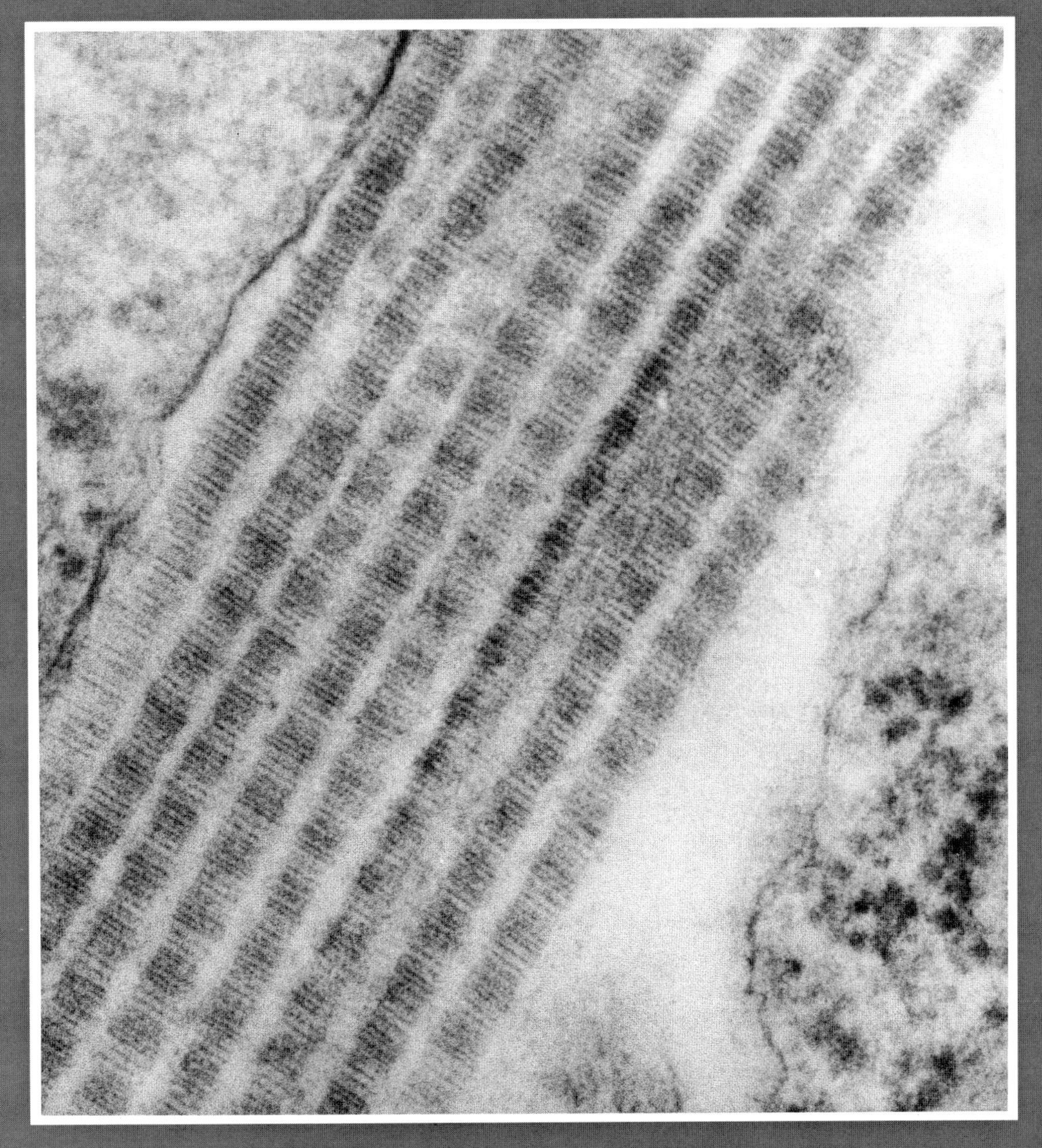

Collagen fibrils in chick embryo tendon

(for more details see 'Fibrillar Collagens' by B.R. Olsen and Y. Ninomiya).

Extracellular Matrix Molecules and their Receptors

The extracellular matrix regulates the development and function of almost all cells in metazoan organisms. Novel extracellular matrix constituents and receptors, discovered in recent years, provide a basis for understanding functions of the extracellular matrix on the molecular level.

The development and normal functioning of all cell types in an organism depend upon interactions with molecules in their environments. The major classes of molecules that regulate cellular development and function include growth and differentiation factors, cell adhesion molecules, and components of the extracellular matrix. For many years, it appeared that the least interesting of these major classes was the extracellular matrix (ECM). It was clear that the ECM contributed to the mechanical integrity, rigidity and elasticity of skin, the vasculature, tendons, lungs, and other organs. This has been confirmed by analysis of phenotypes of mice and humans with alterations in **collagen** genes[1]. Glycosaminoglycans in the ECM were also believed to be important in regulating the size and properties of the extracellular spaces in organs and embryos, regulating macromolecular transport across the extracellular matrix and creating permissive environments for cell migration. Again, the mechanisms of action of these extended, highly charged, sugar chains seemed clear. More recent evidence indicates, however, that the extracellular matrix has more interesting and often more puzzling functions. It now appears that the ECM influences many of the same major steps in development and cellular function that are regulated by growth and differentiation factors. In the nervous system, for example, these include neuronal survival and determination, cell migration and axon guidance) and synapse formation. Outside the nervous system, components of the extracellular matrix have been shown to interact with virtually all cell types in embryonic and adult organisms. They have similarly dramatic effects on development and function of these cells.

This review focuses on the extracellular matrix, its receptors, and their functions. More extended reviews on these topics include those of Reichardt and Tomaselli (l991)[2], Hynes (l990)[3], and Hemler (l990)[4]. A summary of the salient properties of **proteoglycans** is included elsewhere in this volume.

■ COMPOSITION OF THE EXTRACELLULAR MATRIX

The extracellular matrix was originally defined morphologically as extracellular material visible as fibrils or sheets in the electron microscope. It is now defined more broadly to include essentially all secreted molecules that are immobilized outside cells. In the brain, for example, a morphologically visible extracellular matrix is not detected, even though many of the molecules are present that are constituents of visible extracellular matrices in other tissues[2].

Major constituents identified initially in extracellular matrices included collagens, noncollagenous glycoproteins, and proteoglycans. There is tremendous diversity in these molecules. The number of distinct collagens is now greater than 14; there is a similar number of adhesive glycoproteins. Multiple genes and differential splicing further add to diversity in these classes of ECM constituents[3,5-8]. There are, for example, multiple sets of genes which encode isoforms of collagen IV, **tenascin**, **thrombospondin**, and the **laminin** B1 and A chain subunits. Differential splicing generates further diversity in the thrombospondin, tenascin, **fibronectin**, and many other ECM macromolecules. Many of these recently discovered molecules or isoforms have very restricted and developmentally regulated distributions in developing embryos.

Available evidence indicates that this diversity in molecules of the ECM has important functional and developmental consequences. Of particular interest is the recently described heterogeneity in laminin. It was appreciated for many years that laminins isolated from different sources had different structures and antigenic properties[9]. These differences were attributed to the effects of proteases or **laminin binding proteoglycans**. The discovery that much, perhaps all of this diversity is generated by differential gene expression has transformed a previously confusing literature. The key discoveries derived from the cloning of two antigens with interesting distributions, merosin and JS-1 (S-laminin). Sequence analysis has revealed that these are homologues of the laminin A and B-1 chains, respectively[6,7]. Further experiments showed that both of these subunits associate with other laminin subunits to form trimeric isoforms of native laminin[10]. At least four distinct laminin isoforms can be assembled utilizing one of two B1 like subunits, one of two A like subunits, and the B2 subunit. Each of these isoforms has a much more restricted distribution than total laminin. Specific functions relevant for cell differentiation appear to be encoded by each of the laminin subunits. In the case of the A-like chains, expression of the classic A chain appears necessary for normal development of kidney epithelial cells in organ cultures of embryonic kidney anlage[11]. In the case of the B1-like chains, S-laminin appears to interact via a triamino acid sequence - leucine-arginine-glutamate- with specific subsets of neurons, notably motoneurons from the ciliary ganglion and photoreceptors in the retina[7,12,13]. S-laminin is localized specifically *in vivo* to neuromuscular junctions, sites where axons cease elongation and form synapses. S-laminin fusion proteins inhibit axon extension by ciliary neurons *in vitro*, suggesting that laminin isoforms containing this subunit may direct normal synapse formation. These effects are quite different from those of classic laminin, which interacts strongly with almost all classes of neurons and promotes axon extension. Within the embryonic neuroretina, S-laminin is similarly localized specifically to the basal lamina adjacent to developing photoreceptors. In organ cultures of retinal anlage, antibodies to S-laminin specifically inhibit photoreceptor differentiation[13].

Tenascin is a second protein which now appears to be one member of a large family. Tenascin is particularly interesting because antibody injections implicate it as a regulator of cell migration during embryogenesis and because it contains both adhesive and "antiadhesive" domains[14,15]. Diversity in tenascin like proteins is generated in several ways. First, multiple isoforms that differ in the number of FN type III repeats are generated from a single gene by differential splicing[14]. Recent work indicates a domain with "anti-adhesive" activity is encoded in an alternately spliced exon[16]. In addition, there is a second gene encoding an isoform of tenascin whose expression pattern and functional properties have not yet been characterized in detail[17].

The **J1** family of proteins appear to represent an even larger family of structurally similar glycoproteins. J1 antibodies recognize at least four proteins, two of which are differentially spliced isoforms of tenascin. In addition, two other proteins, J1-160 and J1-180 are visualized in the electron microscope as dimeric and trimeric kinked arm rods, similar to, but distinct from tenascin which is a hexamer of kinked arm rods. Thus J1-160 and J1-180 have striking similarities in structure to tenascin, even though they are not products of the tenascin gene. Similar to tenascin, J1-160 and J1-180 have "anti-adhesive" activities on attachment and migration of certain cells. J1-160 and J1-180 are also synthesized by distinct cell types, such as oligodendrocytes, which do not express tenascin. Recently, a brain-derived extracellular matrix protein named **restrictin** has been purified and cloned[18]. Sequence analysis reveals that it is novel, but has extensive similarities with tenascin. It may correspond to J-160 or J-180, which have not yet been cloned.

Probably the single most illuminating studies illustrating the potential functions of the expanding families of extracellular matrix proteins are those on regeneration of vertebrate neuromuscular junctions. Incisive experiments performed a decade ago demonstrated that the extracellular matrix at the neuromuscular junction contains all the information needed to induce synaptic differentiation by both axons and skeletal myotubes *in vivo*[7,19]. The basal lamina at the neuromuscular junction, which covers only ca 0.1% of the myotube surface, is distinct molecularly from the nonsynaptic basal lamina which covers the remainder of the myotube[10]. The synaptic basal lamina contains a collagen IV trimer assembled from the α_3 (IV) and α_4 (IV) chains, while the nonsynaptic basal lamina contains a collagen IV assembled from the α_1 (IV) and α_2 (IV) subunits. Similarly, the laminin isoform in the synapse is assembled primarily from the A, S, and B2 chains, while laminin in the nonsynaptic myotube basal lamina is assembled from the M (merosin), B1 and B2 chains. Additional molecules, such as isoforms of **acetylcholinesterase** containing collagen-like tails, are also localized specifically in the synaptic basal lamina. Inhibition of function of this esterase dramatically perturbs synaptic transmission by prolonging action of the major transmitter, acetylcholine.

The ECM glycoprotein, **agrin**, is specifically localized at the neuromuscular junction and appears to account for much of the ability of the basal lamina to induce synapse formation. Agrin, one of the few extracellular matrix molecules discovered based upon its biological activity, was detected as a factor in biochemical preparations of synaptic ECM that induces clustering of acetylcholine receptors on skeletal myotubes in cell culture[19]. Clustering of this receptor is one of the most dramatic events in neuromuscular synapse formation and regeneration *in vivo*. Purified agrin has been shown to induce clustering of synaptic molecules *in vitro*, such as ECM-associated acetylcholinesterase and membrane-associated acetylcholine receptors. Since agrin is specifically localized to synaptic regions of basal lamina, it is very likely to function similarly *in vivo*. As predicted for an agent that induces differentiation in skeletal myotubes, agrin is synthesized in motoneurons and transported to their terminals in skeletal muscles[19,20]. The predicted amino acid sequence deduced from agrin cDNAs contains domains homologous to the Kazal family of protease inhibitors, epidermal growth factor repeats, and two domains in laminin, domain III, and domain G[20,21]. Comparison of the intron-exon boundaries of domain III in laminin subunits and agrin indicates they are derived from a common ancestral gene, providing a compelling argument for exon shuffling (Sheller, personal communication). Differential splicing generates several isoforms of agrin, at least some of which differ in apparent activity[21,22] (Sheller, personal communication).

It is not known which receptors mediate agrin's unusual biological activities. Agrin has been shown, however, to induce phosphorylation on tyrosine on the acetylcholine receptor β subunit[23]. Treatments that inhibit receptor aggregation prevent tyrosine phosphorylation. Results suggest that the agrin receptor regulates a tyrosine protein kinase or phosphatase that in turn regulates receptor clustering.

Agrin is synthesized primarily by motoneurons and induces differentiation in myotubes. It seems likely that one or more of the synapse-specific collagen IV or laminin subunits, each of which is synthesize by skeletal myotubes, acts similarly to induce synapse development in the terminals of motoneurons. In particular, the triamino acid sequence - leucine, arginine, glutamate - which is present in S-laminin and other ECM glycoproteins, may be a motoneuron-selective site that acts to stop axonal outgrowth at synaptic sites[12].

The above results are important because they demonstrate that the extracellular matrix contains, in at least one instance, all the essential information needed to form a complex structure, the synapse. Heterogeneity in molecular composition of the ECM is extensive and may explain the unique developmental and functional properties of the synaptic ECM. Individual ECM molecules present at the synapse have dramatic inductive effects on both neurons and myotubes that are distinct from simply promoting cell adhesion. Most likely, specific signals, such as tyrosine phosphorylation, are responsible[23]. With appropriate assays, it seems likely that other matrix proteins will prove to have equally important and specific effects on cell differentiation and function.

Until recently, proteoglycans were poorly characterized and categorized primarily according to the nature of their glycosaminoglycan side chains (Proteoglycans by Lander, this volume). Recently, the core proteins of several proteoglycans have been cloned and sequenced. The analysis of these sequences has clarified a previously confusing literature and has shown that this class of molecules is best considered as a group of diverse glycoproteins with functions mediated both by their protein cores and by their carbohydrate side chains[24,25]. The domains found in these molecules include putative, hyaluronic acid binding domains, Ca^{2+}-dependent lectin (sugar-binding) domains, leucine-rich repeats, epidermal growth factor (EGF) repeats, and immunoglobulin like domains. Often many of these domains are found in the same core protein. The very large chondroitin sulphate proteoglycan named **versican**, for example, contains a functional hyaluronic acid binding domain, two EGF repeats, a lectin-like domain and a complement-regulatory protein-like domain. The basement membrane proteoglycan **perlecan** contains domains similar to the ligand-binding domain in the low density lipoprotein receptor, laminin domain III-like regions, Ig-like domains, laminin G domain-like regions and EGF repeats[26]. The natures of these domains suggest adhesive and mitogenic functions for these proteoglycans.

Unfortunately, equivalent progress has not been made in determining the sequences and heterogeneity of glycosaminoglycan side chains associated with individual core proteins. We have known for decades that carbohydrate structure and sequence is important in regulating associations of glycosaminoglycans with proteins. In particular, interactions of heparin with thrombin have been extensively studied[27]. It has also been appreciated that tremendous diversity can potentially be encoded by carbohydrate side chains, particularly in glycosaminoglycans where these chains are modified by N- and O-sulphation and N-acetylation. Sequence determination remains difficult, however, making it virtually impossible to analyze the complete structures of individual proteoglycans. In principle, the fidelity of chain sequence and modification must limit the specificity of information encoded by glycosaminoglycan chains expressed on individual proteoglycans. Cloning and expression of the genes encoding the enzymes that catalyze glycosaminoglycan synthesis may facilitate studies on fidelity. Methods of analyzing the interactions between glycosaminoglycans and proteins are also limited, but new, more sensitive methods are being developed that should help rectify this deficiency[28].

Comparatively slow progress in proteoglycan biochemistry has been frustrating because it now seems that individual proteoglycans have quite important biological functions. Heparan-sulphate proteoglycans, such as **syndecan**, can function as receptors, binding cells to ECM glycoproteins[29,30]. Proteoglycans can also modulate the binding of other receptors. A chondroitin sulphate proteoglycan with weak affinity for fibronectin, for example, has been shown to reduce binding of fibronectin to both cells and the purified fibronectin receptor (the **integrin** $\alpha_5\beta_1$), probably by steric hindrance[31]. Finally, as will be discussed below, proteoglycans bind and modulate the activities of many growth and differentiation factors. In individual examples, these interactions are mediated by both glycosaminoglycan and protein moieties of individual proteoglycans[25].

In an additional, important class of interactions, the extracellular matrix is now known to bind to several growth factors, including basic fibroblast growth factor (FGF), transforming growth factor-beta (TGF-β), granulocyte-macrophage colony stimulating factor (GM-CSF), platelet-derived growth factor, and interleukin-3[25]. It seems likely that the matrix captures most of these growth factors, as they are released by cells, restricting their diffusion and range of action. Many of these growth factors contain extremely basic domains able to interact with glycosaminoglycans. Differential splicing of such a domain in PDGF-A regulates its association with the cell surface and ECM[32]. Interactions of TGF-β with the proteoglycan **decorin** inhibits its activity[33]. Interactions with a second, cell surface associated proteoglycan, beta-glycan, appears to modulate subsequent binding of TGF-β to signal-transducing receptors[34,35]. Interactions between basic FGF and heparan sulphate chains of proteoglycans have been shown to protect it from proteolysis and to regulate its diffusion[36]. Most dramatically, recent evidence indicates that the binding of basic FGF to its receptor requires prior binding of the ligand to heparan sulphate[37]. Thus interactions with the extracellular matrix localize and regulate the activities of many growth factors.

New classes of molecules have also been detected in the extracellular matrix in recent years. Several cell adhesion molecules have now been shown to have isoforms that are immobilized in the extracellular matrix. These include **NCAM**, **myelin-associated glycoprotein** and **L1**[38,39]. Isoforms of NCAM and L1, respectively, have been described that lack membrane anchoring domains. It seems likely that the extracellular matrix provides an alternative anchor for these adhesion molecules. Finally, several matrix constituents have been shown to bind to proteases or protease inhibitors, again regulating the diffusion and activities of these important enzymes[40,41].

DOMAINS IN ECM CONSTITUENTS

Many extracellular matrix glycoproteins are unusually large molecules with extended conformations spanning distances of several hundred nanometers. Analyses of cDNA and genomic sequences indicate that they can be considered as protein chaemeras, containing domains assembled by exon shuffling. As illustrated in Figure 1,

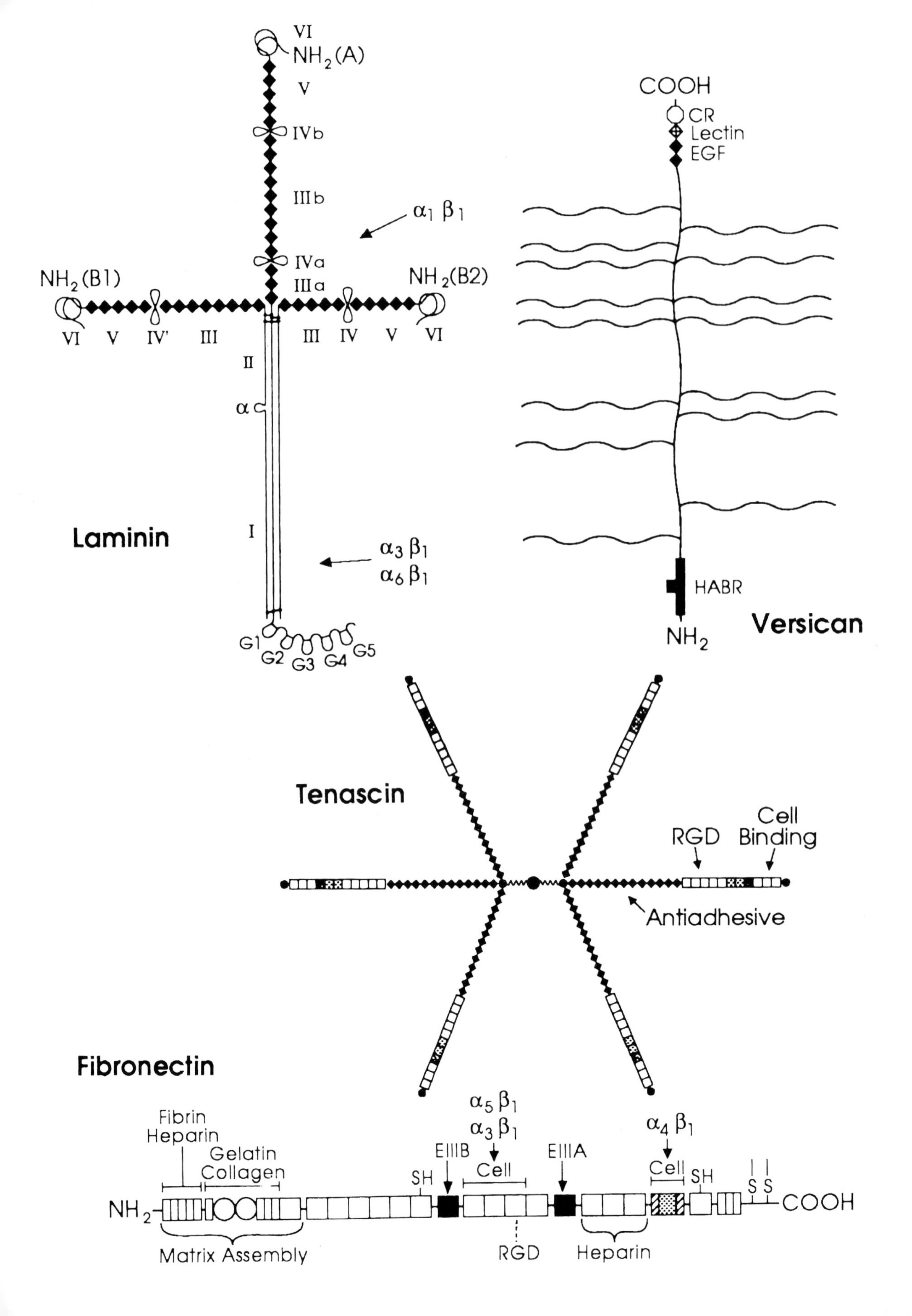
Laminin
VI
NH_2(A)
V
IVb
IIIb
$\alpha_1 \beta_1$
IVa
IIIa
NH_2(B1)
NH_2(B2)
VI V IV' III
III IV V VI
II
α
I
$\alpha_3 \beta_1$
$\alpha_6 \beta_1$
G1 G2 G3 G4 G5
Versican
COOH
CR
Lectin
EGF
HABR
NH_2
Tenascin
RGD
Cell Binding
Antiadhesive
Fibronectin
Fibrin Heparin
Gelatin Collagen
$\alpha_5 \beta_1$
$\alpha_3 \beta_1$
EIIIB
Cell
EIIIA
$\alpha_4 \beta_1$
Cell
SH
SH
S S
NH_2
COOH
Matrix Assembly
RGD
Heparin

these include distinct structural modules, such as immunoglobulin, fibronectin type III, epidermal growth factor, and numerous other domains. The names suggest specific functions, and indeed multiple functional domains can be identified in most extracellular matrix proteins. These include binding sites for cells, other extracellular matrix glycoproteins, proteoglycans, glycosaminoglycans, glycolipids, growth factors, and proteases or protease inhibitors. In many cases, it has been possible to localize specific functions, such as cell binding, to specific domains[42]. In several instances recognition domains have been shown to consist, at least in part, of peptide sequences as short as 3-10 amino acids. One of these sequences, arginine-glycine-asparate (RGD), functions as a cell attachment site in several different extracellular matrix glycoproteins, including fibronectin, thrombospondin, **fibrinogen**, **von Willebrand's Factor**, and **vitronectin**[43].

Once a specific function for a domain in one protein has been demonstrated, it has seemed possible that similar domains have similar functions wherever they are found. By this criteria, most of the domain structures in sequenced extracellular matrix proteins are potential mediators of protein-protein interactions or cell adhesion. As one example, the fibronectin type III repeat, one of which contains the sequence RGDS as a major cell attachment site in fibronectin, is found in many extracellular matrix proteins and cell adhesion molecules, sometimes in multiple copies. Currently, at least four of 18 such domains in fibronectin, one of seven in tenascin, and one of six in thrombospondin have been shown to mediate cell attachment[14,44-47]. It seems likely that more of these domains will prove to have adhesive functions when tested with appropriate cells. Adding to potential diversity, exons encoding fibronectin type III domains are differentially spliced in isoforms of RNAs encoding both fibronectin and tenascin[3,14]. One of the differentially spliced type III domains in fibronectin has been shown to contain an important cell adhesion site. In contrast, a differentially spliced type III repeat in tenascin interacts with cells, but disrupts adhesive interactions promoted by other ECM glycoproteins[16].

A second prominent domain implicated in cell adhesion, the immunoglobulin domain, exists in several copies in perlecan, a major basal lamina heparan sulphate proteoglycan[26]. Similar domains are functionally important structures in virtually all Ca^{2+}-independent adhesion molecules, such as NCAM, the **ICAMs** and other neural or immune system-associated adhesion molecules. Many of these cell adhesion molecules also contain fibronectin type III repeats. Thus, neither class of domain distinguishes cell membrane-associated from extracellular matrix-associated adhesive molecules.

Domains homologous to epidermal growth factor, named EGF repeats, are found in many extracellular matrix glycoproteins and proteoglycans, including laminin, **entactin**, thrombospondin, tenascin, aggrecan and versican. In many proteins, EGF repeats have been shown to

Figure 1. Examples of extracellular matrix glycoproteins and proteoglycans. Schematic structures for the adhesive glycoproteins tenascin, fibronectin, and laminin and for the proteoglycan versican are presented. Fibronectin is a dimer, but only one subunit is depicted. The sulphhydryl residues near its carboxyl terminus mediate its dimerization. The positions of FN type I [▯], FN type II [○], and FN type III [□] repeats are indicated in fibronectin. FN type III repeats are also found in tenascin where their positions are marked by the same symbol [□]. Shaded FN type III repeats in fibronectin and tenascin are encoded by differentially spliced exons and are thus missing from some forms of these glycoproteins. The positions of domains homologous to those in the EGF precursor, named EGF repeats [◆] are indicated in tenascin, laminin, and versican. The positions of domains homologous to functional domains in lectins [◆], complement regulatory proteins [○], and hyaluronic acid binding proteins [▄█▄] are indicated in versican. Note that many of these domains are also found in receptors, some of which are illustrated in Figure 2. In tenascin, the positions of the RGD sequence, Lorraine the major cell binding domain and the region implicated in the anti-adhesive activity of this protein are shown. Similarly, in fibronectin, the positions of the RGD sequence and domains mediating interactions with fibrin, heparin, gelatin, collagen and cells are indicated. Both the N-terminal region and the major cell attachment site containing the RGD sequence are required for assembly of fibronectin into the extracellular matrix. The regions of fibronectin recognized by the integrins $\alpha_5\beta_1$, $\alpha_3\beta_1$ and $\alpha_4\beta_1$ are indicated by arrows. Note that the integrin $\alpha_4\beta_1$ recognizes a sequence located in a differentially spliced exon that is not present in all forms of fibronectin. In laminin, homologous domains present in the A, B1 and B2 chains are indicated by roman numerals. Loops indicate the locations of putative, globular domains, many of which have been visualized in the electron microscope. Domains I and II in the long arm of laminin's cruciform structure represent regions where the three subunits associate in a triple coiled rod. The major proteolytic fragments of laminin E3, E8, E1, and E1-4 described in the text of this review contain approximately the following domains: E3 (G4 and G5), E8 (G1, G2, G3 and most of I), E1 (part of II, III, IV, IV', IIIa, IVa, IIIb, IVb), E1-4 (part of II, III, IIIa, IIIb, IV, IV', IVa, IVb,, V and VI). As indicated, the binding domains of the integrins $\alpha_1\beta_1$, $\alpha_3\beta_1$ and $\alpha_6\beta_1$ have been roughly mapped using these fragments. The site utilized by the integrin $\alpha_2\beta_1$ is close to that of $\alpha_3\beta_1$ and $\alpha_6\beta_1$. The wavy lines in versican indicate the positions of several chondroitin sulphate side chains. All of these glycoproteins contain N-linked carbohydrate, but this is not shown.

mediate protein-protein interactions, making it seem likely that many of these repeats function similarly in extracellular matrix constituents. Particular interest has been directed to the possibility that some repeats may function similarly to EGF, i.e. interact with receptors that initiate intracellular signalling cascades. Consistent with this possibility, tenascin contains an anti-adhesive site localized to its EGF repeat region that inhibits cell flattening and motility on more adhesive substrates[14]. Thrombospondin, **SPARC**, and laminin also have similar anti-adhesive activities[16,48]. The domain(s) responsible for these effect have not yet been mapped, but may include EGF repeats. These results are difficult to explain without postulating effects on intracellular second messenger systems.

In addition to effects on adhesion, many matrix proteins containing EGF-like domains directly regulate cell proliferation. Thrombospondin, tenascin and laminin promote proliferation of smooth muscle cells, fibroblasts, and Schwann cells, respectively[49]. In the case of soluble laminin, the mitogenic activity has been localized within a fragment containing multiple EGF repeats that does not contain the major cell attachment site. A variant cell line lacking EGF receptors does not respond to soluble laminin, suggesting that the EGF receptor may mediate some of these mitogenic actions. Despite experimental efforts, however, there is no biochemical evidence that laminin can interact directly with this receptor. To summarize, argument by analogy suggests that EGF repeats in matrix proteins may act similarly to EGF and other mitogenic growth factors. At least some seem able directly modulate intracellular second messengers.

RECEPTORS FOR ECM CONSTITUENTS

Many classes of molecules are believed to function as receptors for extracellular matrix constituents. These include transmembrane glycoproteins, proteoglycans, gangliosides and other glycolipids.

Virtually all ECM glycoproteins have been shown to interact with integrins, a large family of receptor heterodimers to be discussed below. Integrin receptor function is required in most cell adhesion assays, perhaps because integrins can interact with and reorganize the cytoskeleton, strengthening weak primary interactions[50]. It also seems clear, though, that additional cell surface molecules interact with the ECM. As examples, the Hermes antigen (**CD44**) is a hyaluronic acid binding receptor present on many cells that has been shown to associate with both ECM and cell surface associated ligands[51]. Recent results suggest that one ligand in the ECM is fibronectin[52]. Versican is an intrinsic membrane-associated proteoglycan that can mediate cellular interactions with fibronectin, thrombospondin, and several collagens [29,30]. The galactase-specific lectin laminin binding protein (CBP35) binds laminin with high affinity[53]. A cell surface enzyme, **galactosyltransferase**, also mediates motility responses of neural crest cells and neuronal growth cones to laminin[54]. In addition, gangliosides and other glycolipids, differentially expressed macromolecules present on the surfaces of all cells, have been shown to interact directly with ECM glycoproteins, such as laminin, fibronectin, and thrombospondin[55]. They can also interact with and modulate integrin functions directly[56,57].

Integrins are noncovalently associated heterodimeric glycoprotein complexes expressed on the surfaces of most cell types that function as receptors for essentially all of the major extracellular matrix glycoproteins (agrin and tenascin may be exceptions). Each heterodimer consists of one α chain noncovalently associated with a β subunit. Multiple genes encode families of both α and β subunits. At last count, cDNAs encoding 8 distinct β subunits and 13 distinct α subunits have been characterized in vertebrates. These numbers seem almost certain to increase. Differential splicing, particularly of exons encoding cytoplasmic domains of both α and β subunits further extends the diversity of these subunits. Integrins have been classified into subfamilies according to the β subunit. Originally, groups of α subunits were thought to associate with one particular β subunit. It has recently become clear, however, that some α subunits, most notably α_{VN} (vitronectin), can associate with more than one β subunit[4,58]. Of the 104 potential $\alpha\beta$ heterodimers that could in principle be assembled combinatorially from 8β and 13α subunits, only 19 have been actually detected[4,58,59].

As summarized in Figure 2, which depicts the 19 identified integrins and their extracellular ligands, integrins bind virtually all major constituents of the extracellular matrix. Integrin ligands in the extracellular matrix include several collagens, fibronectin, laminin, thrombospondin, vitronectin, von Willebrand's factor, and fibrinogen. Some evidence suggests that at least some cells also utilize an integrin to bind tenascin[60]. The possible roles of integrins in mediating cellular interactions with novel ECM constituents, such as agrin, have not yet been investigated definitively. A second major class of integrin ligands is cell surface associated members of the Ig superfamily. These include ICAM-1, ICAM-2, and **VCAM**, each of which has been implicated in lymphocyte homing. It seems likely that additional members of this class of integrin ligands will be discovered, particularly in tissues, such as the nervous system, where cell to cell interactions appear more prominent than cell-matrix interactions. Finally, the lymphocyte integrin $\alpha_{mac\text{-}1}\ \beta_2$ has been shown to interact with the complement factor C3bi and the coagulation factor X[61]. Each interaction leads to local generation of proteases. If similar interactions that produce protease activation can be demonstrated for other integrins, they could be important for regulation of cellular migratory and invasive behaviours.

In addition to binding extracellular ligands, integrins also transmit signals to cells. One major type of signalling is almost certainly mediated through the cytoskeleton. For example, binding by many integrins promotes formation of focal contacts in which integrins associate with the termini of F-***actin*** filaments. Based on protein-protein interactions detected *in vitro* and colocalization studies *in vivo*, ***talin***, ***α-actinin***, and ***vinculin*** are believed to link integrins with the cytoskeleton[62]. The cytoplasmic domains of integrin β subunits are strong candidates to regulate these associations. Deletion of the cytoplasmic domain of the β_1 subunit, for example, eliminates the ability of receptors containing the β_1 subunit to associate with focal contacts[63,64]. In addition, not all β subunits form receptors that associate with focal contacts. For example, $\alpha_6\beta_1$ can be localized in focal contacts, but $\alpha_6\beta_4$ is localized instead to hemidesmosomes[65]. Similarly, the $\alpha_V\beta_3$ heterodimer is found in focal contacts, while $\alpha_V\beta_5$ is not[66]. The functional consequences of differential localization can be dramatic. The receptors $\alpha_V\beta_3$ and $\alpha_V\beta_5$, for example, both mediate cell attachment to vitronectin, but only the former is able to induce focal contacts, cell flattening, and cell migration[66]. Differences in sequence between β-subunit cytoplasmic domains seem likely to account for these different localization patterns.

Cytoskeletal associations may also be regulated by intracellular signals. The cytoplasmic domains for several integrin β-subunits have been shown to be substrates for protein kinases *in vivo*. Phosphorylation appears to regulate the association of the β_1-subunit with the cytoskeleton *in vitro*[67].

Binding of antibodies or ligands to integrins also can regulate other cytoplasmic events, not all of which can be attributed to influences on cell shape or adhesion. For example, some monoclonal antibodies specific for the integrin $\alpha_V\beta_3$ can initiate activation of platelets[68]. Fragments of fibronectin and anti-integrin antibodies can induce expression of genes encoding proteases[69]. These effects are not related to visible changes in cell shape or cytoskeletal organization, but do require receptor clustering. In recent experiments, integrin receptor clustering by fibronectin has been shown to increase the cytoplasmic pH by modulating the activity of the Na+/H+ antiporter[70]. Intriguingly, the ECM protein tenascin (also named cytotactin) which prevents fibroblasts from flattening on fibronectin-coated surfaces, also inhibits pH increases induced by growth factors[71]. Activity of the Na^+/H^+ antiporter has been shown to be regulated by protein kinase C. Integrin crosslinking has been shown to increase formation of PIP_2, hydrolysis of which by phospholipase C-*gamma* releases the protein kinase C activator diacylglycerol[72]. Mechanisms by which integrin binding alters phosphoinositide turnover are not clear. At present there is no evidence that integrins associate directly with enzymes in this pathway, analogous to associations observed between receptor tyrosine kinases and phospholipase C-*gamma*. Cytoplasmic kinases, including *c-src*, however, are recruited to integrin-rich adhesion plaques[62]. In platelets, there is also evidence that the integrin $\alpha_{IIb}\beta_3$ can associate with an additional membrane protein, CD9, which also participates in platelet activation[73].

Intracellular metabolism can also alter the structure and function of extracellular domains of integrins. Physiological stimuli, modulators of second messenger systems, and ligand binding to other non-integrin receptors regulates appearance of conformation-dependent epitopes and ligand binding activity[74-76]. For the integrin, $\alpha_{IIb}\beta_3$, signalling requires the α_{IIb} cytoplasmic domain[75]. Intracellular metabolism also regulates surface concentration of some integrins. In neutrophils, large fractions of the integrins $\alpha_{mac}\beta_2$ and $\alpha_{gp150}\beta_2$ are sequestered in intracellular vesicles[77]. Physiological stimuli induce exocytosis, increasing integrin concentrations on the cell surface. Integrins have been shown to cycle rapidly between intracellular and surface compartments in fibroblasts, being preferentially inserted at leading edges of motile cells[78]. Regulation and targeting of integrin insertion may be one mechanism by which growth factors, chemotropic agents, and other molecules affect cell motility.

■ CONCLUSIONS AND PERSPECTIVES

Recent realization of the number and categories of constituents of the ECM has increased our appreciation of the diversity and importance of the ECM in regulating embryonic development and physiological functioning of the adult organism. Focusing on traditional constituents of the ECM, new collagens, adhesive glycoproteins and proteoglycans are discovered each year. These include completely new proteins such as restrictin and agrin, new members of previously identified families, such as merosin and S-laminin, and growing arrays of isoforms generated by differential splicing and other post-transcriptional mechanisms. It seems safe to predict that large numbers of similar discoveries will be made in the next few years. The new additions seem likely to have more restricted distributions. As illustrated by agrin, although, many will probably have quite important functions.

Impressive, but uneven progress has been made in the past few years in characterization of proteoglycans. Molecular biology has increased our appreciation of the structures and functions of the protein moieties of these macromolecules[24]. Our understanding of carbohydrate structure and function remains deficient. New methods seem necessary to pursue this topic more efficiently[28].

New classes of molecules, including growth factors and cell adhesion molecules, have been localized to the extracellular matrix in the past few years. It seems safe to predict that the numbers of such molecules localized there will continue to increase. A major function of the ECM is clearly to modulate the diffusion, activity, stability and other properties of these classes of molecules[25].

The classical constituents of the ECM tend to be very large molecules with extended conformations, assembled by exon shuffling. Even if one postulates that all important matrix constituents will soon be identified, one will still have to identify the functions of the individual modules in these proteins. Fibronectin, for example, has 32 spatially distinct modules. Other ECM proteins also have large numbers of domains. Each of these seems capable in principle of mediating interactions with cells or other macromolecules. Increasing numbers of these modules in proteins such as fibronectin or thrombospondin have been shown to have specific functions[45,46]. A high degree of evolutionary conservation between species suggests that most regions of characterized ECM glycoproteins have important functions[3]. In *Drosophila*, genetics has been extraordinarily useful in dissecting functions of individual domains in molecular giants, such as *notch*[79]. Analysis of mutants in invertebrates has resulted in

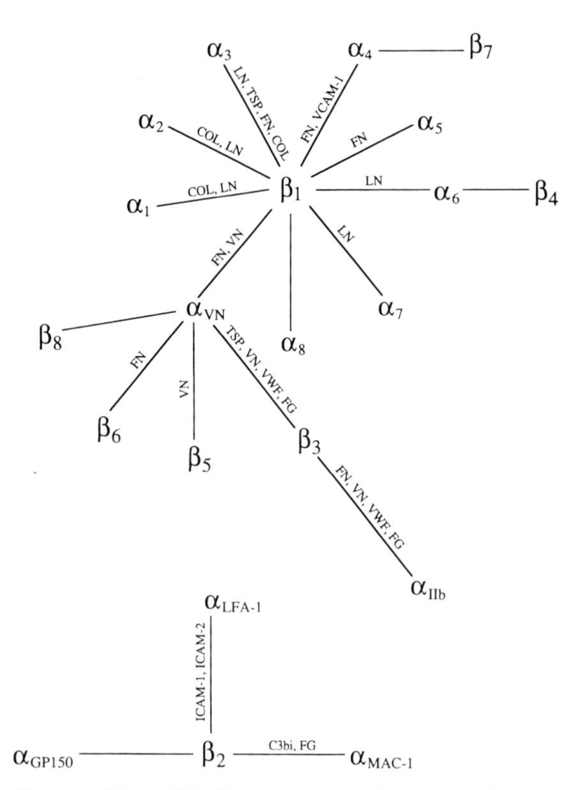

Figure 2. Ligand binding properties of integrin $\alpha\beta$ heterodimers. This figure depicts the heterodimers formed between the 8 known β and 13 known α integrin subunits. It is possible that additional associations between these subunits will be detected in the future, but only 19 of 104 potential heterodimers appear to be formed. The major ligands recognized by each heterodimer are also indicated. Ligands have not yet been identified for every integrin. Additional ligands will almost certainly be discovered.

discovery of novel matrix proteins and receptors[80]. Although a daunting prospect, a genetic approach in vertebrates seems essential to provide a comprehensive insight into the functions of ECM domains and proteins.

The comparatively recent discovery of integrins as prominent receptors for the extracellular matrix is a dramatic advance, but should not obscure our ignorance about other aspects of cell-matrix interactions. First, while the recent rate of discovery of novel integrin subunits makes it virtually certain that additional subunits remain to be discovered, it is hard to understand why there are so many, given their redundancies in ligand binding. Figure 2 reveals more integrin receptors than ligands. Possible explanations include the need to create specificity in expression or in signalling. Secondly, virtually nothing is known about how adhesion affects cytoskeletal structure and cytoplasmic signalling systems. We know equally little about mechanisms by which cytoplasmic signals regulate integrin structure and function. Finally, an abundance of evidence indicates that there must be additional classes of receptors. Integrins are not strong candidates to mediate the "anti-adhesive" effects of tenascin, thrombospondin, and other matrix molecules. Integrins also seem unlikely to mediate the motility inhibiting activity of S-laminin or the synapse-inducing activity of agrin at the neuromuscular junction[12,19]. It remains to be seen whether known macromolecules that bind the ECM, such as the Hermes antigen (CD44), proteoglycans, or gangliosides, can mediate such signals. It is quite likely, though, that novel receptors, less easily detected in cell adhesion assays, remain to be discovered.

■ REFERENCES

1. Engel, J. and Prockop, D.J. (1991) Annu. Rev. Biophys. Biophys. Chem. 20, 137-152.
2. Reichardt, L.F. and Tomaselli, K.J. (1991) Annu. Rev. Neurosci. 14, 531-570.
3. Hynes, R.O. (1990) Fibronectins. New York, Springer-Verlag.
4. Hemler, M.E. (1990) Ann. Rev. Immunol. 8, 365-400.
5. Vuorio, E. and de Crombrugghe, B. (1990) Annu. Rev. Biochem. 59, 837-872.
6. Ehrig, K., Leivo, I., Argraves, W.S., Ruoslahti, E. and Engvall, E. (1990) Proc. Natl. Acad. Sci. (USA) 87, 3264-3268.
7. Sanes, J.R., Hunter, D.D., Green, T.L. and Merlie, J.P. (1990). Cold Spring Harbor Symp. Quant. Biol. 55, 419-430.
8. Bornstein, P., Devarayalu, S., Li, P., Disteche, C.M. and Framson, P. (1991) Proc. Natl. Acad. Sci. (USA) 88, 8636-8640.
9. Edgar, D., Timpl, R. and Thoenen, H. (1988) J. Cell Biol. 106, 1299-1306.
10. Sanes, J.R., Engvall, E., Butkowski, R., Hunter, D.D. (1990) J. Cell Biol. 111, 1685-1699.
11. Klein, L.M., Lavker, R.M., Matis, W.L. and Murphy, G.F. (1988) Proc. Natl. Acad. Sci. (USA) 86, 8972-8976.
12. Hunter, D.D., Cashman, N., Morris-Valero, R., Bulock, J.W., Adams, S.P. and Sanes, J.R. (1991) J. Neurosci. 11, 3960-3971.
13. Hunter, D.D., Murphy, M.D., Olssen, C.V. and Brunken, W.J. (1992) Neuron 8, 399-413.
14. Spring, J., Beck, K., Chiquet-Ehrismann, R. (1989) Cell 59, 325-334.
15. Erickson, H.P. and Bourdon, M.A. (1989) Ann. Rev. Cell Biol. 5, 71-92.
16. Murphy-Ullrich, J.E., Lightner, V.A., Aukhil, I., Yan, Y.Z., Erickson, H.P. and Hook, M. (1991) J. Cell Biol. 115, 1127-1136.
17. Morel, Y., Bristow, J., Gitelman, S.E. and Miller, W.L. (1989) Proc. Natl. Acad. Sci. (USA) 86, 6582-6586.
18. Norenberg, U., Wille, H., Wolff, J.M., Frank, R. and Rathjen, F.G. (1992) Neuron 8, 849-863.
19. McMahan, U.J. (1990) Cold Spring Harbor Symp. Quant. Biol. 55, 407-418.
20. Rupp, F., Payan, D.G., Magill-Solc, C., Cowan, D.M. and Scheller, R.H. (1991) Neuron 6, 811-823.
21. Tsim, K.W.K., Ruegg, M.A., Escher, G., Kroger, S. and McMahan, U.J. (1992) Neuron 8, 677-689.
22. Ruegg, M.A., Tsim, K.W.K., Horton, S.E., Kroger, S., Escher, G., Gensch, E.M. and McMahan, U.J. (1992) Neuron 8, 691-699.
23. Wallace, B.G., Qu, Z., and Huganir, R.L. (1991) Neuron 6, 869-878.
24. Ruoslahti, E. (1989) Biol. Chem. 264, 13369-13372.
25. Ruoslahti, E. and Yamaguchi, Y. (1991) Cell 64, 867-869.
26. Kallunki, P. and Tryggvason, K. (1992) J. Cell Biol. 116, 559-571.
27. Bray, B., Lane, D.A., Freyssinet, J.M., Pejler, G. and Lindahl, U. (1989) Biochem. J. 262, 225-232.

28.Lee, M.K. and Lander, A.D. (1991) Proc. Natl. Acad. Sci. (USA) 88, 2768-2772.
29.LeBaron, R.G., Hook, A., Esko, J.D., Gay, S. and Hook, M. (1989) J. Biol. Chem. 264, 7950-7956.
30.Saunders, S., Jalkanen, M., O'Farrell, S. and Bernfield, M. (1989) J. Cell Biol. 108, 1547-1556.
31.Hautanen, A., Gailit, J., Mann, D.M. and Ruoslahti, E. (1989) J. Biol. Chem. 264, 1437-1442.
32.Raines, E.W. and Ross, R. (1992) J. Cell Biol. 116, 533-543.
33.Yamaguchi, Y., Mann, D.M. and Ruoslahti, E. (1990) Nature 346, 281-284.
34.Andres, J.L., Stanley, D., Cheifetz, S. and Massague, J. (1989) J. Cell Biol. 109, 3137-3145.
35.Wang, X.-F., Lin, H.Y., Ng-Eaton, E., Downward, J., Lodish, H.F. and Weinberg, R.A. (1991) Cell 67, 797-805.
36.Saksela, O. and Rifkin, D.B. (1990) J. Cell Biol. 110, 767-775.
37.Yavon, A., Klagsbrun, M., Esko, J.D., Leder, P. and Ornitz, D.M. (1991) Cell 64, 841-848.
38.Sanes, J.R., Schachner, M. and Covault, J. (1986) J. Cell Biol. 102, 420-431.
39.Fahrig, T., Landa, C., Pesheva, P., Kuhn, K. and Schachner, M. (1987) EMBO J. 6, 2875-2883.
40.Silverstein, R.L., Harpel, P.C. and Nachman, R.L. (1986) J. Biol. Chem. 261, 9959-9965.
41.Tomasini, B.R. and Mosher, D.F. (1991) Prog. Hemost. Thromb. 10, 269-305.
42.Pierschbacher, M.D. and Ruoslahti, E. (1984) Nature 303, 31-33.
43.D'Souza, S.E., Ginsberg, M.H. and Plow, E.F. (1991) Trends Biochem. Sci. 16, 246-250.
44.Guan, J.-L. and Hynes, R.O. (1990) Cell. 60, 53-61.
45.Mould, A.P. and Humphries, M.J. (1991) EMBO J. 10, 4089-4095.
46.Bowditch, R.D., Halloran, C.E., Aota, S., Obara, M., and Plow, E.F. (1991) J. Biol. Chem. 266, 23323-23328.
47.Lawler, J., Weinstein, R. and Hynes, R.O. (1988) J. Cell Biol. 107, 2341-2361.
48.Calof, A.L. and Lander, A.D. (1991) J. Cell Biol. 115, 779-794.
49.Engel, J. (1989) FEBS Lett. 251, 1-7.
50.Lotz, M.M., Burdsal, C.A., Erickson, H.P. and McClay, D.R. (1989) J. Cell Biol. 109, 1795-1805.
51.St. John, T., Meyer, J., Idzerda, R. and Gallatin, W.M. (1990) Cell 60, 45-52.
52.Jalkanen, S. and Jalkanen, M. (1992) J. Cell Biol. 116, 817-825.
53.Woo, H.-J., Shaw, L.M., Messier, J.M., and Mercurio, A.M. (1990) J. Biol. Chem. 265, 7097-7099.
54.Begovac, P.C. and Shur, B.D. (1990) J. Cell Biol. 110, 461-470.
55.Roberts, D.D., Rao, C.N., Magnani, J.L., Spoitalnik, S.L., Liotta, L.A. and Ginsburg, V. (1985) Proc. Natl. Acad. Sci. (USA) 82, 1306-1311.
56.Cheresh, D.A., Pytela, R., Pierschbacher, M.D., Klier, G.F., Ruoslahti, E. and Reisfeld, R.A. (1987) J. Cell Biol. 105, 1163-1173.
57.Santoro, S.A. (1989) Blood 73, 484-489.
58.Moyle, M., Napier, M.A., and McLean, J.W. (1991) J. Biol. Chem. 266, 19650-19658.
59.Kramer, R.H., Vu, M.P., Cheng, Y.-F., Ramos, D., Timpl, R. and Waleh, N. (1991) Cell Regulation 2, 805-817.
60.Bourdon, M.A. and Ruoslahti, E. (1989) J. Cell Biol. 108, 1149-1155.
61.Altieri, D.C., Etingen, O.R., Fair, D.S., Brunck, T.K., Geltosky, J.E., Hajjar, D.P. and Edgington, T.S. (1991) Science 254, 1200-1202.
62.Burridge, K., Fath, K., Kelly, T., Nuckolls, G. and Turner, C. (1988) Ann. Rev. Cell Biol. 4, 487-525.
63.Solowska, J., Guan, J.-L., Marcantonio, E.E., Trevithick, J.E., Buck, C.A. and Hynes, R.O. (1989) J. Cell Biol. 109, 853-861.
64.Hayashi, Y., Haimovich, B., Reszka, A., Boettiger, D. and Horwitz, A. (1990) J. Cell Biol. 110, 175-184.
65.Stepp, M.A., Spurr-Michaud, S., Tisdale, A., Elwell, J. and Gipson, I.K. (1990) Proc. Natl. Acad. Sci. (USA) 87, 8970-8974.
66.Wayner, E.A.; Orlando, R.A. and Cheresh, D.A. (1991) J. Cell Biol. 113, 919-929.
67.Tapley, P., Horwitz, A., Buck, C., Duggan, K. and Rohrschneider, L. (1989) Oncogene 4, 325-333.
68.Jennings, L., Phillips, D.R. and Walker, W.S. (1985) Blood 65, 1112-1119.
69.Werb, Z., Tremble, P.M., Behrendtsen, O., Crowley, E. and Damsky, C.H. (1989) J. Cell Biol. 109, 877-889.
70.Schwartz, M.A., Lechene, C. and Ingber, D.E. (1991) Proc. Natl. Acad. Sci. (USA) 88, 7849-7853.
71.Crossin, K.L. (1991) Proc. Natl. Acad. Sci. (USA) 88, 11403-11407.
72.McNamee, H., Ingber, D.E. and Schwartz, M.A. (1991) J. Cell Biol. 115, 245a.
73.Slupsky, J.R., Seehafer, J.G., Tang, S.-C., Masellis-Smith, A. and Shaw, A.R.E. (1989) J. Biol. Chem. 264, 12289-12293.
74.Dransfield, I., Cabanas, C., Craig, A. and Hogg, N. (1992) J. Cell Biol. 116, 219-226.
75.O'Toole, T.E., Mandelman, D., Forsyth, J., Shattil, S.J., Plow, E.F. and Ginsberg, M.H. (1991) Science. 254, 845-847.
76.Dustin, M.L. and Springer, T.A. (1989) Nature 341, 619-624.
77.Bainton, D.F., Miller, L.J., Kishimoto, T.K., Springer, R.A. (1987) J. Exp. Med. 166, 1641-1653.
78.Bretscher, M.S. (1989) EMBO J. 8, 1341-1348.
79.Campos-Ortega, J.A. and Jan, Y.N. (1991) Annu. Rev. Neurosci. 14, 399-420.
80.Hedgecock, E.M., Culotti, J.G. and Hall, D.H. (1990) Neuron 4, 61-85.

Louis F. Reichardt:
Department of Physiology and Howard Hughes Medical Institute,
University of California, School of Medicine,
San Francisco, CA 94143-0724, USA

Abbreviations: col, collagen; c3bi, complement fragment 3bi; FG, fibrinogen; FN, fibronectin; ICAM, intercellular adhesion molecule; LN, laminin; TSP, thrombospondin; VCAM, vascular cell adhesion molecule; VN, vitronectin; VWF, von Willebrand factor.

Proteoglycans

Proteoglycans are a set of ubiquitous proteins found on cell surfaces, within intracellular vesicles, and incorporated into extracellular matrices. They have been most extensively studied in vertebrates, but have also been detected in invertebrates, including sea urchins and Drosophila. Unlike other proteins that are grouped into families on the basis of amino acid similarities, the proteoglycans are defined by a common type of posttranslational modification: the glycosaminoglycan (GAG) moiety. As described below, GAGs are polysaccharides that are structurally very different from the N- and O-linked oligosaccharides found on most cell surface and secreted proteins.

There is no single structural feature common to all proteoglycans, other than the presence of GAGs. It has become clear over the last several years that proteoglycans can consist of polypeptide chains as small as 10 kDa or as large as 400,000 kDa; they can be soluble or insoluble; they can be membrane-spanning, lipid-tailed, or secreted; they can contain only a single GAG chain, or well over a hundred. What then justifies grouping the proteoglycans together as a "class" of molecules? The reasons are both practical and theoretical. The practical reason is that GAGs tend to dominate the biochemical behaviours of the molecules to which they are attached, giving proteoglycans similar fractionation properties. Among other things, GAGs are responsible for the extraordinarily poor resolution of proteoglycans by SDS-PAGE, a feature that, more than any other, has hindered the identification of proteoglycans. The theoretical, and more important, reason for treating proteoglycans as a class, is the conviction among many who work with proteoglycans that the GAGs are often the business ends of these molecules. This conviction stems both from an appreciation of the unique biophysical properties of GAGs, as well as from an increasing awareness that GAGs are likely ligands for a diverse and growing set of interesting cell surface, secreted, and extracellular matrix proteins. At present, however, there is a wide gulf between what is known about the properties of GAGs and the actual biological functions of proteoglycans. Indeed, most of the proposed functions of proteoglycans are still rather speculative, and the abundance, ubiquity and diversity of proteoglycans suggest that many of the functions of these molecules have yet to be suspected.

GLYCOSAMINOGLYCAN STRUCTURE

Given the fact that GAGs are among the largest of protein-bound polysaccharide moieties, one might suspect that GAGs exhibit an almost indecipherable degree of complexity. Yet GAG structures are surprisingly easy to understand: GAGs are linear polymers; there are no branches such as are found in N-linked oligosaccharides. Except for the short linkage region by which GAGs are attached to the serine residues of protein cores, each GAG is synthesized from only two monosaccharides, strung together in strictly alternating fashion. There are only three such disaccharide repeat units that can be polymerized onto proteins in this fashion, giving rise to three basic "parent polymers" from which all protein-bound GAGs are fashioned. Following polymerization of these simple chains (the lengths of which are variable), certain types of enzymatic modifications are carried out on the sugars themselves.

Thus, early in their biosynthetic histories, proteoglycans start out with simple, structurally homogeneous GAG chains. It is the subsequent modification of these chains, therefore, that generates most of the complexity in GAGs. This is not because the number of modifications that can be made is very large, but rather because modifications are made only sporadically throughout any GAG chain. Thus, while there are only a handful of different modifications - namely addition of O-sulphate groups at various positions, replacement of glucosamine N-acetyl groups by N-sulphate, and an isomerization step that involves the fifth carbon in D-glucuronic acid and converts that sugar into its epimer L-iduronic acid - each disaccharide in a GAG may contain any (or none) of these modifications. As a result, GAGs may be spoken of as having linear sequences - the units of which are disaccharides, each of which has been modified in a particular way. Theoretically, such sequences can encode considerable information - just as the sequences in DNA do - provided that cellular mechanisms for "reading" that information exist. Yet, before researchers can determine whether the sequence information in GAGs is biologically meaningful, they must themselves be able to read that information. Unfortunately, just as little was known about the meaning of sequences in nucleic acids before the advent of DNA sequencing and chemical synthesis, little is currently known about the meaning of GAG sequences, which can only be sequenced and synthesized by slow and cumbersome techniques. One notable success in this area is the identification of the antithrombin-binding sequence found on heparin (see below).

GLYCOSAMINOGLYCAN NOMENCLATURE

GAG nomenclature has evolved over many years, during much of which time neither GAG composition nor biosynthesis were well understood. Biochemical features of GAGs such as the disaccharide repeat unit, susceptibility to digestion by certain enzymes or chemicals, and overall charge, have traditionally formed the main criteria by which GAGs are classified. As one might expect, information about the sequences in which sugar modifications occur is not reflected in current GAG nomenclature.

The GAGs *heparin* and *heparan sulphate* are both derived from the parent polymer [D-glucuronic acid beta (1->4) D-N-acetyl glucosamine alpha (1->4)]$_n$, the disaccharides of which are then subjected to any or all of five different chemical modifications (epimerization of glucuronate to iduronate, N-deacetylation/N-sulphation, and O-sulphation at position 2 of iduronate and positions 3 and/or 6 of glucosamine). What distinguishes heparin is how heavily it is modified compared with heparan sulphate (it is an unfortunate historical accident that the less modified, and therefore the less sulphated species, was the one to acquire the term "sulphate" in its name). The purely quantitative nature of the distinction between heparin and heparan sulphate is nicely illustrated by comparing the actions of two GAG-degrading bacterial enzymes, heparinase and heparitinase on these GAGs. Heparinase cleaves only at disaccharides containing certain modifications; heparitinase makes the identical cleavage, but only at disaccharides that lack particular modifications. The result is that both enzymes degrade both heparin and heparan sulphate, but heparinase cleaves heparin into small fragments while heparitinase does not, and heparitinase cleaves heparan sulphate into small fragments while heparinase does not. It should be added that, while heparan sulphate is found in virtually all tissues and can be attached to a variety of different core proteins, heparin is now known to be derived from a single proteoglycan made by a single cell type, the mast cell. Under the circumstances, it might be more logical to rename heparin "mast cell heparan sulphate", although such a change in nomenclature seems unlikely to happen.

Chondroitin sulphate and dermatan sulphate are also GAGs that derive from the same starting polymer [D-glucuronic acid beta (1->3)D-N-acetyl galactosamine beta (1->4)]$_n$. Both may be modified by sulphation at position 2 of the uronic acid and positions 4 or 6 of the amino sugar. N-sulphation does not occur. The distinction between chondroitin sulphate and dermatan sulphate involves epimerization of glucuronic acid to iduronic acid; if this modification is found with some frequency, the GAG is called dermatan sulphate; if not, it is called chondroitin sulphate. Since it is not necessary that epimerization occur at every disaccharide position for a GAG to be called dermatan sulphate, it follows that stretches of sequence within dermatan sulphate will potentially be identical to those in chondroitin sulphate. As with heparin and heparan sulphate, bacterial enzymes can be used to examine the differences between chondroitin and dermatan sulphates. The enzyme chondroitinase AC cleaves only at glucuronate-containing disaccharides, while the enzyme chondroitinase ABC cleaves at glucuronate or iduronate; accordingly, dermatan sulphates (at least those that contain extensive conversion of glucuronate to iduronate) will be relatively resistant to chondroitinase AC, but not ABC. The rather unique "ABC" terminology used with chondroitinases reflects an older nomenclature in which chondroitin sulphate A referred to chondroitin sulphate in which 4-O-sulphation of galactosamine is prevalent (also called chondroitin 4-sulphate); chondroitin sulphate B to chondroitin sulphate in which epimerization of glucuronic acid to iduronic acid is prevalent (i.e. dermatan sulphate), and chondroitin sulphate C to chondroitin sulphate in which 6-O-sulphation of galactosamine is prevalent (also called chondroitin 6-sulphate).

Keratan sulphate differs from its parent polymer, [D-galactose beta (1->4) D-N-acetyl glucosamine beta (1->3)]$_n$, only by O-sulphation at position 6 of glucosamine and/or position 6 of galactose. Keratan sulphate is unique in that it can be synthesized not only as an O-linked sugar attached to serine (as are other GAGs), but also an an N-linked sugar. This latter structure comes about because the parent polymer for keratan sulphate is simply a repetition of the lactosamine disaccharide that is found on many N-linked oligosaccharide structures. Indeed, many cells are known to produce long polylactosamine chains (also known as lactosaminoglycans) at the ends of N-linked structures. Addition of sulphate moieties to these chains renders them structurally identical to keratan sulphate synthesized by the more "traditional" route.

Hyaluronic acid (**Hyaluronan**) differs from other GAGs in that it is (apparently) not synthesized attached to protein, and is not further modified from the structure of its parent polymer, [D-glucuronic acid beta (1->3) D-N-acetyl glucosamine beta (1->4)]$_n$.

PROTEOGLYCAN STRUCTURE AND NOMENCLATURE

Until recently, proteoglycans were named by the type of GAG chain found on them, the tissue source from which they had been isolated (e.g. "the basement membrane heparan sulphate proteoglycan"), and/or distinguishing biochemical features (e.g. "the large aggregating chondroitin sulphate proteoglycan of cartilage").With the cloning and sequencing of the cDNAs of polypeptide components of proteoglycans, these molecules are now generally referred to by the one-word names given to their protein cores (frequently these names end with the suffix "-can"). Although the new names represent a much needed improvement, the fact that they no longer convey any information about GAGs is somewhat of a drawback. Because different cell types are known to attach different GAGs to the same protein cores, one name can now potentially be used to describe structurally very different proteoglycans. It remains to be seen whether a useful nomenclature can be devised that imparts relevant information about both cores and GAGs.

As the following chapters show, several proteoglycan core proteins can now be classified into families on the basis of similar amino acid motifs. For example, the small, interstitial matrix proteoglycans, **decorin**, **biglycan**, **fibromodulin** and lumican have stretches of leucine-rich repeats similar to those found in leucine-rich glycoprotein and *Drosophila* chaoptin. These proteoglycans also share the ability to bind to and inhibit the fibrillogenesis of **collagen**. The large, aggregating chondroitin sulphate proteoglycans **aggrecan** and **versican** also form a closely-related family; these two molecules contain similar

Proteases and Anti-Proteases	Growth Factors	Extracellular Matrix	Cell Adhesion Molecules	Lipoprotein Uptake and Metabolism
Antithrombin III	Acidic FGF	Laminin	NCAM	Apo B
Protease nexin-1	Basic FGF	Fibronectin	Myelin-Associated Glycoprotein	Apo E
Heparin cofactor II	hst/K-fgf	Merosin	Amyloid β-Protein Precursor	Lipoprotein Lipase
Thrombin	FGF-5	Vitronectin		Hepatic Triglyceride Lipase
Tissue plasminogen activator	PDGF	Thrombospondin		Pancreatic Triglyceride Lipase
Urokinase plasminogen activator	GM-CSF	Tenascin		Cholesterol Esterase
Factor IXa	IL-3	von Willebrand Factor		
Factor XIa	TGF-β	Collagens (including types I through VI)		
Chymase		Retinal purpurin		
		p30		
		Acetylcholinesterase		

Table 1. Some glycosaminoglcyan-binding proteins. Proteins listed are known to bind, under physiologic conditions of salt and pH, to glycosaminoglycans, or to proteoglycans in a glycosaminoglycan-dependent manner. Other glycosaminoglycan-binding proteins include platelet factor 4, histidine rich glycoprotein, coat proteins of certain viruses (e.g. HIV, Herpes Simplex), and L-type voltage-gated calcium channels. For review, see Jackson et al., 1991.

hyaluronate-binding domains as well as domains related to those found in the LEC-CAM or **Selectin** family of lymphocyte homing receptors. The cell surface proteoglycan **CD44** (also known as H-CAM and the Hermes antigen) also contains a hyaluronic acid binding domain structurally related to those found in aggrecan and versican.

There is some suggestion that families of related cell surface proteoglycans may also exist. **Syndecan** and **fibroglycan**, for example, show extensive similarity in their transmembrane and cytoplasmic domains. **Glypican**, a fibroblast proteoglycan linked to the membrane via covalent attachment to phosphatidylinositol, may be related to lipid-linked proteoglycans that have been detected in other tissues, such as brain[1]. The recently cloned NG2 proteoglycan, a transmembrane proteoglycan found on melanoma and certain glial cells, shows no similarities to syndecan, glypican, nor any other proteoglycan identified so far[2], suggesting that the number of unrelated types of cell surface proteoglycans may be large.

FUNCTIONS OF PROTEOGLYCANS

As mentioned earlier, the functions of proteoglycans are not well understood. It is convenient to divide known functions into those mediated by the core proteins with little or no participation by GAG chains, and those mediated by the GAG chains themselves. Good examples of core protein-mediated functions include the ability of the small interstitial proteoglycans (decorin, biglycan, fibromodulin) to modulate collagen fibrillogenesis; the ability of the TGF-beta binding proteoglycan to bind TGF-beta[3]; and the ability of aggrecan, versican and CD44/H-CAM to bind hyaluronic acid.

GAG dependent functions can be loosely subdivided into two classes: the biophysical and the biochemical. The former term refers to functions that depend on the unique biophysical properties of GAGs - the ability to fill space, bind and organize water molecules, and repel negatively charged molecules. The large quantities of chondroitin sulphate and keratan sulphate found on aggrecan, for example, are thought to play an important role in the hydration of cartilage. In contrast, the heparan sulphate on the kidney glomerular basement membrane proteoglycan appears to play a role in filtration, impeding the passage of anionic serum proteins into the urine.

The other, more biochemical functions of GAGs, are those that are mediated by specific binding of GAGs to other macromolecules, notably proteins. In recent years an enormous amount of information has been gathered on the binding of GAGs to proteins. This information has suggested numerous ways in which proteoglycans might participate in cell and tissue development and physiology. Some of the known GAG-protein interactions are outlined in Table 1. For some of these proteins, all that is known is that they bind to GAG affinity columns (e.g. heparin-agarose) under physiological conditions of salt and pH. For others, affinity constants have been measured (K_ds tend to be in the range of 10^{-6} to 10^{-9} M) and for still others, direct evidence for interaction with proteoglycans has been obtained (e.g. copurification, proteoglycan dependent binding to cells, etc.). For the most part, GAG binding proteins fall into five classes:

1. Secreted proteases and anti-proteases

The best understood GAG-mediated function has been worked out with this group of molecules. The protease inhibitor antithrombin III (ATIII) binds tightly to heparin and certain heparan sulphates; several of the proteases that are inactivated by ATIII (e.g. thrombin, factor IXa, factor XIa) also bind these GAGs. In the absence of GAGs, the kinetics with which ATIII inactivates these proteases are very slow; in the presence of appropriate GAGs, these reactions are accelerated by as much as 2000-fold. The explanation of this phenomenon is twofold: First most GAG chains are sufficiently long that both protease and protease inhibitor can bind to the same chain, as a result of being confined to the same limited space, the likelihood of the two proteins then binding to each other is increased enormously. Second, heparin appears also to have an effect on protein conformation that contributes to improving ATIII's binding kinetics. It is clear that these effects of GAGs on blood coagulation enzymes account for the potent anticoagulant effects of heparin *in vivo*. More recent evidence suggests that endogenous heparan sulphate of vascular endothelial cells may play a critical role in producing a nonthrombogenic surface along blood vessel walls[4]. It is also suspected that two additional heparin binding plasma proteins (platelet factor 4 and histidine-rich glycoprotein) play a physiological role in regulating the accessibility of heparin and heparan sulphate to blood-borned proteases and protease inhibitors.

2. Polypeptide growth factors

Members of the fibroblast growth factor family, as well as several other growth factors, bind avidly to heparin or heparan sulphate. There is evidence that binding to endogenous GAGs entraps these molecules in extracellular matrices from which they may later be released. GAGs can alter the conformation, proteolytic susceptibility, and biologic potency of some of these proteins. Recent data suggest that cell surface heparan sulphate must be present on a cell in order for bFGF to exert its effects on that cell[5]. This is an intriguing observation in view of strong evidence that proteoglycans are not themselves the signal-transducing receptors for bFGF. The data raise the possibility that, as with ATIII, GAGs perform an ancillary, or catalytic role in the functions of growth factors.

3. Extracellular matrix proteins

Most of the large, multi-domain extracellular matrix proteins that have been characterized contain at least one GAG binding site, and in some cases contain several distinct ones. Most of these proteins bind GAGs of the heparin/heparan sulphate class, but one of them, **tenascin**, binds chondroitin sulphate and, under some conditions, copurifies from tissues with a chondroitin sulphate proteoglycan. Interactions between these proteins and secreted proteoglycans are likely to play an important role in matrix assembly. In addition, there is evidence that cell surface proteoglycans, such as syndecan, can mediate cell adhesion to these proteins.

4. Cell-cell adhesion molecules

Most of what is known about the interactions of cell-cell adhesion molecules with GAGs comes from the study of the homophilic adhesion molecule NCAM. There is evidence that **NCAM** interacts with cell surface heparan sulphate proteoglycans, and that this interaction is required for its function. NCAM has recently been shown to have a distinct heparin binding domain, the genetic delection of which results in a molecule incapable of mediating cell adhesion[6]. Whether other cell adhesion molecules are functionally dependent on GAGs remains to be tested (most such molecules have not yet been tested for binding to GAGs).

5. Proteins involved in lipoprotein uptake

Several of the molecules involved in the uptake and catabolism of blood-borne lipoproteins have been shown to bind heparin. It has been suggested that the heparan sulphate on the hepatocyte cell surface serves the dual function of facilitating the binding of lipoproteins (via interactions with apolipoproteins B and E) and immobilizing soluble enzymes, such as lipoprotein lipase and hepatic triglyceride lipase, on the cell surface.

Currently, there is much interest in assessing the specificity of GAG protein interactions. Are the individual sequences of sugar modification that occur in GAGs of any importance in determining protein-GAG affinity, or do proteins merely recognize overall, sequence-independent features of GAGs (e.g. local charge density)? For only one protein, antithrombin III, is the answer known, and in this case at least, the carbohydrate sequence is critical. In particular, for ATIII to exhibit significant affinity for a GAG of the heparan sulphate class, the GAG chain must contain at least one pentasaccharide in which a strictly specified pattern of modifications (including the relatively uncommon 3-O-sulphation of glucosamine) must be present. There are suggestions that other protease inhibitors homologous to ATIII (heparin cofactor II, protease nexin I) exhibit different, but also very specific, affinities for GAG structures. Little is known, however, about the degree with which other GAG binding proteins exhibit sequence specific binding.

■ SUMMARY

The proteoglycans are clearly a large, interesting, and important class of molecules. Current intense efforts to define the structures of proteoglycan cores, the sequences on the GAG chains they bear, and molecules with which they interact, are likely to lead rapidly to a better understanding of the roles that proteoglycans play *in vivo*.

For further information on proteoglycans, the reader is referred to the review articles listed below, from which much of the information presented above was drawn. Additional articles cited in the text are also listed below; these articles contain information not covered in either the review articles, or in the individual descriptions of proteoglycans on the following pages.

■ REFERENCES

Review Articles:

Bjork, I. and Lindahl U. (1982) Mol. Cell. Biochem. 48, 161-182.

Gallagher, J.T. (1989) Curr. Opinions Cell Biol. 1, 1201-1218.

Jackson, R.L., Busch, S.J. and Cardin, A.D. (1991) Physiol. Rev. 71, 481-539.

Ruoslahti, E. (1989) J. Biol. Chem. 264, 13369-13372.

Other References:

1. Herndon, M.E. and Lander, A.D. (1990) Neuron 4, 949-961.
2. Nishiyama, A., Dahlin, K.J., Prince, J.T., Johnstone, S.R. and Stallcup, W.B. (1991) J. Cell Biol. 114, 359-371.
3. Andres J.L., Stanley, K., Cheifetz, S. and Massague, J. (1989) J. Cell Biol. 109, 3137-3145.
4. Marcum, J.A., Reilly, C.F. and Rosenberg, R.D. (1987) In Biology of Proteoglycans, T.N. Wight and R.P. Mecham, Editors. Academic Press, San Diego, CA. pp. 301-343.

5. Yayon, A., Klagsbrun, M., Esko, J.D., Leder, P. and Ornitz, D.M. (1991) Cell 64, 841-848.
6. Reyes, A.A., Akeson, R., Brezina, L. and Cole, G.J. (1990) Cell Regulation 1, 567-576.

Arthur D. Lander:
Department of Brain and Cognitive Sciences,
Department of Biology,
Massachusetts Institute of Technology,
Cambridge, MA 02139, USA

Aggrecan

Aggrecan[1] is the core protein of the large aggregating keratan-sulphate/chondroitin-sulphate proteoglycan found in cartilaginous tissues[2-4]. An extracellular matrix structural molecule, aggrecan localizes in cartilage by virtue of a high affinity binding to hyaluron-link protein complex, and provides a strongly hydrated space filling gel due to the large number of polyanionic glycosaminoglycan chains covalently attached to the protein core.

Aggrecan consists of a highly-substituted monomeric core protein of 210-250 kDa (Figure 1), as deduced from its cDNA sequence; core protein sizes vary with different species and also due to alternative exon splicing[1] (Figure 2). The mature proteoglycan monomer bears several types of covalent substituents attached to the core protein: approximately 100-150 keratan-sulphate and the more abundant chondroitin-sulphate glycosaminoglycan chains, along with N- and O-linked oligosaccharides. The mature molecule is about 2,500 kDa[2-4]. The glycosaminoglycans are attached to repetitive sequences in the middle two-thirds of the molecule, including several types of repeats containing serine-glycine, the linkage site for chondroitin-sulphate; preceding these serine-glycine repeats is a region rich in proline, glutamic acid, and serine, where keratan-sulphate sites are clustered, possibly in a set of hexameric repeats[1,5]. Complex disulphide containing globular structures, G1-G3, are found at the ends of the molecule. G1 and G2 are at the N-terminus, and each contains homologous repeated domains B and B' of 100 amino acids each, which are also found in **link protein**[6]. G1, but not G2, is thought to bind hyaluronic acid and link protein[7]. G3 is at the C-terminus, and is composed of three structural domains: an EGF like repeat, a lectin-like portion, and a sequence homologous to the complement regulatory protein (CRP) motif. The lectin-like domain appears to be present in all forms of the molecule, while the EGF and CRP like domains are alternatively spliced. Functions for G2 and the various forms of G3 have not been determined, although there is evidence that the lectin-like sequence of G3 can bind to fucose and galactose[8]. While aggrecan is largely confined to cartilage, there are reports of biochemically and immunologically related molecules in several tissues including tendon, sclera, aorta and bone[3,9,10], and a chicken cDNA probe was used to detect aggrecan mRNA in eye tissues[11]. Another proteoglycan, **versican**, of related structure, appears to be more widely expressed[12]. These two proteoglycans share similarities not only with link protein, but with **CD44** hermes antigen (G1 and G2); and a family of LEC-CAMs or **selectins** (G3)[13]. As one of the major structural proteins of cartilage, aggrecan has been widely investigated for involvement in several genetic diseases of skeletal growth, but no definitive link has been established; the possible role of this molecule in degenerative joint conditions is also being studied.

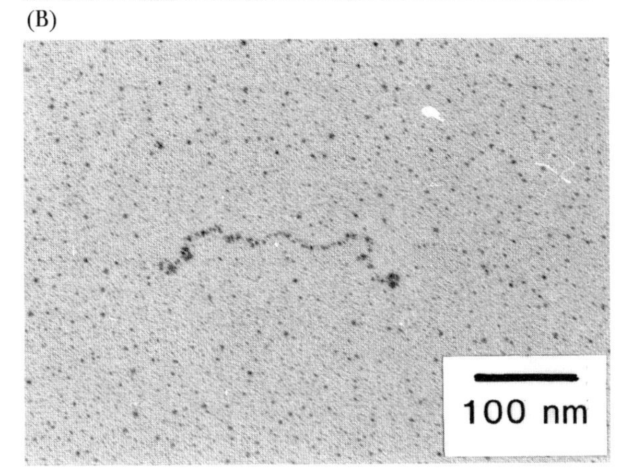

Figure 1. Rotary shadow electron micrographs of aggrecan proteoglycan monomer (A), and the aggrecan core protein (B) (Pictures provided by Drs. M. Morgelin and J. Engel).

■ PURIFICATION

Aggrecan has generally been purified as the proteoglycan, utilizing its high buoyant density in CsCl conferred by the numerous negative charges of the attached glycosaminoglycans[14]. Cartilage is dispersed in 4 M guanidine-HCl to reversibly denature the hyaluronan binding structure, and the extracts are then centrifuged in CsCl either in 4 M guanidine, to sediment the monomer; or after reassociation in 0.5 M guanidine to sediment the aggregate with hyaluronan and link protein. Anion exchange and molecular sieve chromatography have also been applied, exploiting the high charge and large hydrodynamic volume of the proteoglycan[15].

Aggrecan Structural Domains

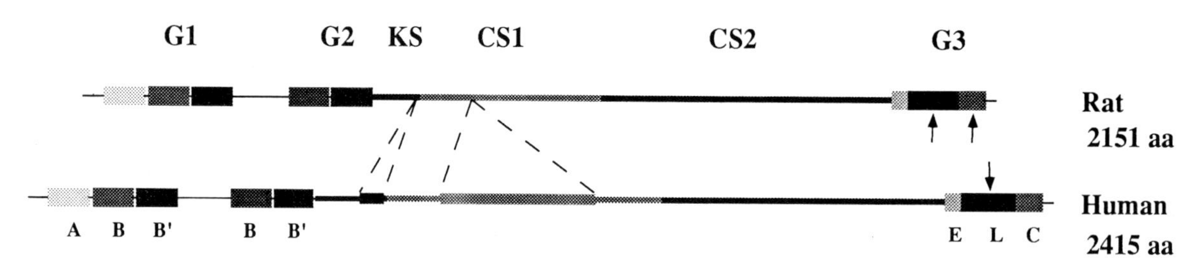

Figure 2. Schematic of aggrecan structural domains deduced from the rat (upper, 2151 amino acids) and the human (lower, 2415 amino acids) cDNA sequences. Dashed lines indicate repeated sequences inserted in the human sequence, relative to the rat; arrows indicate which of the (E)GF, (L)ectin, and (C)RP subdomains of G3 are preferentially spliced into aggrecan in the two species.

■ ACTIVITIES

Aggrecan is a structural molecule, with its high concentration of fixed charges creating a hyperosmotic environment in the extracellular space, resulting in a rigid, reversibly deformable gel which can withstand pressure. The G1 domain binds irreversibly under native conditions to both the unbranched polymer hyaluronan, and to link protein, creating a huge aggregate of up to 100 monomers arrayed along the strand[2]. Other structural domains (G2, G3) have no known functions, despite suggestive homologies.

■ ANTIBODIES

Numerous polyclonal and monoclonal immunological reagents have been prepared to both the protein and carbohydrate portions of the proteoglycan[16], and have been applied in studies of its biosynthesis, and tissue distribution[10]. One monoclonal antibody to a core protein epitope (1C-6) is available from the Developmental Hybridoma Bank (Johns Hopkins University, Baltimore, MD).

■ GENES

Full length cDNAs have been reported for the rat (GenBank J03485)[17] and human (GenBank M55172)[1] species of aggrecan, while partial cDNA clones for chicken[18,19], bovine (GenBank Y00319, J05028)[20,5] and human (GenBank J05062)[21] also have been described. A five kB genomic clone containing the six exons encoding the chicken G3 domain has been sequenced[22].

■ REFERENCES

1. Doege, K., Sasaki, M., Kimura, T. and Yamada, Y. (1991) J. Biol. Chem. 266, 894-902.
2. Hascall, V. and Hacall, G. (1981) In Cell Biology of Extracellular Matrix (Hay, E., ed.) pp. 39-63, Plenum, NY.
3. Hassell, J.R., Kimura, J.H. and Hascall, V.C. (1986) Ann. Rev. Biochem. 55, 539-567.
4. Gallagher, J. (1989) Curr. Op. Cell Biol. 1, 1201-1218.
5. Antonsson, P., Heinegard, D. and Oldberg, A. (1989) J. Biol. Chem. 264, 16170-16173.
6. Doege, K., Hassell, J.R., Caterson, B. and Yamada, Y. (1986) Proc. Natl. Acad. Sci. (USA) 83, 3761-3765.
7. Mörgelin, M., Paulsson, M., Hardingham, T.E., Heinegård, D. and Engel, J. (1988) Biochem. J. 253, 175-185.
8. Halberg, D., Proulx, G., Doege, K., Yamada, Y. and Drickamer, K. (1988) J. Biol. Chem. 263, 9486-9490.
9. Heinegård, D., Franze, A., Hedbom, E. and Sommarin, Y. (1986) Ciba Found. Symp. 124, 69-88.
10. Kimura, J.H., Shinomura, T. and Thonar, E.J.-M.A. (1987) Meth. Enz. 144, 372-393.
11. Tsonis, P. and Goetinck, P. (1988) Exp. Eye. Res. 46, 753-764.
12. Zimmerman, R. and Ruoslahti, E. (1989) EMBO J. 8, 2975-2981.
13. Coombe, D. and Rider, C. (1989) Immunol. Today 1, 289-291.
14. Hascall, V. and Sajdera, S. (1989) J. Biol. Chem. 244, 2384-2396.
15. Heinegård, D. and Sommarin, Y. (1987) Methods Enzymol. 144, 319-372.
16. Caterson, B., Christner, J.E., Baker, J.R. and Couchman, J.R. (1985) Fed. Proc. 44, 386-393.
17. Doege, K., Sasaki, M., Horigan, E., Hassell, J.R. and Yamada, Y. (1987) J. Biol. Chem. 262, 17757-17767.
18. Sai, S., Tanaka, T., Kosher, R.A. and Tanzer, M.L. (1986) Proc. Natl. Acad. Sci. (USA) 83, 5081-5085.
19. Krueger Jr., R.C., Fields, T.A., Mensch Jr., J.R. and Schwartz, N.B. (1990) J. Biol. Chem. 265, 12088-12097.
20. Oldberg, Å., Antonsson, P. and Heinegård, D. (1987) Biochem. J. 243, 255-259.
21. Baldwin, C.T., Reginato, A.M. and Prockop, D.J. (1989) J. Biol. Chem. 264, 15747-15750.
22. Tanaka, T., Har-el, R. and Tanzer, M. (1988) J. Biol. Chem. 263, 15831-15835.

■ *Kurt J. Doege:*
Shriners Hospital,
Department of Biochemistry and Molecular Biology,
Oregon Health Sciences University,
Portland, Oregon, USA

Agrin

Agrin[1] is a high molecular weight protein extracted from basal lamina fractions of the synapse rich electric organ of the marine ray, Torpedo californica. When it is applied to cultured myotubes, it causes them to form on their surface patches of proteins that have a high concentration of acetylcholine receptors (AChR), acetylcholinesterase (AChE) and several other proteins that are aggregated in the postsynaptic apparatus of neuromuscular junctions in vertebrate skeletal muscles[1-3]. Several lines of evidence, based primarily on the use of agrin probes and activity assays, indicate that agrin in vertebrates is synthesized in the cell bodies of motor neurons in the central nervous system (CNS) and transported along their axons to muscle where it is released from the axons' terminals to bind to myofibre basal lamina in the synaptic cleft and to an agrin receptor in the myofibres' plasma membrane triggering the myofibres to form postsynaptic apparatus[1,4]. Agrin is a member of a family of extracellular matrix proteins (the agrin protein family) that are highly similar in structure but differ in function and distribution[5].

There are 150 kDa and 95 kDa forms of agrin in electric organ extracts. Both have the same activities in myotube cultures and both are recognized by several different monoclonal antibodies (mAbs)[1]. They may be proteolytic fragments of a larger protein cleaved either during extraction or during *in situ* posttranslational modification[6]. cDNA that encodes an agrin homologue (see below) has been cloned from a marine ray (*Discopyge ommata*) CNS library[6]. It is partial but predicts a protein having a stretch of amino acids similar to the N-terminal amino acid sequence of the 95 kDa protein. The ray cDNA has been used to clone from a chick brain library a cDNA that codes for the C-terminal half of chick agrin[7]. It encompasses a region homologous to the 95 kDa form of ray agrin, and, when constructs of the cDNA are transfected into COS-7 cells, the cells secrete a protein that is recognized by anti-agrin mAbs and induces cultured myotubes to form AChR/AChE aggregates identical to those induced by agrin extracted from ray in electric organ, both in their appearance and in the manner of their formation. Full length chick cDNA has been assembled by piecing together primer extension cDNAs and linking them to cDNA coding the the C-terminal half of agrin[7]. The molecular weight of agrin like protein deduced from this assembly is ~205 kDa. It has several structural motifs common to other extracellular matrix and secreted proteins[7] (Figure –). The motifs are based on the number and spacing of cysteine residues of which the N-terminal half of the protein is particularly rich. Some of the cysteine rich domains are similar to domains found in EGF (EGF repeats), some are similar to domains in follistatin, osteonectin and pancreatic secretory trypsin inhibitors of the Kazal family, and one is similar to a domain in **laminin** and s-laminin. Flanking an EGF like repeat near the C-terminus (Figure) are stretches of 4 and 11 amino acids which are required for AChR aggregating activity[5,7]. It is not yet known whether amino acid residues in these stretches compose the active site of agrin or confer a shape on the protein that exposes the active site. There is evidence that the cysteine rich domains and the stretches of 4 and 11 amino acids are conserved in agrin across species[7] and that the molecular weight of agrin is similar in chick, ray and rat[6-8].

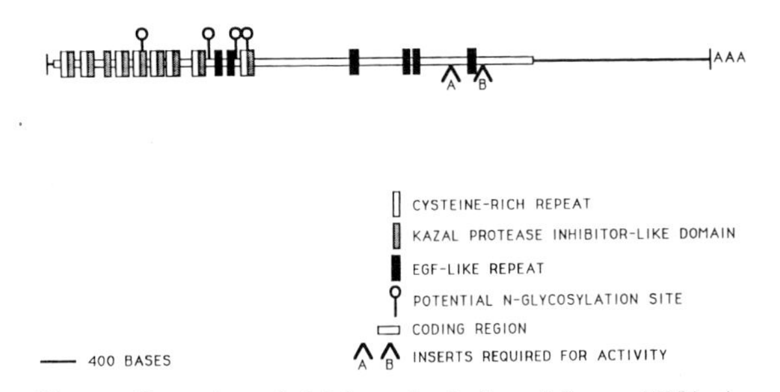

Figure: Domains of chick agrin deduced from cDNA. A, 4 amino acids: B, 11 amino acids.

Chick cDNAs that encode the C-terminal half of two isoforms of agrin have been isolated[5]. The amino acid sequence encoded by each is identical to the C-terminal half of agrin except that one (agrin-related protein-1, ARP-1) lacks the 11 amino acid stretch and the other (agrin-related protein-2, ARP-2) lacks the 11 and 4 amino acid stretches that are required for AChR aggregating activity. Indeed, when constructs of the cDNAs of the two ARPs are transfected into COS cells, the cells secrete proteins recognized by anti-agrin mAb but inactive in AChR aggregation assays. Northern blot analysis indicates that full length agrin and ARPs have similar molecular weights. PCR experiments demonstrate that agrin is expressed in the cell bodies of motor neurons in the CNS as predicted by the agrin hypothesis[7]. Such experiments also indicate that the ARPs are present in both the nervous system and muscle and that, in muscle, myofibres and Schwann cells synthesize ARPs while motor axon terminals are the sole source of agrin. cDNAs that encode the C-terminal half of ARP-1 have also been isolated from ray[6] and rat (ARP-1 referred to as agrin)[8,9]. Transfection experiments using a full length rat ARP-like cDNA suggest that the N-terminal half of agrin and/or ARPs can have some AChR aggregating activity under certain conditions[9,5]. In *in situ* hybridizations, probes that recognize both agrin and ARP transcripts label motor neurons[6-8] and other types of neurons in the CNS. They also label at low levels the CNS white matter which contains transcripts of glial cells and endothelial cells but not neurons[6].

PURIFICATION

Agrin has been purified from a basal lamina-enriched fraction of *T. californica* electric organ insoluble in isotonic saline and detergent at pH 7.5[1]. It is extracted from this fraction by treatment with 0.2M bicarbonate buffer at pH 9. Subsequent purification involves affinity chromatography using Cibacron Blue 3GA-Agarose and immunoaffinity chromatography using anti-agrin mAbs. Repeated extraction of the insoluble fraction with the bicarbonate buffer can yield up to 10 times the amount of activity obtained with the first extraction[1].

ACTIVITIES

AChR-aggregating activity is routinely assayed on chick myotube cultures by fluorescence microscopy after labelling the AChRs with rhodamine conjugated α-bungarotoxin[3]. Agrin causes a 3-20 fold increase in the number of AChR aggregates. It has little or no effect on myotube size, total number of AChRs on the myotube surface or the rate of AChR degradation[10]. The AChR-aggregating activity is dose-dependent and due, at least in part, to the lateral migration of AChRs present in the plasma membrane at the time extracts are applied. The increase in number of AChR aggregates is first seen 2 hours after adding agrin to the medium and it is maximal by 24 hours. Agrin also causes the formation of aggregates of AChE (membrane and **collagen**-tail forms), butyrylcholinesterase, heparan sulphate proteoglycan and ***43 kDa AChR-associated protein***[2,3]. Each accumulates with a time course similar to that of the AChR aggregates and the aggregates tend to be coextensive. Within 30 minutes of adding agrin to chick myotube cultures, there is an increase in tyrosine phosphorylation of the β subunit of AChRs, suggesting that agrin-induced tyrosine phosphorylation of the β subunit may play a role in regulating the formation of AChR aggregates[11].

ANTIBODIES

mAbs against *T. californica* agrin stain basal lamina in the synaptic cleft at neuromuscular junctions of ray, frog and chick skeletal muscles[12]. The only neuronal cell bodies in the CNS in which staining has been detected are those of the motor neurons that innervate skeletal muscle (and in ray, electric organ)[13]. The antibodies also stain the basal lamina of capillaries in the CNS. In chick the extrasynaptic basal lamina of slow muscle fibers stains as does the basal lamina of smooth muscle fibers of blood vessels in skeletal muscles, the basal lamina of Schwann cells in nerves, the glomerular basal lamina in kidney and the myofiber basal lamina of cardiac muscle[12,14]. The mAbs immunoprecipitate AChR/AChE-aggregating activity (agrin) extracted from the CNS of ray, frog and chick[13]. Proteins immunoprecipitated by anti-agrin mAbs from chick kidney and embryonic heart are inactive in standard AChR aggregation assays for agrin[15] suggesting that they are ARPs[5]. Antiserum directed against ray agrin inhibits the motor neuron-induced aggregation of AChRs on myotubes in chick motor neuron-myotube cocultures[16].

GENES

Southern blot analysis of chick genomic DNA using probes common to agrin, agrin-related protein-1 and agrin-related protein-2, revealed a single band[7]. Analysis of intron/exon boundaries showed that the 4 and 11 amino acid stretches in agrin that are absent from the agrin-related proteins are encoded by separate exons. Thus, agrin and the agrin-related proteins are generated from the agrin gene by alternative splicing of common precursor mRNA.

REFERENCES

1. Nitkin, R.M., Smith, M.A., Magill, C., Fallon, J.R., Yao, Y.-M.M., Wallace, B.G. and McMahan, U.J. (1987) J. Cell Biol. 105, 2471-2478.
2. Wallace, B.G. (1986) J. Cell Biol. 102, 783-794.
3. McMahan, U.J. and Wallace, B.G. (1989) Dev. Neurosci. 11, 227-247.
4. McMahan, U.J. (1990) Cold Spring Harbor Symp. Quant. Biol. 50, 407-418.
5. Ruegg, M.A., Tsim, K.W.K., Horton, S.E., Kröger, S., Escher, G., Gensch, E.M. and McMahan, U.J. (1992) Neuron, 8, 691-699.
6. Smith, M.A., Magill-Solc, C., Rupp, F., Yao, Y.-M.M., Schilling, J.W., Snow, P. and McMahan, U.J. (1992) submitted for publication.
7. Tsim, K.W.K., Ruegg, M.A., Escher, G., Kröger, S. and McMahan, U.J. (1992) Neuron, 8, 677-689.
8. Rupp, F., Payan, D.G., Magill-Solc, C., Cowan, D.M. and Scheller, R.H. (1991) Neuron 6, 811-823.
9. Campanelli, J.T., Hoch, W., Rupp, F., Kreiner, T. and Scheller, R.H. (1991) Cell 67, 909-916.
10. Godfrey, E.W., Nitkin, R.M., Wallace, B.G., Rubin, L.L. and McMahan, U.J. (1984) J. Cell Biol. 99, 615-627.
11. Wallace, B.G., Qu, Z. and Huganir, R.L. (1991) Neuron 6, 869-878.
12. Reist, N.E., Magill, C. and McMahan, U.J. (1987) J. Cell Biol. 105, 2457-2469.
13. Magill-Solc, C. and McMahan, U.J. (1988) J. Cell Biol. 107, 1825-1833.
14. Godfrey, E.W., Deitz, M.E., Morstad, A.L., Wallskog, P.A. and Yorde, D.E. (1988) J. Cell Biol. 106, 1263-1272.
15. Godfrey, E.W. (1991) Exp. Cell Res. 195, 99-105.
16. Reist, N.E., Werle, M.J. and McMahan, U.J. (1992) Neuron, 8, 677-89.

U.J. McMahan:
Department of Neurobiology,
Stanford University School of Medicine,
Stanford, CA 94305, USA

Biglycan (BGN)

Biglycan (BGN)[1] is a small proteoglycan whose primary gene product is found associated with the cell surface or pericellular matrix of a variety of cells including specific subsets of developing mesenchymal (skeletal muscle, bone and cartilage), endothelial (blood vessels), and epithelial (keratinocytes cells)[2]. Other names for this proteoglycan includes; PG-1, PG-I, DS-PGI, PG-S1 and DS-I. BGN is composed of two chondroitin (CS) or dermatan sulphate (DS) chains on a 38 kDa core protein that is predominantly made of 12 tandem 24 amino acid repeat structures, each characterized by ordered hydrophobic residues. Similar tandem repeat structures have been used throughout evolution when a protein is destined to bind another protein or perhaps a cell surface[1,3]. The function of biglycan is unknown, but, like the homologous proteoglycan, decorin, it may bind to transforming growth factor-β (TGF-β)[4].

The chondroitin sulphate containing biglycan is most commonly isolated from fetal or young bone[5,6], while the dermatan sulphate containing form is isolated from articular cartilage[7]. The BGN is heterogeneous with respect to the size of the glycosaminoglycan chains which results in a broad band on SDS-PAGE centered anywhere from 200-350 kDa. Although BGN may differ in size between tissues and developmental stage, it is almost always larger than the other small proteoglycan, **decorin**, when decorin is also present. (BGN may occasionally be present with a single CS/DS chain, thus making it the same size as decorin.) Removal of the glycosaminoglycan chains with the enzyme chondroitin ABC-lyase results in a 45 kDa band. The gene for BGN is on the human X chromosome (Xq27ter)[8] and its mRNA encodes a 42.5 kDa preproprotein. The human (cDNA)[1], bovine protein[9] and rat[10] sequences have been reported. BGN probably contains three disulphide bonds. Unlike its close relatives, decorin and **fibromodulin**, purified BGN does not bind to **collagen** fibrils *in vitro,* nor is it found associated with classic collagen bundles in tissues. BGN (both protein and mRNA) is expressed in a range of specialized cell types in developing human tissues including bone, cartilage, blood vessel endothelial cells, skeletal myofibrils, renal tubular epithelia, and differentiating keratinocytes[2]. Generally, the BGN is immunolocalized to the cell surface or pericellular matrices, but in a tissue such as bone, the protein is detected in the matrix proper. This localization in the extracellular matrix may be due to the adsorption of the BGN to hydroxylapatite crystals after having been shed from the osteoblasts. Localization of BGN by immunoelectron microscopy has not yet been performed. BGN may bind TGF-β *in vitro*[4], and thus may modulate the biological effects of this growth factor in the extracellular environment. The human BGN gene has been cloned and partially sequenced.

PURIFICATION

BGN containing chrondroitin sulphate chains can be purified from fetal or young bone by a series of extractions procedures and protein chromatography[5]. Bone is milled to a fine powder, extracted with denaturing buffers to remove blood and cellular proteins, and the residue extracted with demineralization buffers. Standard ion exchange and molecular sieve chromotography is performed in denaturing solvents to suppress the endogenous protease activity associated with the mineralized bone matrix. BGN is too hydrophobic to be chromatographed on standard reverse phase columns using methanol or acetonitrile gradients. Dermatan sulphate containing BGN can be isolated in good yield from articular cartilage using the same method as above (less the demineralizations step) plus a reverse phase chromatography step using octyl-Sepharose and a detergent gradient[7]. The isolation of undenatured BGN in good yield has not been reported.

ACTIVITIES

BGN has no known activities but may bind to TGF-β *in vitro.*

ANTISERA

BGN is poorly antigenic in rabbit and goat. The only available antisera to date are against a synthetic peptide modeled after amino acids 11-24 of the human BGN sequence (LF-15 and LF-51) which detect both human and monkey BGN but not any other species[2,5,8] and a similar set of peptide antisera for cow BGN (LF-96+97) (L. Fisher unpublished data). Both antisera work in immunolocalization and immunoblot analyses.

GENES

Full length cDNA for human BGN (GenBank J04599) is available in Bluescript plasmid (plasmid P16)[1]. The human gene has been cloned from the X chromosome and has been published[11]. The gene is approximately 7 kB and contains eight exons.

REFERENCES

1. Fisher, L.W., Termine, J.D. and Young, M.F. (1989) J. Biol. Chem. 264, 4571-4576.
2. Bianco, P., Fisher, L.W., Young, M.F., Termine, J.D. and Robey, P.G. (1990) J. Histochem. Cytochem. 38, 1549-1563.
3. Patthy, L. (1987) J. Mol. Biol. 198, 567-577.
4. Yamaguchi, Y., Mann, D.M. and Rouslahti, E. (1990) Nature 346, 281-284.

5.Fisher, L.W., Hawkins, R.G., Tuross, N. and Termine, J.D. (1987) J. Biol. Chem. 262, 9702-9708.
6.Fisher, L.W., Termine, J.D., Dejter Jr., S.J., Whitson, S.W., Yanagishita, M., Kimura, J.R., Hascall, V.C., Kleinman, H.K., Hassell, J.R. and Nilsson, B. (1983) J.Biol. Chem. 258, 6588-6594.
7.Choi, H.U., Johnson, T.L., Subhash, P., Tang, L.-H., Rosenberg, L.C. and Neame, P.J. (1989) J. Biol. Chem. 264, 2876-2884.
8.Fisher, L.W., Lindner, W., Young, M.F. and Termine, J.D. (1989) Connect. Tissue Res. 21, 43-50.
9.Neame, P.J., Choi, H.U. and Rosenberg, L.C. (1989) J. Biol. Chem. 264, 8653-8661.
10.Dreher, K.L., Asundi, V., Matzura, D. and Cowan, K. (1990) Eur. J. Cell Biol. 53, 296-304.
11.Fisher, L.W., Heegaard, A.-M., Vetter, U., Vogel, W., Just, W., Termine, J.D. and Young, M.F. (1991) J. Biol. Chem. 266, 14371-14377.

Larry W. Fisher:
Bone Research Branch,
NIDR, NIH,
Bethesda, MD, USA

Bone Sialoprotein (BSP)

Bone sialoprotein (BSP) is a heavily glycosylated extracellular matrix protein[1] that supports cell attachment on bacterial plastic through the Arg-Gly-Asp (RGD) cell attachment mechanism[2-4]. The ~75 kDa BSP appears to be relatively specific to the skeleton[1].

Bone sialoprotein (also BSP-II) constitutes 10-15% of the noncollagenous proteins in the mineral compartment of developing bone. In man[5], cow[1], and rat[6], BSP is a protein containing about 50% carbohydrate, while in rabbit bone it is a keratan sulphate proteoglycan[7]. BSP contains no disulphide bonds. Both the rat[8] and human[9] cDNA showed a Arg-Gly-Asp tripeptide known to often support cell attachment through the **integrin** receptor[2]. Indeed, rat BSP binds to the **vitronectin** type receptor[4], and human BSP supports cell attachment through the RGD mechanism *in vitro*[3]. Thus BSP has many of the properties of the cell-attachment domain of **fibronectin** or vitronectin but in a considerably smaller gene product. The original observation[9] that the RGD site is flanked by tyrosine residues that fit the consensus requirement for tyrosine sulfation has been supported by the purification of a heavily sulphated rat BSP isolated from a an osteosarcoma cell line[10]. The modulation of the sulphation of these tyrosine sulphates may play a role in the cell attachment and migration processes.

BSP has been localized in the developing human using both immunolocalization and *in situ* methods[11]. BSP appears to be limited to the skeleton (osteoblasts, osteocytes, osteoclasts and hypertrophic chondrocytes) and the trophoblasts of developing placenta. These localization results suggest that BSP is specifically used for cell-matrix and/or cell-cell interactions.

PURIFICATION

BSP can be easily purified from developing bone by the use of standard biochemical techniques[1,5-7], although these generally result in denatured product. A rat osteosarcoma cell line, UMR-106-01, has been observed to have subclones (UMR-106-01-BSP) that can produce mg/l quantities of BSP under serum free conditions[10]. These transformed cells make a heavily sulphated form of the BSP, however the addition of sulphation blocking drugs such as chlorate may allow the investigator to titrate the sulphation levels desired.

ACTIVITIES

BSP can support cell attachment on bacterial plastic and this activity can be blocked by addition of peptides containing the RGD sequence[6-8]. The biological significance of this is under debate.

ANTIBODIES

Polyclonal antisera to human[5] and cow[1] has been reported. The human BSP polyclonal antiserum is reported to crossreact with monkey and dog. Recently, a rat BSP polyclonal antiserum has been made by the Bone Research Branch. Synthetic peptide antisera to the RGD-region of human BSP have also been made by the same laboratory, although the monospecificity of these have not been reported.

GENES

Full length rat[8] and human[9] cDNA are available with GenBank accession numbers J04215 and J05213 respectively. The human BSP gene has been mapped to chromosome 4[9].

REFERENCES

1.Fisher, L.W., Whitson, S.W., Avioli, L.V. and Termine, J.D. (1983) J. Biol. Chem. 258, 12723-12727.
2.Rouslahti, E. and Pierschbacher, M.D. (1987) Science 238, 491-497.
3.Somerman, M.J., Fisher, L.W., Foster, R.A. and Sauk, J.J. (1988) Calcif. Tissue Int. 43, 50-53.
4.Oldberg, A., Franzen, A., Heinegård, D., Pierschbacher, M. and Ruoslahti, E. (1988) J. Biol. Chem. 263, 19433-19436.

5.Fisher, L.W., Hawkins, G.R., Tuross, N. and Termine, J.D. (1987) J. Biol. Chem. 262, 9702-9708.
6.Franzen, A. and Heinegård, D. (1985) In Chemistry and Biology of Mineralized Tissues (Butler, W.T., ed.) pp 132-141, EBSCO, Birminham, AL.
7. Kinne, R.W. and Fisher, L.W. (1987) J. Biol. Chem. 262, 10206-10211.
8.Oldberg, A., Franzen, A. and Heinegård, D. (1988) J. Biol. Chem. 263, 19430-19432.
9.Fisher, L.W., McBride, O.W., Termine, J.D. and Young, M.F. (1990) J. Biol. Chem. 265, 2347-2351.
10.Midura, R.J., McQuillan, D.J., Benham, K.J., Fisher, L.W. and Hascall, V.C. (1990) J. Biol. Chem. 265, 5285-5291.
11.Bianco, P., Fisher, L.W., Young, M.F., Termine, J.D. and Robey, P.G. (1991) Calcif. Tissue Int. 49, 421-426..

■ *Larry W. Fisher:*
Bone Research Branch,
NIDR, NIH,
Bethesda, Maryland, USA

Cartilage Matrix Protein (CMP)

Cartilage matrix protein (CMP) is a noncollagenous protein of cartilage that was first described in extracts of bovine tracheal cartilage[1]. The molecular weight of the intact protein is 148 kDa as determined by sedimentation equilibrium centrifugation. Upon reduction of the disulphide bonds, the protein behaves as a single subunit of 54 kDa. Initial characterization of the protein indicated that the cysteine content was high and the carbohydrate content was relatively low.

The amino acid sequence of CMP has been deduced from chicken[2,3] and human[4] cDNA and genomic DNA sequences. CMP contains two homologous repeats (CMP1 and CMP2) which are separated by an EGF-like domain. Domains with homology to the CMP repeats have been reported in **von Willebrand factor**[5-7] (vWF), complement factors B[8] and C2[9], the α-chains of the β2 **integrins** (Mac-1[10,11], p150,95[12], **LFA-1**[13]), VLA-2[13] and type VI **collagen**[14,15]. (Figure 1) These CMP domains have also been referred to as I-domains or A-domains. The protein has been reported to be present in tracheal, nasal septum, xiphisternal, auricular and epiphyseal cartilage but absent in extracts of articular cartilage, the anulus fibrosus and the nucleus pulposus of the intervertebral disc.

There is a single copy of the CMP gene in the genome of chicken[2,3] and human[4]. The human CMP gene has been mapped[4] to chromosome 1p35. Both the chicken and human genes consist of eight exons and seven introns. The RNA splice junctions of the seventh intron (Intron G) of the chicken and human CMP genes do not conform to

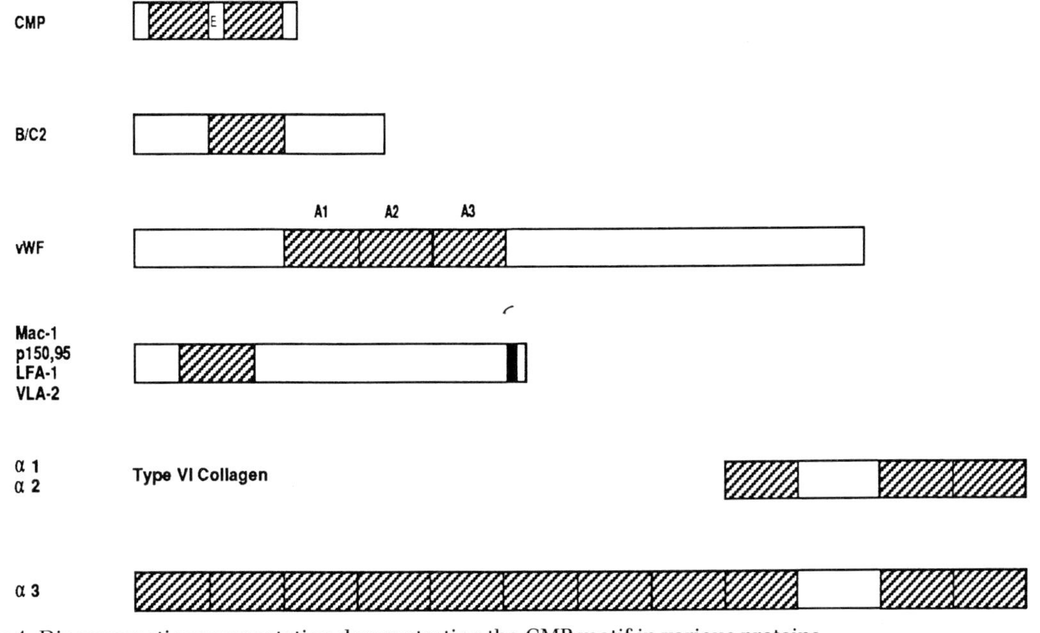

Figure 1. Diagrammatic representation demonstrating the CMP motif in various proteins.

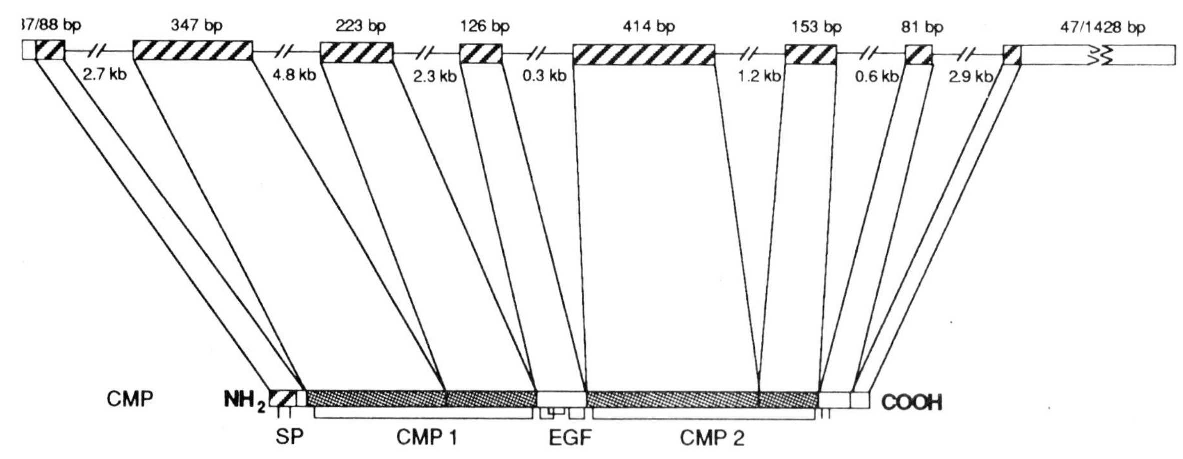

Figure 2. Diagrammatic representation showing the relationship between CMP and the CMP gene. SP, putative signal peptide; CMP1 and CMP2, homologous repeats; EGF, EGF like domain. The vertical lines indicate cysteine residues. All disulphide bonds are assumed.

concensus splice sequences, suggesting a novel type of splicing mechanism in cartilage. The relationship between the structure of CMP and the CMP gene in the chicken is shown in Figure 2. A similar organization exists for the human gene. Each of the CMP domains is encoded by two exons whereas the EGF like domain is encoded by a single exon. The exon-intron junction within the CMP domains is at a different position within the coding regions of each of the two domains[3]. It is of interest that the exonic composition of the CMP-like domains of factor B, p150,95, and vWF show no distinct pattern. This domain in factor B is encoded by five exons and the A3 domain of vWF and p150,95 are each encoded by four exons. Domains A1 and A2 of vWF are encoded by one single large exon.

The temporal expression of the gene for CMP is independent of the expression of type II collagen, **link protein** and **aggrecan** core protein in the developing limb bud of the chick embryo. The CMP gene is also expressed in the notochord and in the early somites of the chick embryo[16].

■ PURIFICATION

CMP can be extracted from cartilage using dissociative solvents (4 M guanidine-HCl) followed by CsCl density gradient centrifugation, gel chromatography and precipitation from low ionic strength solutions[1].

■ ACTIVITIES

The function of CMP is not known. CMP binds to collagen and has been shown to be a component of the collagen fibril of cartilage[17].

■ ANTIBODIES

Rabbit polyclonal antisera against bovine CMP have been generated[18]. Monoclonal antibodoes against chicken CMP have been used as has a rabbit polyclonal antiserum against a synthetic peptide covering a sequence of the CMP2 domain[17].

■ GENES

The complete nucleotide sequences are available for chicken[3] (EMBL X12346-X12354) and human[4] (EMBL J05666 and J05667) CMP.

■ REFERENCES

1.Paulsson, M. and D. Heinegård (1981) Biochem J. 197, 367-375.
2.Argraves, S.W., Deák, F., Sparks, K.J., Kiss, I. and Goetinck, P.F. (1987) Proc. Natl. Acad. Sci (USA) 84, 464-468.
3.Kiss, I., Deák, F., Holloway, R.G., Delius, H., Mebust, K.A., Frimberger, E., Argraves, W.S., Tsonis, P.A., Winterbottom, N. and Goetinck, P.F. (1989) J. Biol. Chem. 264, 8126-8134.
4.Jenkins, R.N., Osborne-Lawrence, S.L., Sinclair, A.K., Eddy Jr., R.L., Byers, M.G., Shows, T.B. and Duby, A.D. (1990) J. Biol. Chem. 265, 19624-19632.
5.Sadler, J.E., Shelton-Inloes, B.B., Sorace, J.M. and Titani, K. (1986) Cold Spring Harbor Symp. Quant. Biol. 51, 515-523.
6.Shelton-Inloes, B.B., Titani, K. and Sadler, J.E. (1986) Biochemistry 25, 3164-3171.
7.Titani, K., Kumar, S., Takio, K., Ericsson, L.H., Wade, R.D., Asnida, K., Walsh, K.A., Chopek, M.W., Sadler, J.E. and Fujikawa, K. (1986) Biochemistry 25, 3171-3184.
8.Mole, J.E., Anderson, J.K., Davison, E.A. and Woods, D.E. (1984) J. Biol. Chem. 259, 3407-3412.
9.Bentley, D.R. (1986) Biochem. J. 239, 339-345.
10.Pytela, R. (1988) EMBO J. 7, 1371-1378.
11.Corbi, A.L., Kishimoto, T.K., Miller, L.J. and Springer, T.A. (1988) EMBO J. 263, 12403-12411.
12.Corbi, A.L., Miller, L.J., O'Connor, K., Larson, R.S. and Springer, T.A. (1987) EMBO J. 6, 4023-4028.
13.Takada, Y. and Hemler, M.E. (1989) J. Cell Biol. 109, 397-407.
14.Koller, E., Winterhalter, K.H. and Trueb, B. (1989) EMBO J. 8, 1073-1077.
15.Bonaldo, P. and Colombatti, A. (1989) J. Biol. Chem. 264, 20235-20239.

16.Stirpe, N.S. and Goetinck, P.F. (1989) Development 107, 22-33.
17.Winterbottom, N., Tondravi, M.M., Harrington, T.L., Klier, F.G., Vertel, B.M. and Goetinck, P.F. (1992) Develop, Dynamics 193, 266-276.
18.Paulsson, M. and Heinegård, D. (1982) Biochem. J. 207, 207-213.

Paul F. Goetinck:
Cutaneous Biology Research Center,
Massachusetts General Hospital East,
Building 149, 13th Street, Charlestown, MA 02129, USA

Cat-301 Proteoglycan

The Cat-301 proteoglycan is a high molecular chondroitin sulphate proteoglycan (CSPG) distributed on the extracellular surface of a subset of neurons in the mature mammalian central nervous system. The Cat-301 CSPG is first expressed late in development, at a time that correlates with the end of the period of developmental synaptic plasticity. Perturbation of the normal pattern of neuronal activity during the early postnatal period attenuates expression of the protein. The function of the Cat-301 CSPG has not yet been established definitively, but the pattern of its expression and regulation suggests a role in stabilizing mature synaptic structure.

The Cat-301 CSPG was originally identified by its expression (detected immunocytochemically with monoclonal antibody Cat-301) on the surface of subsets of neurons in several areas of the central nervous system (CNS) in many mammalian species[1] (Figure 1). Expression of the Cat-301 CSPG is largely restricted to grey matter in the CNS. A characteristic subset of neurons is recognized by monoclonal antibody Cat-301 in each area of the CNS examined to date[2-4]. In the monkey visual system the antibody demarcates a series of functionally related neurons through several orders of subcortical and cortical processing[2] (Figure 2).

The developmental regulation of the Cat-301 CSPG is one of its most intriguing properties. In every system examined, the surface associated Cat-301 staining is first observed relatively late in development, at a time that correlates with the end of the period of synaptic plasticity[5,6]. Perturbation of normal patterns of neuronal activity during the early postnatal period (the critical period) by a variety of pharmacological, surgical and behavioral manipulations produces a marked and irreversible reduction in levels of the Cat-301 CSPG[4-8]. Perturbation of activity in adult animals has no effect on Cat-301 CSPG expression. The Cat-301 CSPG is, thus, the first protein described with a pattern of expression that parallels neuronal maturation during critical periods in CNS development.

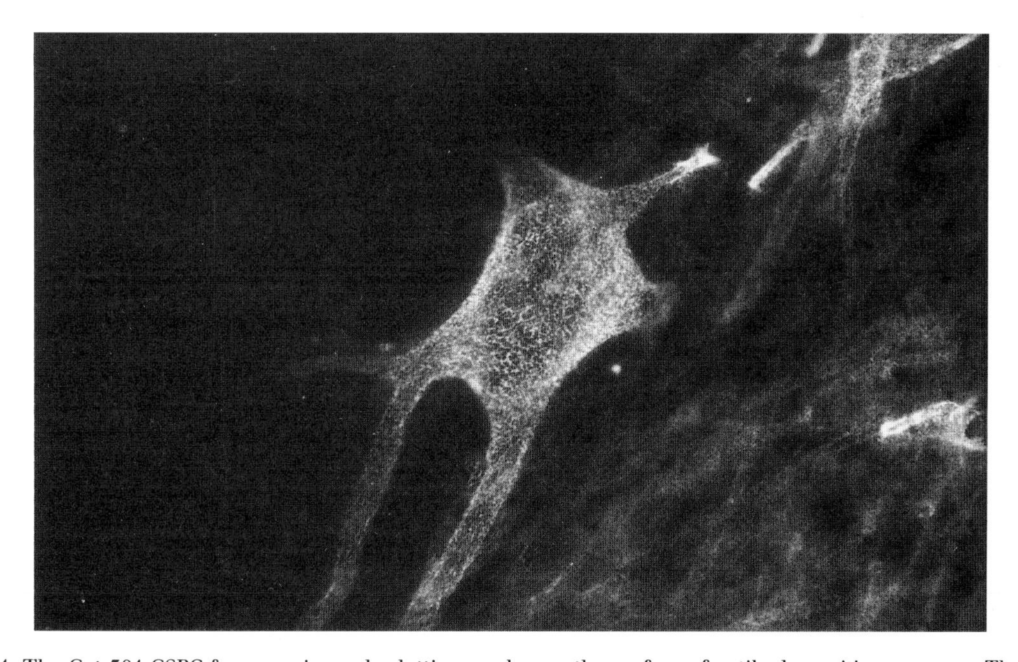

Figure 1. The Cat-301 CSPG forms an irregular lattice-work over the surface of antibody-positive neurons. The small, round areas devoid of Cat-301 staining are the sites of synapses.

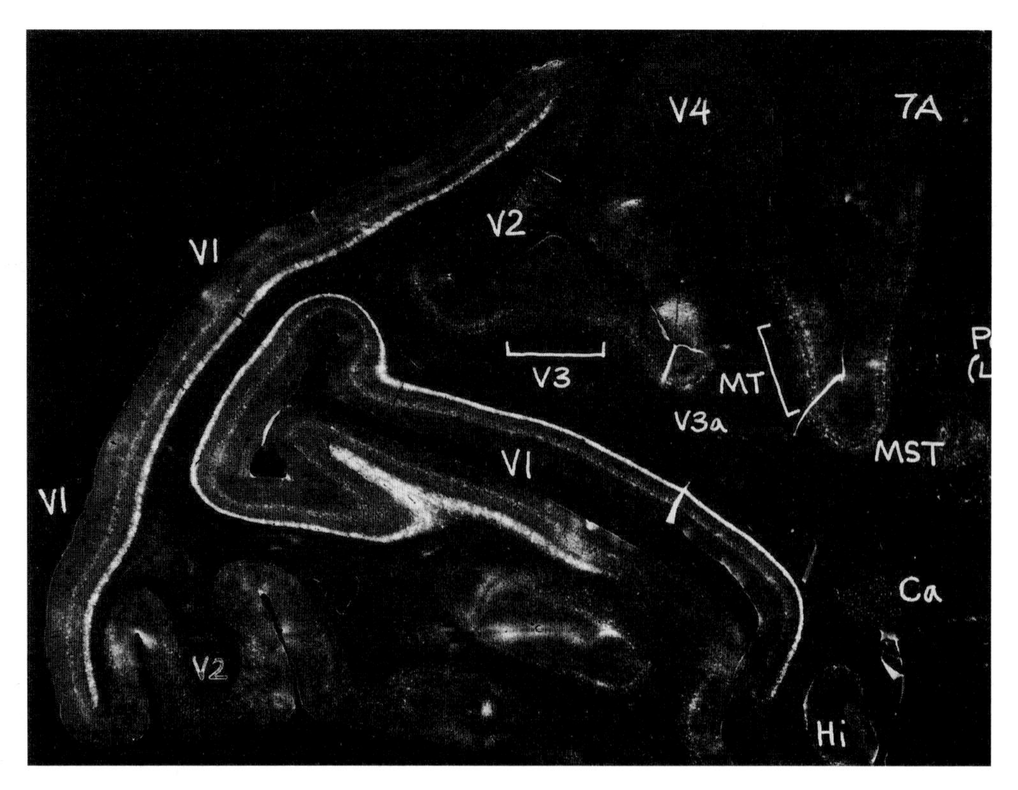

Figure 2. In monkey neocortex, antibody Cat-301 preferentially recognizes neurons in areas associated with processing the motion component of visual stimuli. Cortical areas V1, V2, MT and MST contain antibody-positive neurons. In contrast, area V4 which is largely concerned with processing the color component of visual stimuli contains far fewer antibody-positive neurons.

On immunoblots, the Cat-301 CSPG appears as a 650 kDa polydisperse band, originally identified in brain extracts[9]. Digestion by chondroitinase ABC reduces the size to approximately 550 kDa while digestion with keratanase has no effect. Furthermore, antibodies to keratan sulphate indicate that the Cat-301 proteoglycan lacks modification by keratan sulphate. A second monoclonal antibody, Cat-304, raised using an immunosuppression protocol[7,9,10], recognizes an independent epitope on the Cat-301 CSPG[9].

The Cat-301 proteoglycan is antigenically related to the large, aggregating chondroitin sulphate proteoglycan from cartilage, **aggrecan**[11,12]. Both Cat-301 and Cat-304 recognize an antigen in cartilagenous tissues and also recognize purified aggrecan. On immunoblots purified aggrecan digested with chondroitinase and keratanase has the same apparent molecular weight as the chondroitinase-treated brain Cat-301 CSPG. Preliminary studies indicate that neither the Cat-301 nor the Cat-304 epitope is located in the G1 domain (the hyaluronic acid binding region) of aggrecan. The lack of modification of the brain derived Cat-301 CSPG by keratan sulphate indicates that it is not identical to aggrecan.

PURIFICATION

The Cat-301 proteoglycan has been purified[9,11] from guanidine HCl extracts of cat cortex using CsCl gradients under associative conditions for CSPGs and hyaluronic acid[12] followed by immunoaffinity chromatography using the Cat-301 antibody immobilized on Affigel-HZ. Alternatively, the Cat-301 proteoglycan can be purified using standard purification procedures for the large chondroitin sulphate proteoglycan from cartilage, aggrecan[13].

ACTIVITIES

The biological role of the Cat-301 proteoglycan is not known, but the temporal features of its expression[4-8] and its localization[1,9] suggest a role in the stabilization of mature synaptic structure in the mammalian CNS.

ANTIBODIES

Monoclonal antibody Cat-301 recognizes the brain CSPG in many mammalian species, including cat, hamster, mouse, rat, monkey and human[1,2,5,6]. Monoclonal antibody

Cat-304 recognizes a different epitope on the Cat-301 CSPG[7,9] expressed in cat and cow. Both antibodies appear to be directed to peptide, rather than carbohydrate, epitopes. Both also recognize aggrecan purified from cat and cow cartilage[11].

GENES

None identified at present.

REFERENCES

1. Hockfield, S. and McKay, R. (1983) Proc. Natl. Acad. Sci. (USA) 80, 5758-5761.
2. DeYoe, E.A., Hockfield, S., Garren, H. and Van Essen, D. (1990) Visual Neuroscience 5, 67-81.
3. Sahin, M. and Hockfield, S. (1990) J. Comp. Neurol. 301, 575-584.
4. Sur, M., Frost, D. and Hockfield, S. (1988) J. Neurosci. 8, 874-882.
5. Hockfield, S., Zaremba, S., Kalb, R.G. and Fryer, H. (1990) Cold Spring Harbor Symp. Quant. Biol. 55, 505-514.
6. Kalb, R. and Hockfield, S. (1988) J. Neuroscience 8, 2350-2360.
7. Guimarães, A., Zaremba, S. and Hockfield, S. (1990) J. Neuroscience 10, 3014-3024.
8. Kalb, R.G. and Hockfield, S. (1990) Science 250, 294-296.
9. Zaremba, S., Guimarães, A., Kalb, R.G. and Hockfield, S. (1989) Neuron 2, 1207-1219.
10. Hockfield, S. (1987) Science 237, 67-70.
11. Fryer, H.J.L., Molinaro, L. and Hockfield, S. (1990) Soc. Neurosci. 16, 496.
12. Heinegård, D. and Paulsson, M. (1984) In: Extracellular Matrix Biochemistry, ed. K.A. Piez and A.H. Reddi. Elsevier, New York, pp. 272-328.
13. Hascall, V.C. and Kimura, J.H. (1982) Methods Enzymol. 82, 769-800.

Susan Hockfield:
Section of Neurobiology,
Yale University School of Medicine,
New Haven, CT, USA

CD44

CD44 is an abundantly expressed cell surface glycoprotein. This molecule has been independently described in the literature as a T cell, granulocytic, and brain antigen[1], as a major cell surface protein of 3T3 cells[1] (Pgp-1), as a marker of mouse lymphoid tissue expressed early in thymic development[1], as a receptor for a component of the extracellular matrix (ECMR III)[1], as a surface product regulated by the (In(Lu)) gene[1], as a lymphocyte homing receptor (Hermes)[1], and most recently as the major hyaluronic acid receptor of mammalian cells[2]. CD44 associates with the Triton X-100 insoluble cytoskeleton and may associate with ankyrin[3-5]. CD44 is unusual amongst cell surface proteins in having an unusually low diffusion coefficient[6].

The single chain CD44 molecule includes a 250 amino acid extracellular domain with sequence homology to cartilage **link proteins**, a transmembrane domain and a 72 amino acid cytoplasmic domain. The region of the link protein similarity is well conserved between species (80-90%), while the rest of the external domain shows significant sequence divergence between any pair of species compared (approximately 35% similarity). Some cell lines express alternately spliced forms of CD44 with extra exons of 132 or 162 amino acids located in the external domain of the molecule[7,8]. The murine CD44 transcript is unusual and does not purify with the poly-A containing mRNA from several sources[9].

The most common form of CD44 occurs as an abundant 80-95 kDa glycoprotein. Other forms of CD44 have larger apparent molecular masses ranging to over 250 kDa[10,11]. The 37 kDa core polypeptide chain is extensively decorated with N-linked carbohydrate, chondroitin sulphate and heparin sulphate glycosaminoglycans and O-linked mucin and polylactosamine groups[10]. CD44 is expressed on the surfaces of most mammalian cell types examined, including most rodent and primate hematopoietic cell types, fibroblastoid, neural, and muscle cells. With approximately 10^6 copies on the surface of a mouse 3T3 cell[1], CD44 is one of the major protein species on the cell surface (Figure 1). The major forms on lymphoid cells, 90 kDa and 180-200 kDa, differ primarily due to the presence of chondroitin sulphate on the larger form. CD44 heterogeneity has been characterized at the level of posttranslational modification in macaque lymphocytes, from which five to ten 90 kDa isoforms ranging in pI from 4 to 7 have been identified. These molecules differ in their degree of sialylation and phosphorylation. Human PBLs show a more restricted CD44 charge heterogeneity with a predominant pI of approximately 4.2[12].

PURIFICATION

CD44 has been purified from detergent extracts of mammalian cell lines by affinity chromatography on type VI **collagen**[13], by its affinity for wheat germ agglutinin and by affinity chromatography with monoclonal antibodies[14]. The abundance of this molecule makes purification relatively simple. As the isolated molecule has little relevant biological activity, studies with CD44 are usually achieved by the study of transfected cells.

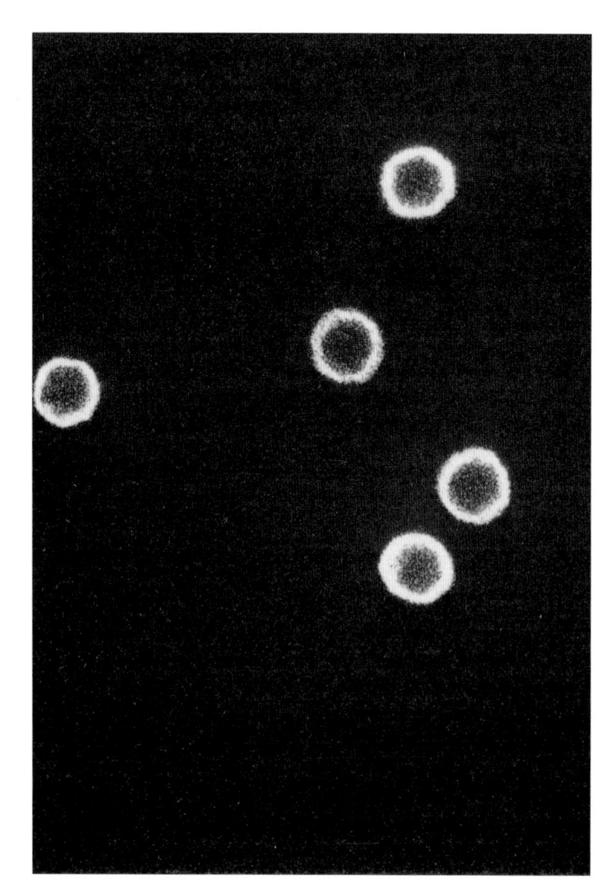

Figure 1. Direct immunofluorescent visualization of peripheral blood lymphocyte CD44 using FITC-conjugated monoclonal antibody shows uniform surface staining.

Figure 2. Capping of human PBL surface CD44 is visualized by indirect immunofluorescence using biotin-conjugated CD44-specific monoclonal antibody and Texas Red-avidin. CD44 cap formation is energy-dependent and cytoskeletally driven in lymphocytes.

■ ACTIVITIES

CD44 has been described as a receptor for a component of the ECM, probably type I or type VI collagen[13]. The external domain has been directly identified as a receptor for hyaluronic acid[2]. In some cell types, CD44 is concentrated in filapodia but is absent from sites of focal adhesion[10]. It also may concentrate at cell contact junctions. In mouse fibroblasts transfected with a CD44 gene, self adhesion between the transfected cells is strongly up-regulated[15].

Inhibition of the adhesion of human lymphocytes to mucosal postcapillary endothelium in an *in vitro* assay by the Hermes-3 CD44 monoclonal antibody implicates CD44 as a receptor used by circulating leukocytes for the postcapillary endothelium of mucosal lymphoid tissue[1] (Figure 2). CD44 may participate in cellular adhesive processes through recognition of hyaluronic acid and/or other ECM components including collagens. The role of CD44 in adhesion to postcapillary high endothelium is generally thought not to be a primary event dictating organ specific binding (Figure 3).

CD44 also functions as an accessory molecule in T lymphocyte activation[16-18]. CD44 monoclonals augment CD2- or CD3-mediated stimulation of T cell proliferation and IL-2 release. In addition, exposure of monocytic cells to CD44 antibodies can result in the release of cytokines, including IL-1 and TNF[19]. Macaque peripheral blood lymphocytes expressing high levels of CD44, whether cycling or not, are primary target for infection by SIV[20].

■ ANTIBODIES

Numerous CD44 monoclonal antibodies exist that recognize the human, nonhuman primate, hamster, and mouse proteins[1]. Two mouse allotypes exist, distinguishable by monoclonal antibodies.

■ GENES

Gene sequences encoding CD44 have been isolated from baboon[14] (GenBank accession M22452), human[21,22] (M24915 and M25078), mouse[9,23,24] (M27129, M27130, M30655, J05163) hamster[2] (M33827), bovine[25] and canine[26] cells.

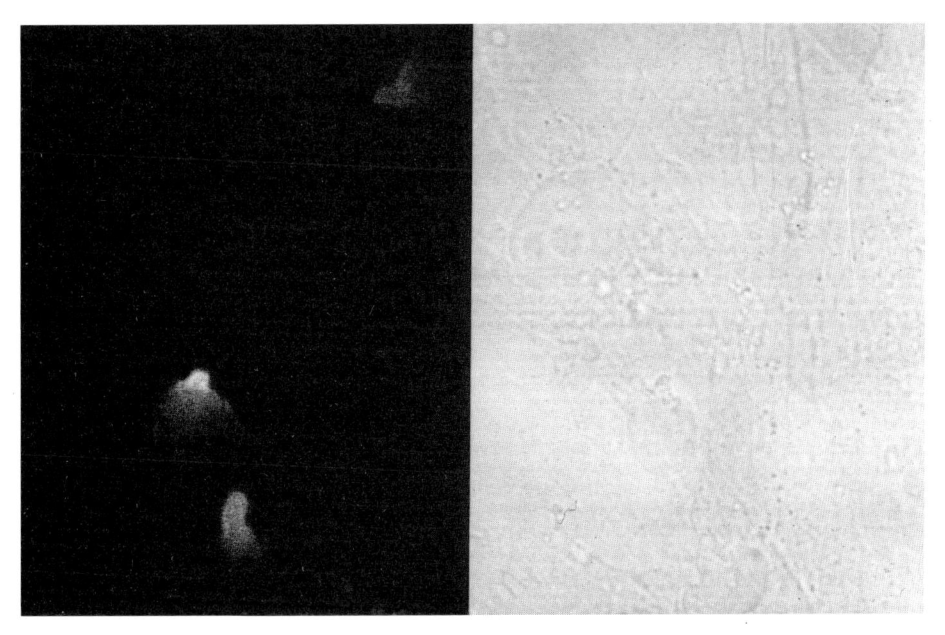

Figure 3. Redistribution of lymphocyte surface CD44 into nozzle-like processes accompanies cell shape change during interaction with endothelial cells (EC). PBL CD44 was directly labelled with FITC-mAb as in Figure 1 prior to co-culture with cultured EC monolayer. Fluorescent image at left shows CD44 concentrated in nonuniform distribution. Transmitted light image at right reveals EC and distorted PBL following 2 hours co-culture.

REFERENCES

1. Haynes, B.F., Telen, M.J., Hale, L.P. and Denning, S.M. (1989) Immunology Today 10, 423-428.
2. Aruffo, A., Stamenkovic, I., Melnick, M., Underhill, C.B. and Seed, B. (1990) Cell 61, 1303-1313.
3. Tarone, G., Ferracini, R., Galetto, G. and Comoglio, P. (1984) J. Cell Biol. 99, 512-519.
4. Carter, W.G. and Wayner, E.A. (1988) J. Biol. Chem. 263, 4193-4201.
5. Kalomiris, E.L. and Bourguigon, L.Y.W. (1988) J. Cell Biol. 106, 319-327.
6. Jacobson, K., O'Dell, D. and August, J.T. (1984) J. Cell Biol. 99, 1624-1633.
7. Dougherty, G.J., Lansdorp, P.M., Cooper, D.L. and Humphries, R.K. (1991) J. Exp. Med. 174, 1-5.
8. Günthert, U., Hofmann, M., Rudy, W., Reber, S., Zöller, M., Haußmann, I. Matzku, S., Wenzel, A., Ponta, H. and Herrlich, P. (1991) Cell 65, 13-24.
9. Nottenburg, C., Rees, G. and St. John, T. (1989) Proc. Natl. Acad. Sci. (USA) 86, 8521-8525.
10. Brown, T.A., Bouchard, T., St. John, T., Wayner, E. and Carter, W.G. (1991) J. Cell Biol. 113, 207-221.
11. Omary, M.B., Trowbridge, I.S., Letarte, M., Kagnoff, M.F. and Isacke, C.M. (1988) Immunogenetics 27, 460-464.
12. Rosenman, S. (1990) personal observations.
13. Wayner, E.A. and Carter, W.G. (1987) J. Cell Biol. 105, 1873-1884.
14. Idzerda, R.L., Carter, W.G., Nottenburg, C., Wayner, E.A., Gallatin, W.M. and St. John, T. (1989) Proc. Natl. Acad. Sci. (USA) 86, 4659-4663.
15. St. John, T., Meyer, J., Idzerda, R.L. and Gallatin, W.M. (1990) Cell 60, 45-52.
16. Denning, S.M., Le, P.T., Singer, K.H. and Haynes, B.F. (1990) J. Immunol. 144, 7-15.
17. Huet, S., Groux, H., Caillou, B., Valentin, H., Prieur, A. and Bernard, A. (1989) J. Immunol. 143, 798-801.
18. Shimizu, Y., Van Seventer, G.A., Siraganian, R., Wahl, L. and Shaw, S. (1989) J. Immunol. 143, 2457-2463.
19. Webb, D.S.A., Shimizu, Y., Van Seventer, G.A., Shaw, S. and Gerrard, T.L. (1990) Science 249, 1295-1297.
20. Willerford, D.M., Gale Jr., M.J., Benveniste, R.E., Clark, E.A. and Gallatin, W.M. (1990) J. Immunol. 144, 3779-3783.
21. Goldstein, L.A., Zhou, D.F., Picker, L.J., Minty, C.N., Bargatze, R.F., Ding, J.F. and Butcher, E.C. (1989) Cell 56, 1063-1072.
22. Stamenkovic, I., Amiot, M., Pesando, J.M. and Seed, B. (1989) Cell 56, 1057-1062.
23. Wolffe, E.J., Gause, W.C., Pelfrey, C.M., Holland, S.M., Steinberg, A.D. and August, J.T. (1990) J. Biol. Chem. 265, 341-347.
24. Zhou, D.F., Ding, J.F., Picker, L.J., Bargatze, R.F., Butcher, E.C. and Goeddel, D.V. (1989) J. Immunol. 143, 3390-3395.
25. Bosworth, B.T., St. John, T., Galatin, W.M. and Harp, J.A. (1991) Mol. Immunol 28, 1131-1135.
26. Sandmeier, B. (1990) personal observations.

Steve Rosenman and Tom St. John:
Programs in Cell Adhesion and Structural Cell Biology,
ICOS Corp., 22021 - 20th Ave. SE,
Bothell, WA 98021, USA

Cholinesterases

The Cholinesterases are a homologous family of enzymes whose only well established physiologic function is the efficient hydrolysis of acetylcholine. Acetylcholinesterase (AChE-E.C. 3.1.1.7) may be distinguished from butyrylcholinesterase (BuChE-E.C. 3.1.1.8) by specificity for acetylcholine over butyrylcholine hydrolysis. AChE is typically synthesized in nerve, muscle and certain hematopoietic cells. In excitable tissues, its synthesis is regulated by tissue specific development and it is localized in synapses at the extracellular face of nerve and muscle. BuChE is synthesized in liver with significant quantities appearing in plasma.

The cholinesterases exist in multiple molecular forms which may be distinguished by their subunit associations and hydrodynamic properties (Figure 1)[1,2]. The catalytic subunits associate with a lipid linked or a **collagen** like structural subunit to form distinct heteromeric species. The collagen containing species consist of tetramers of catalytic subunits, each of which is disulphide linked to a single subunit strand of a triple helical, collagen like unit[1-6]. The asymmetry imparted by the filamentous collagen unit has led to the designation of asymmetric or A forms with the subscript specifying the number of attached catalytic subunits (Figure 2). The collagen containing subunit has noncollagenous sequences at its N- and C-termini[7] and associates with acidic basal laminar components within the synapse[8,9]. The lipid linked subunit is approximately 20 kDa in mass[10], contains covalently attached fatty acids and tethers the enzyme at the outside surface of the cell. Only the hydrophilic catalytic subunits appear to be involved in these heteromeric associations.

The homomeric forms typically exist as dimers and tetramers; occasionally monomeric species are also found. Their hydrodynamic properties have led to the classification of globular or G forms and they may be subdivided into hydrophilic or amphiphilic (hydrophobic) G forms. The amphipathic character of the latter arises from cleavage of a 28 amino acid peptide from the C-terminus and attachment of a glycophospholipid to the newly formed C-terminal amino acid. This posttranslational modification also localizes the enzyme at the outer surface of the cell membrane[11,12]. To date only hydrophilic (globular) and asymmetric forms of BuChE have been identified[1]. The glycophospholipid linked forms of AChE are found in excitable tissues of lower vertebrates, whereas in mammals they appear largely restricted to the hematopoietic system.

■ PURIFICATION

The cholinesterases are routinely purified by affinity chromatography using a conjugated ligand that is inhibitory to the enzyme. Differential extraction in buffer, nonionic detergent and high ionic strength buffer yields the hydrophilic, amphiphilic asymmetric forms of the enzyme, respectively[1,13].

■ ACTIVITIES

Catalytic activity is typically measured by acetylthiocholine hydrolysis detecting liberated thiocholine[1,13]. Alternatively, pH stat or radioenzymatic assays are suitable for measuring acetylcholine hydrolysis. The turnover number of AChE (8.2×10^5 min^{-1}) is the fastest of the serine hydrolases. Rates for BuChE are typically 10-25% of this value using butyrylthiocholine as a substrate. BuChE shows a wider substrate specificity than AChE and may have a role in the detoxification of various plant esters that are ingested. The role of cholinesterases in peptide hydrolysis remains controversial.

■ ANTIBODIES

Both polyclonal and monoclonal antibodies have been prepared to intact cholinesterases and their structural subunits from electric fish and mammalian sources[14,15]. Some unusual carbohydrate epitopes have been detected[16]. Antibodies selective for a particular AChE species have been used to localize different forms of the cholinesterases in intact tissue[17].

■ GENES

Although distinct genes encode AChE and BuChE giving gene products with ~50% overall amino acid identity[18-22], all of the diversity within AChE and BuChE peptide chains is encoded in a single gene. The entire open reading frame is encoded by three exons in Torpedo AChE and mammalian BuChE, while four exons are required in mammalian AChE[23,24]. Alternative splicing of the last exon in the open reading frame gives rise to the distinct C-terminal sequences found in the hydrophilic (asymmetric) and amphiphilic forms of AchE[24,25]. To date, cDNA's for *Torpedo*[18], *Drosophila*[19], mouse[22] and human[26] AChE as well as human[20,21], mouse[22] and rabbit[27] BuChE have been reported. The cholinesterases show no global homology with the serine hydrolases of the trypsin or subtilisin families but are homologous with several carboxyesterases, glutactin, neurotactin, and the C-terminal region of thyroglobulin[2]. All of these proteins have a common disulphide bonding pattern[2].

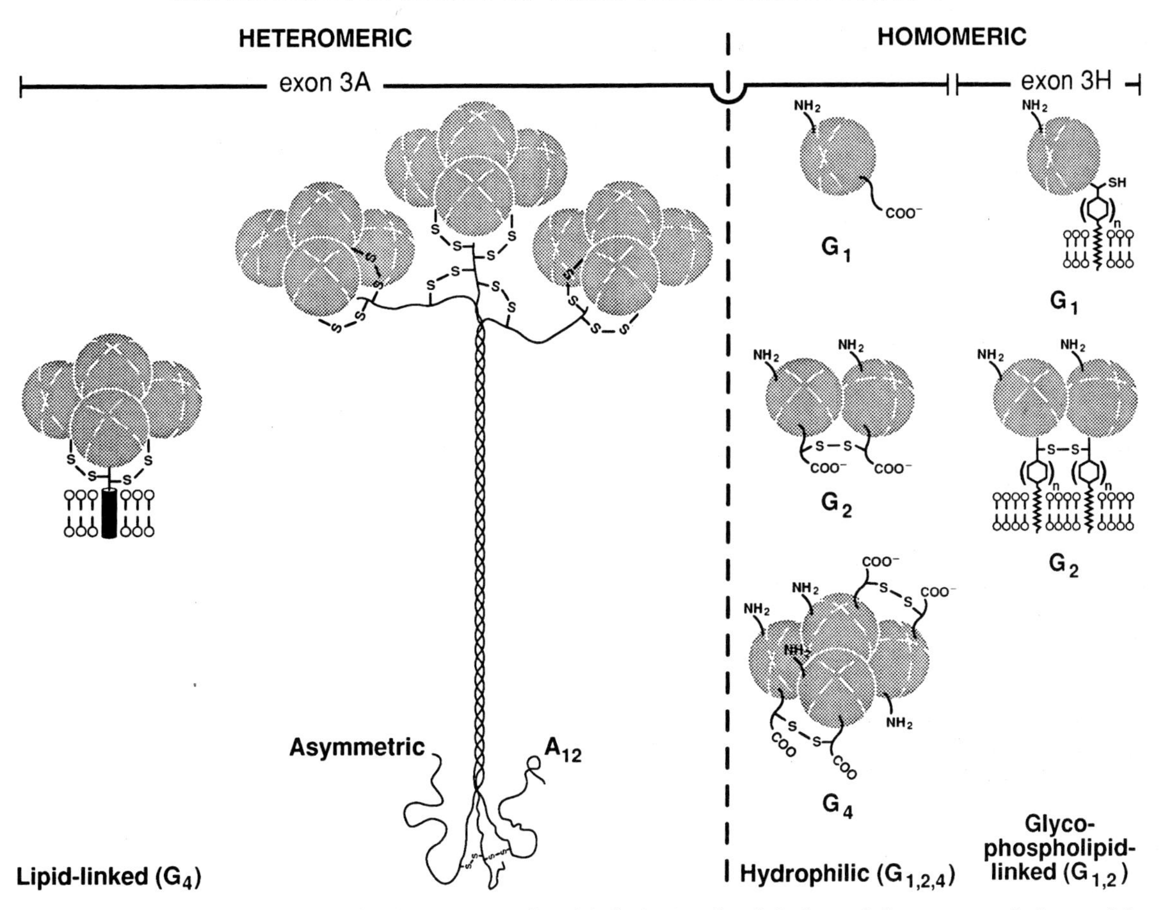

Figure 1. Molecular Species of the Cholinesterases. G and A designate the globular and the asymmetric forms of the enzyme, respectively; the subscript denotes the number of subunits. The catalytic subunit of AChE is encoded by a single gene, the C-terminus of most of the molecular forms known is encoded by an alternatively spliced exon designated 3A; an exception is the glycophospholipid-linked form which is encoded by 3H.

■ REFERENCES

1. Massoulié, J. and Toutant, J.-P. (1988) Handbook Exp. Pharmacol. Springer-Verlag, Berlin, 86, 167-223.
2. Taylor, P. (1991) J. Biol. Chem. 266, 4025-4028.
3. Lwebuga-Mukasa, J., Lappi, S. and Taylor, P. (1975) Biochemistry 15, 1425-1435.
4. Cartaud, J., Bon, S. and Massoulié, J. (1976) Brain Res. 88, 127-130.
5. Anglister, L. and Silman, I. (1978) J. Mol. Biol. 125, 293-311.
6. Rosenberry, T.L. and Richardson, J.M. (1978) Biochemistry 16, 3550-3558.
7. Krejei, E., Coussen, F., Duval, N., Chatel, J.M., Lagay, C., Paype, M., Vandekerckhove, J., Cartaud, J., Bon, S. and Massoulié, J. (1991) EMBO J. 10, 1285-1293.
8. McMahan, U.J., Sanes, J.R. and Marshall, L.M. (1978) Nature 271, 172-174.
9. Brandan, E., Maldonado, M., Garrido, J. and Inestrosa, N.C. (1985) J. Cell Biol. 101, 985-992.
10. Inestrosa, N.C., Roberts, W.L., Marshall, T.L. and Rosenberry, T.L. (1987) J. Biol. Chem. 262, 4441-4444.
11. Silman, I. and Futerman, A.H. (1987) Eur. J. Biochem. 170, 11-22.
12. Roberts, W.L., Kim, B.H. and Rosenberry, T.L. (1987) Proc. Natl. Acad. Sci. (USA) 84, 7817-7821.
13. Rosenberry, T.L. (1975) Adv. Enzymol. 43, 103-218.
14. Brimijoin, S. (1986) Int. Rev. Neurobiol. 28, 363-410.
15. Doctor, B.P., Camp, S., Gentry, M.K., Taylor, S.S. and Taylor, P. (1983) Proc. Natl. Acad. Sci. (USA) 80, 5767-5771.
16. Bon, S., Meflah, K., Musset, F., Grassi, J. and Massoulié, J. (1987) J. Neurochem. 49, 1720-1731.
17. Abramson, S.N., Ellisman, M.N., Deerinck, T.J., Maulet, Y., Gentry, M.K., Doctor, B.P. and Taylor, P. (1989) J. Cell. Biol. 108, 2301-2311.
18. Schumacher, M., Camp, S., Maulet, Y., Newton, M., MacPhee-Quigley, K., Taylor, S.S., Friedman, T. and Taylor, P. (1986) Nature 319, 407-409.
19. Hall, L.M.C. and Spierer, P. (1986) EMBO J. 5, 2949-2954.
20. Lockridge, O., Bartels, C.F., Vaughan, T.A., Wong, C.K., Norton, S.E. and Johnson, L.L. (1987) J. Biol. Chem. 262, 549-557.
21. Prody, C.A., Zevin-Sonkin, D., Gnott, A., Goldberg, O. and Soreq, H. (1987) Proc. Natl. Acad. Sci. (USA) 54, 3555-3559.

Figure 2. Electron micrograph of the asymmetric form of acetylcholinesterase[4].

22. Rachinsky, T.L., Camp, S., Li, Y., Ekström, T.J., Newton, M. and Taylor, P. (1990) Neuron 5, 317-327.
23. Arpagaus, M., Knott, M., Vatsu, K.P., Bartels, C.F., LaDu, B.N. and Lockridge, O. (1990) Biochemistry 29, 124-131.
24. Maulet, Y., Camp, S., Gibney, G., Rachinsky, T.L., Ekström, T.J. and Taylor, P. (1990) Neuron 4, 289-301.
25. Sikorav, J.L., Duval, N., Ansetmet, A., Bon, S., Krejei, E., Legay, C., Osterlund, M., Reimund, B. and Massoulié, J. (1988) EMBO J. 7, 2983-2993.
26. Soreq, H., Ben-Aziz, R., Prody, C.A., Seidman, S., Gnatt, A., Neville, L., Lieman-Hurwitz, J., Lev-Lehman, E., Ginzberg, D., Lapidot-Lifson, Y. and Zakut, H. (1990) Proc. Natl. Acad. Sci. (USA) 87, 9688-9692.
27. Jbilo, O. and Chatonnet, A. (1990) Nucleic Acids Res. 18, 3990.

■ *Palmer Taylor:*
Department of Pharmacology 0636,
University of California,
San Diego, La Jolla, CA 92093, USA

Collagens

The collagens constitute a superfamily of extracellular matrix proteins with a structural role as their primary function. Based on the exon structure of their genes as well as the configuration of the sequence domains of the proteins they can be divided into several families or groups. Within each family, several homologous genes encode polypeptides that have domains with similar sequences. All collagenous proteins, however, have domains with a triple-helical conformation. Such domains are formed by three subunits (α-chains), each containing a $(Gly\text{-}X\text{-}Y)_n$ repetitive sequence motif.

A triple-helical molecular conformation[1], once thought to be unique to collagens, is now known to be present in a variety of proteins that are not classified among the collagens (Figure). For example, the complement component C1q[2], the asymmetric form of acetylcholinesterase[3], surfactant apoproteins[4,5], conglutinin[6], **mannose binding proteins** in serum[7,8], and the macrophage scavenger receptor[9] all contain triple-helical domains. It would not

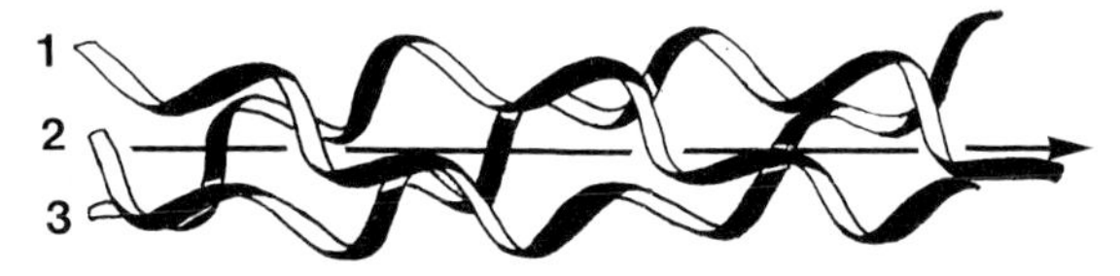

Figure 1. Molecular structure of the triple-helical conformation; three left-handed helices form a right-handed superhelix.

Table: CHROMOSOMAL LOCATION OF COLLAGEN GENES

Gene Locus	Chain Designation	Chromosomal Location	References
Fibrillar collagens			
COL1A	1α1(I)	17q21.3-q22	14
COL1A	2α2(I)	7q21.3-q22	15
COL2A	1α1(II)	12q13-q14	16
COL3A	1α1(III)	2q24.3-q3117,	18
COL5A	1α1(V)	9q34.3	19
COL5A	2α2(V)	2q24.3-q31	17
COL5A	3α3(V)	—	
COL11A1	α1(XI)	1p21	20
COL11A2	α2(XI)	6p212	21
Facit collagens			
COL9A1	α1(IX)	6q12-q14	22
COL9A2	α2(IX)	—	
COL9A3	α3(IX)	—	
COL12A1	α1(XII)	—	
COL14A1	α1(XIV)	—	
Short chain collagens			
COL8A1	α1(VIII)	3q11.1-q13.2	23
COL8A2	α2(VIII)	1p32.3-p34.3	24
COL10A1	α1(X)	6q21-q22	25
Basement membrane collagens			
COL4A1	α1(IV)	13q33-q34	18
COL4A2	α2(IV)	13q33-q34	26-28
COL4A3	α3(IV)	—	
COL4A4	α4(IV)	2q35-2q37.1	29
COL4A5	α5(IV)	Xq22	30
Other collagens			
COL6A1	α1(VI)	21q223	31
COL6A2	α2(VI)	21q223	31
COL6A3	α3(VI)	2q37	31
COL7A1	α1(VII)	3p21.1-p21.3	32
COL13A1	α1(XIII)	10q11-qter	33

be surprising to find similar domains in several additional proteins, where rigid, rod-like molecular structures are required.

What then are the characteristic features of collagens? It has been suggested that for a protein to be classified as a collagen, it must satisfy several criteria[10]:

(a) The protein in the native state must contain at least one triple-helical domain. In **fibrillar collagens** (type I, II, III, V, and XI) each collagen molecule (after complete removal of the propeptides) contains only one such domain which accounts for almost the entire length of the molecule. In other collagens, such as **basement membrane collagens** or **FACIT collagens**, several shorter triple-helical domains are separated by nontriple-helical domains. A protein that contains an amino acid sequence that is sensitive to bacterial collagenase (an enzyme frequently used as a diagnostic tool to identify collagenous sequences), but is not part of a triple-helix, is therefore not a collagen.

(b) The protein must form supramolecular aggregates (fibrils, filaments, or networks), either alone or in conjunction with other extracellular matrix components. This requirement is not always met before a new collagen is given a type designation. In cases where homology with an existing type can be demonstrated, this would seem justifiable since similar molecules are likely to have similar polymerization properties.

(c) The protein contributes to the structural integrity of the extracellular matrix. A protein that contains a triple-helical domain is not a collagen unless its final destination after synthesis is the extracellular compartment or the cell surface. Also, the function of the protein must contribute to the structure of the matrix. Thus, although the macrophage scavenger receptor has an extracellular triple-helical domain and may form an oligomer at the cell surface, it is not classified as a collagen because its function is unrelated to the structure of the extracellular matrix in which the macrophage resides.

The structural roles of collagen are quite obvious for type I collagen in tendons, ligaments, skin, and bone, for type II in hyaline cartilage, and type IV collagens in basement membranes. The structural roles of many **other collagen** types are less clear, and they appear to contribute little to the macroscopic, mechanical properties of various tissues. However, the formation of tissues and organs during embryonic development requires the assembly of matrices of precise three-dimensional structure, and many of the collagens that have been described as minor collagens (in a quantitative sense) may play important roles in this assembly process[11].

Based on a detailed analysis of the exon structures of genes that encode proteins satisfying the above criteria, the collagen superfamily can be divided into several families, as follows[12,13]:

(a) Fibrillar collagens: These include types I, II, III, V, and XI collagen, with molecules forming banded fibrils in various tissues.

(b) Facit collagens: These include types IX, XII, and XIV collagen, with molecules that are associated with fibrils formed by fibrillar collagens.

(c) Short chain collagens: These include types VIII and X collagen, with "short" molecules that form part of unique extracellular matrices.

(d) Basement membrane collagens: These include several different molecules collectively known as type IV collagens. They form the major collagenous component of basement membranes.

(e) Other collagens: This includes several types (VI, VII, XIII) that form specialized structures in a variety of tissues (e.g. microfibrils for type VI, anchoring fibrils for type VII).

Most collagen genes have now been cloned and their chromosomal locations in the human genome determined. The very large number of sequence entries into GenBank/EMBL Data Bank makes it not practical to list the appropriate accession numbers in the pages that follow. The reader is instead referred to the original articles for specific information. The chromosomal locations of the genes are tabulated below.

REFERENCES

1. Rich, A. and Crick, F.H.C. (1961) J. Mol. Biol. 3, 483-506.
2. Reid, K.B.M. (1979) Biochem. J. 179, 367-371.
3. Rosenberry, T.L. and Richardson, J.M. (1977) Biochemistry 16, 3550-3558.
4. Benson, B., Hawgood, S., Schilling, J., Clements, J., Damm, D., Cordell, B. and White, R.T. (1985) Proc. Natl. Acad. Sci. (USA) 82, 6379-6383.
5. Persson, A., Rust, K., Chang, D., Moxley, M., Longmore, W. and Crouch, E. (1989) Biochemistry 28, 6361-6367.
6. Strang, C.J., Slayter, H.S., Lachmann, P.J. and Davis, A.E. (1986) Biochem. J. 234, 381-389.
7. Drickamer, K., Dordal, M.S. and Reynolds, L. (1986) J. Biol. Chem. 261, 6878-6887.
8. Ezekowitz, R.A.B. and Stahl, P.D. (1988) J. Cell Sci. Suppl. 9, 121-133.
9. Kodama, T., Freeman, M., Rohrer, L., Zabrecky, J., Matsudaira, P. and Krieger, M. (1990) Nature 343, 531-535.
10. Ninomiya, Y., Showalter, A.M. and Olsen, B.R. (1984) In The Role of Extracellular Matrix in Development (R.L. Trelstad, ed.) Alan R. Liss, Inc. pp. 255-275.
11. Gordon, M.K. and Olsen, B.R. (1990) Curr. Opinion Cell Biol. 2, 833-838.
12. Vuorio, E. and deCrombrugghe, B. (1990) Annu. Rev. Biochem. 59, 837-872.
13. Jacenko, O., Olsen, B.R. and LuValle, P. (1991) In: Crit. Rev. Eukar. Gene Expression (G.S. Stein, J.L. Stein and J.B. Lian, eds.) CRC Press, 327-353.
14. Huerre, C., Junien, C., Weil, D., Chu, M.-L., Morabito, M., Van Cong, N., Myers, J.C., Foubert, C., Gross, M.-S., Prockop, D.J., Bouè, A., Kaplan, J.-C., de la Chapelle, A. and Ramirez, F. (1982) Proc. Natl. Acad. Sci. (USA) 79, 6627-6630.
15. Junien, C., Weil, D., Myers, J.C., Nguyen, V.C., Chu, M.L., Foubert, C., Gross, M.S., Prockop, D.J., Kaplan, J.C. and Ramirez, F. (1982) Am. J. Hum. Genet. 34, 381-387.
16. Strom, C.M., Eddy, R.L. and Shows, T.B. (1984) Somatic Cell. Mol. Genet. 10, 651-655.
17. Emanuel, B.S., Canizzaro, L.A., Seyer, J.M. and Myers, J.C. (1985) Proc. Natl. Acad. Sci. (USA) 82, 3385-3389.
18. Solomon, E., Hiorns, L.R., Spurr, N., Kurkinen, M., Barlow, D., Hogan, B.L. and Dalgleish, R. (1985) Proc. Natl. Acad. Sci. (USA) 82, 3330-3334.
19. Inazawa, J., Kimura, T., Tanaka, K., Ariyama, T., Nakagawa, H., Misawa, S. and Abe, T. (1992) Genomics, submitted.
20. Henry, I., Bernheim, A., Bernard, M., van der Rest, M., Kimura, T., Jeanpierre, C., Barichard, F., Berger, R., Olsen, B.R., Ramirez, F. and Junien, C. (1988) Genomics 3, 87-90.
21. Kimura, T., Cheah, K.S.E., Chan, D.H., Lui, V.C.H., Mattei, M.-G., van der Rest, M., Ono, K., Solomon, E., Ninomiya, Y. and Olsen, B.R. (1989) J. Biol. Chem. 264, 13910-13916.
22. Kimura, T., Mattei, M.-G., Stevens, J.W., Goldring, M.B., Ninomiya, Y. and Olsen, B.R. (1989) Eur. J. Biochem. 179, 71-78.
23. Muragaki, Y., Mattei, M.-G., Yamaguchi, N., Olsen, B.R. and Ninomiya, Y. (1991) Eur J. Biochem. 197, 615-622.
24. Muragaki, Y., Jacenko, O., Apte, S., Mattei, M.-G., Ninomiya, Y. and Olsen, B.R. (1991) J. Biol. Chem. 266, 7721-7727.
25. Apte, S., Mattei, M.-G. and Olsen, B.R. (1991) FEBS Lett. 282, 393-396.
26. Griffin, A., Emanuel, B.S, Hansen, J.R., Cavenee, W.K. and Myers, J.C. (1987) Proc. Natl. Acad. Sci. (USA) 84, 512-516.
27. Killen, P.D., Francomano, C.A., Yamada, Y., Modi, W.S. and O'Brien, S.J. (1987) Hum. Genet. 77, 318-324.
28. Solomon, E., Hall, V. and Kurkinen, M. (1987) Ann. Hum. Genet. 51, 125-127.
29. Ninomiya, Y. et al. in preparation.
30. Hostikka, S.L., Eddy, R.L., Byers, M.G., Höyhtyä, M., Shows, T.B. and Tryggvason, K. (1990) Proc. Natl. Acad. Sci. (USA) 87, 1606-1610.
31. Weil, D., Mattei, M.-G., Passague, E., NGuyen, V.C., Pribula-Conway, D., Mann, K., Deutzmann, R., Timpl, R. and Chu, M.-L. (1988) Am. J. Hum. Genet. 42, 435-445.
32. Parente, M.G., Chung, L.C, Woodley, D.T., Wynn, K.C., Bauer, E.A., Mattei, M.-G., Chu, M.-L. and Uitto, J. (1991) Proc. Natl. Acad. Sci. (USA) 88, 6931-6935.
33. Shows, T.B., Tikka, L., Byers, M.G., Eddy, R.L., Haley, L.L, Henry, W.M., Prockop,D.J. and Tryggvason, K. (1989) Genomics 5, 128-133.

- *Bjorn Reino Olsen:*
  Department of Anatomy and Cellular Biology, Harvard Medical School, Boston, MA, USA
- *Yoshifumi Ninomiya, Department of Molecular Biology and Biochemistry, Okayama University Medical School, Okayama, Japan*

Basement Membrane Collagens (Type IV)

Type IV collagen is the major collagenous component of basement membranes, forming a network structure with which other basement membrane components (laminin, nidogen, heparan sulfate proteoglycan) interact[1-3]. Currently, five distinct genes are recognized as belonging to the type IV collagen family[4,5]. The exon structure of these genes is different from that of genes belonging to other collagen families; however, the similarity in structure among the type IV genes indicates that they arose by duplications of a common precursor gene.

Type IV collagen molecules, heterotrimers of two α1(IV) and one α2(IV) chain, have long been recognized as the major collagenous component of basement membranes. Each of the two chains is about 1,700 amino acid residues long, containing at least three distinct domains: the N-terminal cysteine-rich (7 S) domain, a central triple-helical domain, and a C-terminal nontriple-helical domain (NC1)[1]. Type IV molecules assemble into a network which is quite different from the banded fibrils formed by **fibrillar collagen** types (Figure). Within the network, separate molecules are covalently crosslinked within laterally associated 7 S domains and associated by end-to-end interactions through their NC1 domains[1,2]. Lateral associations between the triple-helical domains also contribute to the network structure[1].

Recently, three additional type IV collagen chains have been identified and partially characterized[6-8]. These chains, α3(IV), α4(IV), and α5(IV), form additional type IV molecules in specialized basement membranes; the precise chain composition of these molecules and their tissue distributions are not yet completely understood. However, the α3(IV) chain appears to contain the determinants for autoimmune antibodies in Goodpasture's Syndrome[9]. The α5(IV) collagen gene, located on the X-chromosome, is mutated in several cases of X-linked Alport familial nephritis[10].

PURIFICATION

Fragments of type IV collagen can be extracted from basement membranes with pepsin (resulting in triple-helical fragments) or with bacterial collagenase (resulting in non-triple-helical domains)[1,11]. Intact type IV collagen composed of α1(IV) and α2(IV) chains can be isolated by acetic acid extraction of murine EHS-tumour tissue[1], and is commercially available (Collaborative Research, Inc.). Pepsinized material is also commercially available.

ANTIBODIES

A variety of antibodies are available. These include antibodies against the 7 S domain and the NC1 domain, as well as antibodies against pepsin fragments[1]. Both polyclonal and monoclonal antibodies against type IV collagen are commercially available from several sources, including Chemicon, Collaborative Research, Developmental Studies Hybridoma Bank, Southern Biotechnology Associates, Inc., Upstate Biotechnology, Inc., and DMI/Pasteur Institute in Lyon.

ACTIVITIES

Type IV collagen can interact with cells indirectly through laminin. Strong binding of type IV collagen to laminin is mediated by **nidogen**[12] (**entactin**[13]), a glycoprotein of about 150 kDa which binds tightly to **laminin**[14] and has binding sites also for type IV collagen and cells[3]. In addition, direct low affinity interaction between laminin and type IV collagen is possible[2,15]. Type IV collagen also binds to heparin and heparan sulphate proteoglycan[2,16-18], and it has been reported that heparin can inhibit type IV collagen polymerization[16]. Many cell types adhere to type IV collagen[1,19], and peptides from within type IV sequences can inhibit this adhesion[20]. A major cell binding site within the 390 nm-long $[\alpha1(IV)]_2\alpha2(IV)$ heterotrimers is localized about 100 nm from the N-terminus of the molecule, and this triple-helical binding site interacts specifically with α1b1 and α2b1 integrins on cells[21].

GENES

The complete primary structures of mouse and human α1(IV) and α2(IV) chains have been deduced from cDNA sequences and mouse and human α1(IV) and α2(IV) genomic clones have been extensively characterized[4,5,22,23]. In both species the two genes are arranged head-to-head, separated by a bidirectional promoter; the human genes are located on chromosome 13[24-27]. The human α5(IV) gene, located on the X-chromosome, has been extensively characterized[8,10]. The protein encoded by the α1(IV) collagen gene in *Drosophila* is quite similar to the vertebrate type IV collagen chains[28], but the gene has fewer exons and is smaller than the corresponding vertebrate genes[28]. In *C. elegans* the clb-1 and clb-2 genes are homologous to the vertebrate α1(IV) and α2(IV) collagen genes[29]. Interestingly, mutations in the α1(IV) gene in *C. elegans* result in temperature-sensitive lethality during late embryogenesis[30].

REFERENCES

1. Glanville, R.W. (1987) In: Structure and Function of Collagen Types (R. Mayne and R.E. Burgeson, eds.) Academic Press Inc. pp. 43-79.
2. Yurchenko, P.D. and Schittny, J.C. (1990) FASEB J. 4, 1577-1590.
3. Timpl, R. (1989) Eur. J. Biochem. 180, 487-502.

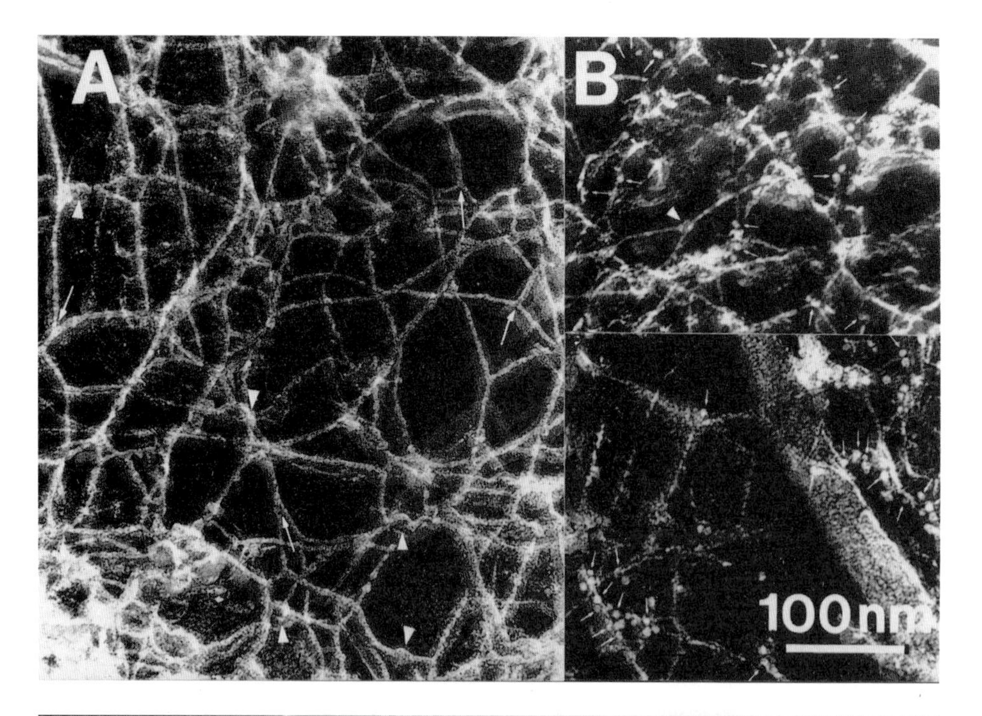

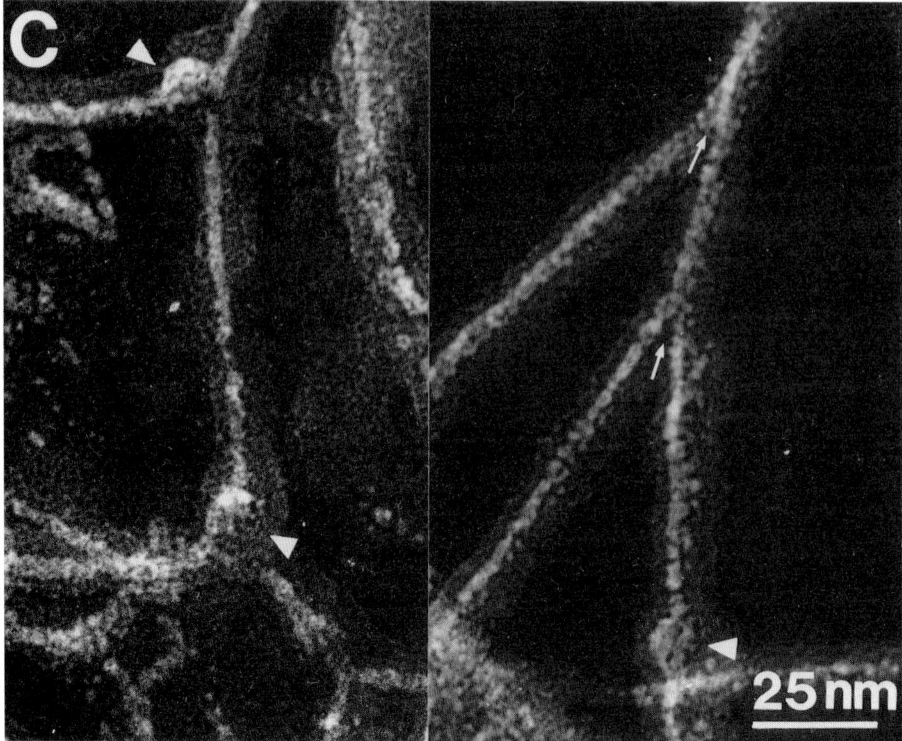

Figure. Amniotic basement membrane collagen network. (A) Panoramic view. (B) Network labelled with antitype IV collagen antibody directly coupled to 5-nm gold spheres. (C) Detail of network revealing 10-12 nm globular domains and branching filaments. (From Yurchenco (1990)[31]).

4.Vuorio, E. and deCrombrugghe, B. (1990) Annu. Rev. Biochem. 59, 837-872.
5.Jacenko, O., Olsen, B.R. and LuValle, P. (1991) In: Crit. Rev. Eukar. Gene Expression (G.S. Stein, J.L. Stein and J.B. Lian, eds.) CRC Press, pp. 327-353.
6.Butkowski, R.J., Langeveld, J.P.M., Wieslander, J., Hamilton, J. and Hudson, B.G. (1987) J. Biol. Chem. 262, 7874-7877.
7.Saus, J., Wieslander, J., Langeveld, J.P.M., Quinones, S. and Hudson, B.G. (1988) J. Biol. Chem. 263, 13374-13380.
8.Hostikka, S.L., Eddy, R.L., Byers, M.G., Höyhtyä, M., Shows, T.B. and Tryggvason, K. (1990) Proc. Natl. Acad. Sci. (USA) 87, 1606-1610.

9.Butkowski, R.J., Wieslander, J., Kleppel, M., Michael, A.F. and Fish, A.J. (1989) Kidney Int. 35, 1195-1202.
10.Barker, D., Hostikka, S.L., Chou, J., Chow, L.T., Oliphant, A.R., Gerken, S.C., Gregory, M.C., Skolnick, M.H., Atkin, C.L. and Tryggvason, K. (1990) Science 248, 1224-1227.
11.Hudson, B.G., Chung, A. and Burgeson, R. (1991) In: Extracellular Matrix Macromolecules - A Practical Approach (M.A. Haralson and J.R. Hassell, eds.) IRL Press, in press.
12.Timpl, R., Dziadek, M., Fujiwara, S., Nowack, H. and Wick, G. (1983) Eur. J. Biochem. 137, 455-465.
13.Carlin, B., Jaffe, R., Bender, B. and Chung, A.E. (1981) J. Biol. Chem. 256, 5209-5214.
14.Paulsson, M., Aumailley, M., Deutzmann, R., Timpl, R., Beck, K. and Engel, J. (1987) Eur. J. Biochem. 166, 11-19.
15.Charonis, A.S., Tsilibary, E.C., Yurchenko, P.D. and Furthmayr, H. (1985) J. Cell Biol. 100, 1848-1853.
16.Tsilibary, E.C., Koliakos, G.G., Charonis, A.S., Vogel, A.M., Reger, L.A. and Furcht, L.T. (1988) J. Biol. Chem. 263, 19112-19118.
17.Koliakos, G.G., Kouzi-Koliakos, K., Furcht, L.T., Reger, L.A. and Tsilibary, E.C. (1989) J. Biol. Chem. 264, 2313-2323.
18.Fujiwara, S., Wiedemann, H., Timpl, R., Lustig, A. and Engel, J. (1984) Eur. J. Biochem. 143, 145-157.
19.Aumailley, M. and Timpl, R. (1986) J. Cell Biol. 103, 1569-1575.
20.Tsilibary, E.C., Reger, L.A., Vogel, A.M., Koliakos, G.G., Anderson, S.S., Charonis, A.S., Alegre, J.N. and Furcht, L.T. (1990) J. Cell Biol. 111, 1583-1591.
21.Vandenberg, P., Kern, A., Reis, A., Luckenbill-Edds, L., Mann, K. and Kühn, K. (1991) J. Cell Biol. 113, 1475-1483.
22.Sandell, L.J. and Boyd, C.D. (1990) In: Extracellular Matrix Genes (L.J. Sandell and C.D. Boyd, eds.) Academic Press Inc. pp. 1-56.
23.Blumberg, B. and Kurkinen, M. (1990) In: Extracellular Matrix Genes (L.J. Sandell and C.D. Boyd, eds.) Academic Press Inc. pp. 115-135.
24.Kaytes, P., Wood, L., Thereault, N., Kurkinen, M. and Vögeli, G. (1988) J. Biol. Chem. 263, 19274-19277.
25.Burbelo, P.D., Martin, G. and Yamada, Y. (1988) Proc. Natl. Acad. Sci. (USA) 85, 9679-9682.
26.Pöschl, E., Pollner, R. and Kühn, K. (1988) EMBO J. 7, 2687-2695.
27.Soininen, R., Huotari, M., Hostikka, S.L., Prockop, D.J. and Tryggvason , K. (1988) J. Biol. Chem. 263, 17217-17220.
28.Blumberg, B. and Kurkinen, M. (1990) In: Extracellular Matrix Genes (L.J. Sandell and C.D. Boyd, eds.) Academic Press, Inc. pp. 115-135.
29.Guo, X. and Kramer, J.M. (1989) J. Biol. Chem. 264, 17474-17582.
30.Guo, X., Johnson, J.J. and Kramer, J.M. (1991) Nature 349, 707-709.
31.Yurchenko, P.D. (1990) Ann. N.Y. Acad. Sci. 580, 195-213.

■ *Bjorn Reino Olsen:*
Department of Anatomy and Cellular Biology, Harvard Medical School, Boston, MA, USA

■ *Yoshifumi Ninomiya, Department of Molecular Biology and Biochemistry, Okayama University Medical School, Okayama, Japan*

FACIT Collagens (Types IX, XII, XIV)

FACIT collagens[1] are a group of proteins that share a common collagen domain structure and may serve as molecular linking devices between fibrillar collagens and other extracellular matrix macromolecules[2]. Their structure is strikingly different from that of other collagens in that FACIT collagens contain two, three, or more relatively short triple-helical domains connected by nontriple-helical sequences. For one FACIT collagen, type IX, two different forms of transcripts are derived from different promoters and are expressed in a tissue specific manner during embryonic development[3].

The FACIT (fibril associated collagens with interrupted triple-helices) group includes at least three types of collagen, IX, XII, and XIV, composed of five distinct polypeptide chains. cDNA clones encoding additional chains are currently being characterized, and it is likely that the group will prove to include many more types. The domain structure of the first molecule of this group to be described, type IX collagen, was predicted by cloning and sequencing of a cDNA encoding the chicken α1(IX) chain[4]. Type IX molecules are heterotrimers consisting of α1(IX), α2(IX), and α3(IX) chains of 84, 160 and 68 kDa (as determined with collagen standards), respectively[5]. As shown in Figure 1, they contain three triple-helical (COL) domains interrupted by nontriple-helical (NC) regions. Most of the NC domains contain cysteinyl residues forming disulphide bridges between subunits. One of the subunits, α2(IX), serves as a proteoglycan core protein and contains a glycosaminoglycan side chain attached to a seryl residue in the NC3 domain[5].

Type IX collagen molecules, expressed in hyaline cartilage, are associated with the surface of collagen fibrils such that two of the triple-helical domains are located at or close to the fibril surface, while an N-terminal globular domain is located in the perifibrillar space at the tip of a triple-helical arm[6] (Figure 2). Immunofluorescence, *in situ* hybridization, and biochemical studies have shown that this FACIT collagen is also present in embryonic chick cornea and in the vitreous body[1]. Interestingly, different tissues contain different forms of type IX molecules. These forms are translation products of two distinct mRNAs generated by alternative transcription of the α1(IX) collagen gene[3,7]. In chondrocytes, the majority of the α1(IX) transcripts are synthesized from an upstream transcription start site leading to the formation of a mRNA that codes for a polypeptide with a 266 (for chicken; 286 for human and mouse) residue, N-terminal, globular domain. In embryonic chick cornea and probably also in the vitreous body, the majority of the transcripts are synthesized from a downstream, alternative start site, leading to the formation of a mRNA encoding an α1(IX) with an alternative

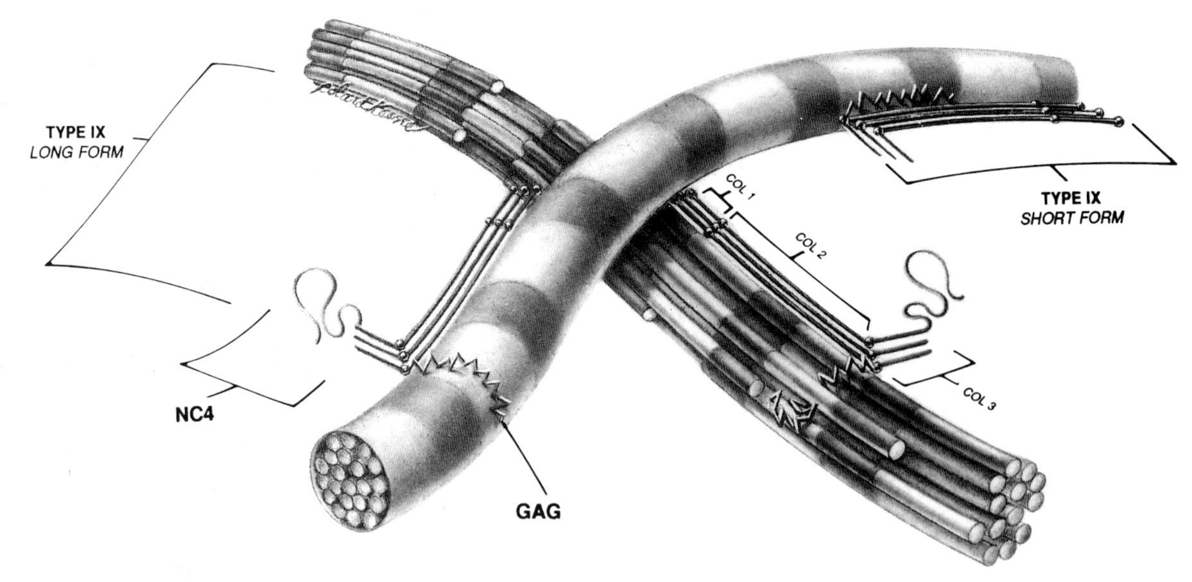

Figure 1. Diagram of type II containing fibrils with type IX collagen molecules on the surface. (From Jacenko et al. (1991)[34]).

signal peptide sequence and lacking the N-terminal globular domain[3,8,9].

Types XII and XIV collagen are homologous, but distinct, homotrimeric molecules of ~220 kDa subunits. Each subunit contains two triple-helical (COL) domains separated by a NC region of more than 40 amino acid residues, a relatively short (< 100 amino acid residues) nontriple-helical C-terminal region, and a very large (> 1500 amino acid residues) N-terminal nontriple-helical domain. Within the native molecules the COL domains of the three subunits form a triple-helical tail attached to a central globule from which three nontriple-helical arms or finger-like structures project. The central globule and the arms are composed of the N-terminal NC domains[10-15] (Figure 3).

PURIFICATION

Type IX collagen can be purified from the medium of chondrocyte cultures or from cartilage tissue extracts[16]. Triple-helical fragments of the molecule have been purified from pepsin extracts of cartilage[16]. Types XII and XIV collagens have been purified from neutral salt extracts of skin and tendons[11,14] and as triple-helical fragments by pepsin extraction[11,12].

ACTIVITIES

Type IX collagen is associated with type II containing fibrils[6]; the association is stabilized by covalent crosslinking[17,18]. This arrangement suggests that type IX may serve

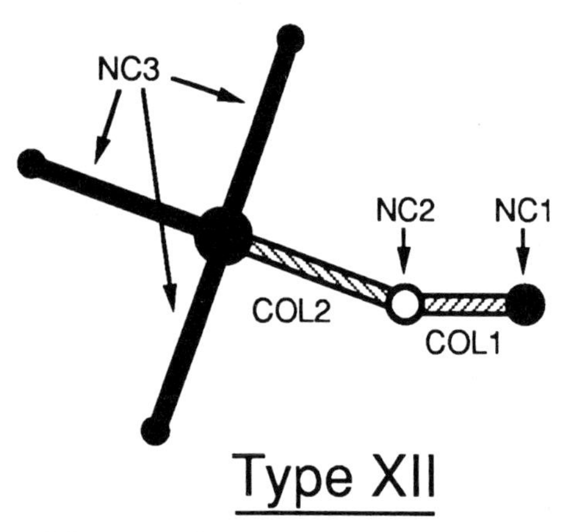

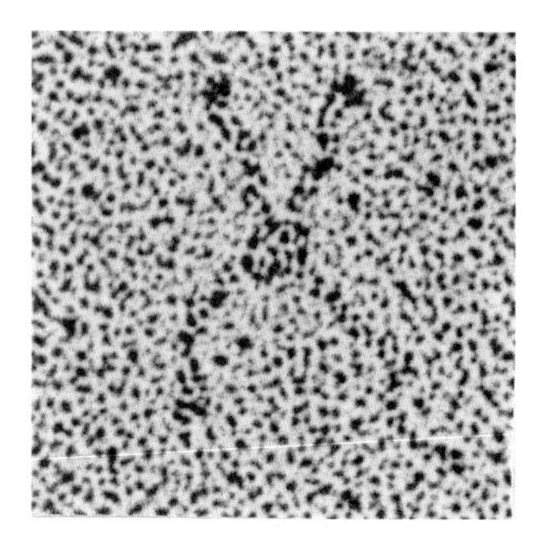

Figure 3. Diagram of the structure of a type XII collagen molecule (left) and rotary shadowing image (right). (Courtesy of Dr. M. van der Rest).

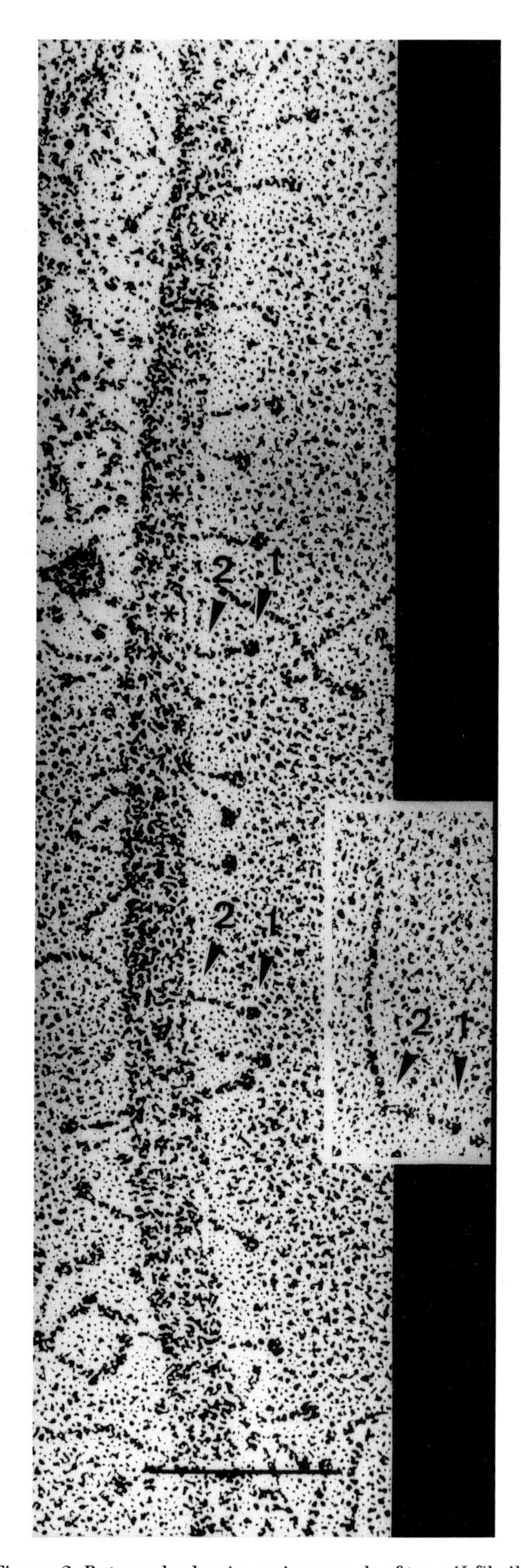

Figure 2. Rotary shadowing micrograph of type II fibril with type IX molecules on the surface. (From Vaughan et al. (1988)[6]).

as a molecular coupling device between type II fibrillar collagen and other extracellular matrix components. Based on the colocalization of types XII and XIV with type I collagen in tissues[15,19] and the partial sequence similarity between types XII, XIV, and IX collagen[10,13], it is thought that types XII and XIV collagen associate with type I collagen fibrils in a similar fashion as type IX with type II fibrils[1]. This model is supported by electron microscopical studies[15].

■ ANTIBODIES

Polyclonal antibodies against synthetic peptides deduced from nucleotide sequences[20] and polyclonal[21-25] as well as monoclonal[26] antibodies against protein fragments have been described for type IX collagen. The location of the epitopes for the monoclonal antibodies 2C2 and 4D6[26] were determined by rotary shadowing to be the C-terminus of the COL2 domain and the NC4 domain, respectively.

A monoclonal antibody against a synthetic peptide derived from cDNA sequences recognizes the chicken α1(XII) chain by Western blotting, and has been used for immunohistochemical studies[19]. Monoclonal antibodies against bovine type XII (TL-A) and type XIV (TL-B) collagen are also available[14,15].

■ GENES

Complete cDNA sequences are available for chicken α1(IX), α2(IX), and human α1(IX)[27,28]. Partial α1(IX) cDNA sequences from rat[29], bovine[27], and mouse[7,27] are also reported. The chicken α1(IX) and α2(IX) genes[27,30-32] and fragments of the mouse[7] and human[7,28] α1(IX) genes have been isolated and characterized. The human α1(IX) gene is located on chromosome 6[29]. Portions of the chicken α1(XII)[10,33] and α1(XIV)[13] cDNAs and genes have been isolated.

■ REFERENCES

1. Gordon, M.K. and Olsen, B.R. (1990) Curr. Opinion Cell Biol. 2, 833-838.
2. Olsen, B.R. (1989) Connect. Tiss. Res. 23, 115-121.
3. Nishimura, I., Muragaki, Y. and Olsen, B.R. (1989) J. Biol. Chem. 264, 20033-20041.
4. Ninomiya, Y. and Olsen, B.R. (1984) Proc. Natl. Acad. Sci. (USA) 81, 3014-3018.
5. van der Rest, M. and Mayne, R. (1987) In: Structure and Function of Collagen Types (R. Mayne and R.E. Burgeson, eds.) Academic Press, Inc., pp. 195-221.
6. Vaughan, L., Mendler, M., Huber, S., Bruckner, P., Winterhalter, K.H., Irwin, M.H. and Mayne, R. (1988) J. Cell Biol. 106, 991-997.
7. Muragaki, Y., Nishimura, I., Henney, A., Ninomiya, Y. and Olsen, B.R. (1990) Proc. Natl. Acad. Sci. (USA) 87, 2400-2404.
8. Svoboda, K.K., Nishimura, I., Sugrue, S.P., Ninomiya, Y. and Olsen, B.R. (1988) Proc. Natl. Acad. Sci. (USA) 85, 7496-7500.
9. Yada, T., Suzuki, S., Kobayashi, K., Kobayashi, M., Hoshino, T., Horie, K. and Kimata, K. (1990) J. Biol. Chem. 265, 6992-6999.
10. Gordon, M.K., Gerecke, D.R., Dublet, B., van der Rest, M. and Olsen, B.R. (1989) J. Biol. Chem. 264, 19772-19778.

11.Dublet, B., Oh, S., Sugrue, S.P., Gordon, M.K., Gerecke, D.R., Olsen, B.R. and van der Rest, M. (1989) J. Biol. Chem. 264, 13150-13156.
12.Dublet, B. and van der Rest, M. (1991) J. Biol. Chem. 266, 6853-6858.
13.Gordon, M.K., Castagnola, P., Dublet, B., Linsenmayer, T.F., van der Rest, M. and Olsen, B.R. (1991) Eur. J. Biochem. 201, 333-338.
14.Lunstrom, G.P., Morri, S.P., McDonough, A.M., Keene, D.R. and Burgeson, R.E. (1991) J. Biol. Chem. 113, 963-969.
15.Keene, D.R., Lunstrum, G.P., Morris, N.P., Stoddard, D.W. and Burgeson, R.E. (1991) J. Cell Biol. 113, 971-978.
16.Mayne, R. and van der Rest, M. (1991) In: Extracellular Matrix Molecules - A Practical Approach (M.A. Haralson and J.R. Hassell, eds.) IRL Press, in press.
17.Eyre, D.R., Apon, S., Wu, J.-J., Ericson, L.H. and Walsh, K.A.. (1987) FEBS Lett. 220, 337-341.
18.van der Rest, M. and Mayne, R. (1988) J. Biol. Chem. 263, 1615-1618.
19.Sugrue, S.P., Gordon, M.K., Seyer, J., Dublet, B., van der Rest, M. and Olsen, B.R. (1989) J. Cell Biol. 109, 939-945.
20.Konomi, H., Seyer, J.M., Ninomiya, Y. and Olsen, B.R. (1986) J. Biol. Chem. 261, 6742-6746.
21.Duance, V.C., Shimokomaki, M. and Bailey, A.J. (1982) Biosci. Rep. 2, 223-227.
22.Ricard-Blum, S., Hartmann, D.J., Herbage, D., Payen-Meyran, C. and Ville, G. (1982) FEBS Lett. 146, 343-347.
23.Evans, H.B., Ayad, S., Abedin, M.Z., Hopkins, S., Morgan, K., Walton, K.W., Weiss, J.B. and Holt, P.J.L. (1983) Ann. Rheum. Dis. 42, 575-581.
24.Hartmann, D.J., Magloire, H., Ricard-Blum, S., Joffre, A., Couble, M.-L., Ville, G. and Herbage, D. (1983) Collagen Relat. Res. 3, 349-357.
25.Müller-Glauser, W., Humbel, B., Glatt, M., Sträuli, P., Winterhalter, K.H. and Bruckner, P. (1986) J. Cell Biol. 102, 1931-1939.
26.Irwin, M.H., Silver, S.H. and Mayne, R. (1985) J. Cell Biol. 101, 814-823.
27.Ninomiya, Y., Castagnola, P., Gerecke, D., Gordon, M., Jacenko, O., LuValle, P., McCarthy, M., Muragaki, Y., Nishimura, I., Oh, S., Rosenblum, N., Sato, N., Sugrue, S., Taylor, R., Vasios, G., Yamaguchi, N. and Olsen, B.R. (1990) In: Extracellular Matrix Genes (L.J. Sandell and C.D. Boyd, eds.) Academic Press, Inc., pp. 79-114.
28.Muragaki, Y., Kimura, T., Ninomiya, Y. and Olsen, B.R. (1990) Eur. J. Biochem. 192, 703-708.
29.Kimura, T., Mattei, M.-G., Stevens, J.W., Goldring, M.B., Ninomiya, Y. and Olsen, B.R. (1989) Eur. J. Biochem. 79, 71-78.
30.Lozano, G., Ninomiya, Y., Thompson, H. and Olsen, B.R. (1985) Proc. Natl. Acad. Sci. (USA) 82, 4050-4054.
31.McCormick, D., van der Rest, M., Goodship, J., Lozano, G., Ninomiya, Y. and Olsen, B.R. (1987) Proc. Natl. Acad. Sci. (USA) 84, 4044-4048.
32.Vasios, G., Nishimura, I., Konomi, H., van der Rest, M., Ninomiya, Y. and Olsen, B.R. (1988) J. Biol. Chem. 263, 2324-2329.
33.Gordon, M., Gerecke, D. and Olsen, B.R. (1987) Proc. Natl. Acad. Sci. (USA) 84, 6040-6044.
34.Jacenko, O., Olsen, B.R. and LuValle, P. (1991) In: Crit. Rev. Eukar. Gene Expression (G.S. Stein, J.L. Stein and J.B. Lian, eds.) CRC Press, pp. 327-353.

■ *Bjorn Reino Olsen:*
Department of Anatomy and Cellular Biology,
Harvard Medical School,
Boston, MA, USA

■ *Yoshifumi Ninomiya:*
Department of Molecular Biology and Biochemistry,
Okayama University Medical School,
Okayama, Japan

Fibrillar Collagens (Types I-III, V, XI)

Collagen types I, II, III, V, and XI participate in the formation of fibrils with molecules packed in quarter-staggered arrays[1,2]. Encoded by homologous genes that contain multiple exons (over 50), fibrillar collagens evolved to provide organisms with supramolecular scaffolds for mechanical support and to provide the proper environment for cellular migration, attachment and differentiation.

Fibrillar collagens include five different molecular types (I, II, III, V, and XI) containing polypeptide subunits (α-chains) encoded by at least nine distinct genes. Each α-chain contains over 300 repeats of the triplet sequence -Gly-X-Y-, flanked by short nontriplet containing sequences, telopeptides, at each end. About 50% of the prolyl residues in the triplet domain are posttranslationally converted to 4-hydroxyproline by an enzyme, prolyl-4-hydroxylase (EC1.14.11.2), located in the rough endoplasmic reticulum[3,4]. The active enzyme is a tetramer of two nonidentical subunits, $\alpha_2\beta_2$, and it hydroxylates prolyl residues that occupy the Y-position in the repeating triplets. The β-subunit is identical to the enzyme protein disulphideisomerase; the prolyl-4-hydroxylase tetramer has itself disulphideisomerase activity. In addition to prolyl hydroxylation, some lysyl residues in the Y-position are hydroxylated by lysyl hydroxylase[5]. The sequential action of galactosylhydroxylysyltransferase and glucosylgalactosylhydroxylysyltransferase add mono- and disaccharides to some hydroxylysyl residues[6].

The hydroxylated α-chains form stable triple-helices at 37°C. These triple-helical molecules are 300 nm rigid rod-like structures with a diameter of 1.5 nm. The molecular weight is about 300,000[7]. Fibrillar collagens are produced by a variety of cells, mostly (but not only) of mesenchymal origin. After posttranslational modifications (hydroxylation, glycosylation) and triple-helix assembly, precursor procollagen molecules containing amino and carboxyl propeptide extensions flanking the central triple-helical domain, are secreted from the cell (Figure 1). During extracellular processing of procollagen to collagen, the propeptides are removed from the major, collagen triple-helical domain by specific endoproteinases[6].

Supramolecular aggregates of fibrillar collagen triple-helices usually contain more than one type of collagen[8],

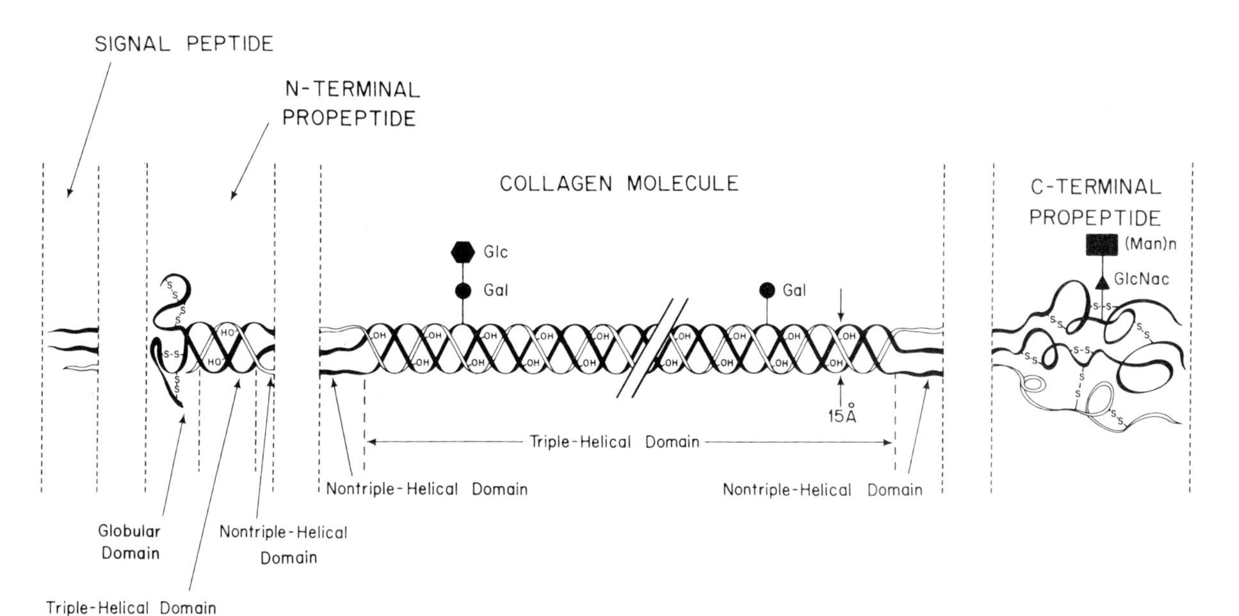

Figure 1. Diagram of the domains of a fibrillar procollagen molecule. (From Olsen (1991)[6]).

and such heterotypic fibrils are arranged in different patterns in different tissues; parallel fibril bundles in tendon, criss-crossing layers in cornea, and spiral arrangements in lamellar bone (Figure 2). Heterotypic fibrils containing types I, III, and/or V collagens are expressed in a number of tissues of mesenchymal origin such as skin, tendon, ligaments, and bone, whereas fibrils with types II and XI are found predominantly in hyaline cartilage and the vitreous body of the eye. It is believed that the presence of different fibrillar collagen types within the same fibril plays a role in determining the fibril diameter. Fibrillar collagens can interact with cells directly via specific cell surface receptors or indirectly via other extracellular matrix components[9]. By such interactions fibrillar collagens influence cell behavior and differentiation during embryonic development.

Type I collagen molecules, consisting of two $\alpha1(I)$ and one $\alpha2(I)$ chain are the products of two genes located on the human chromosomes 17 and 7. The genes contain more than 50 exons[2], like all the fibrillar collagen genes (Figure 3). In fact, nucleotide sequence comparisons among fibrillar collagen genes have shown that the arrangement of exons is practically identical in the region coding for triple-helical domains[10]. This conservation of structure is independent of species or type of fibrillar collagen. It is likely that the triple-helical domain exons evolved by repeated duplications of an exon unit of 54 bp[2]. Mutations in the type I genes (substitutions, deletions) account for the majority of cases of osteogenesis imperfecta and for certain types of the Ehlers-Danlos syndrome[11,12]. Mutations in the $\alpha1(II)$ gene appear to be the basis for many disorders in the spondyloepiphyseal dysplasia group[13,14], while several mutations in the $\alpha1(III)$ gene have been described to result in Ehlers-Danlos type IV[15].

Promoter elements have been investigated in several fibrillar collagen genes. Functional transcription regulatory elements have been identified in the first introns of several genes, including types I and II genes[1].

PURIFICATION

To purify fibrillar collagens, pepsin is commonly used to dissociate collagen triple-helical domains (these are pepsin resistant) from other extracellular matrix molecules. Repeated differential salt precipitations at neutral pH as well as in acid conditions are used to purify each fibrillar collagen type[16]. Types II, V, and XI collagen require higher salt concentrations than types III and I to precipitate[16-20]. Fibrillar procollagen molecules are usually purified from media of cultured cells[7,20,21]. Addition of protease inhibitors and avoidance of acidic pH prevent the action of endogenous proteolytic enzymes that remove the propeptides, resulting in the isolation of intact precursor molecules.

ACTIVITIES

Fibrillar collagens polymerize to form fibrils that serve as stabilizing scaffolds in extracellular matrices[22,23]. Within the fibrils, the 300 nm long rod like molecules overlap with their ends about 30 nm and are arranged in quarter-staggered arrays. The fibrils have therefore a periodic structure. Each period is 67 nm long and consists of a "hole" zone with more loosely packed molecules and an overlap zone with more densely packed molecules[7]. These zones can be easily visualized by negative staining and electron microscopy. When fibrils are positively stained, a periodic cross-striation pattern is observed, reflecting the distribution of clusters of charged amino acid residues

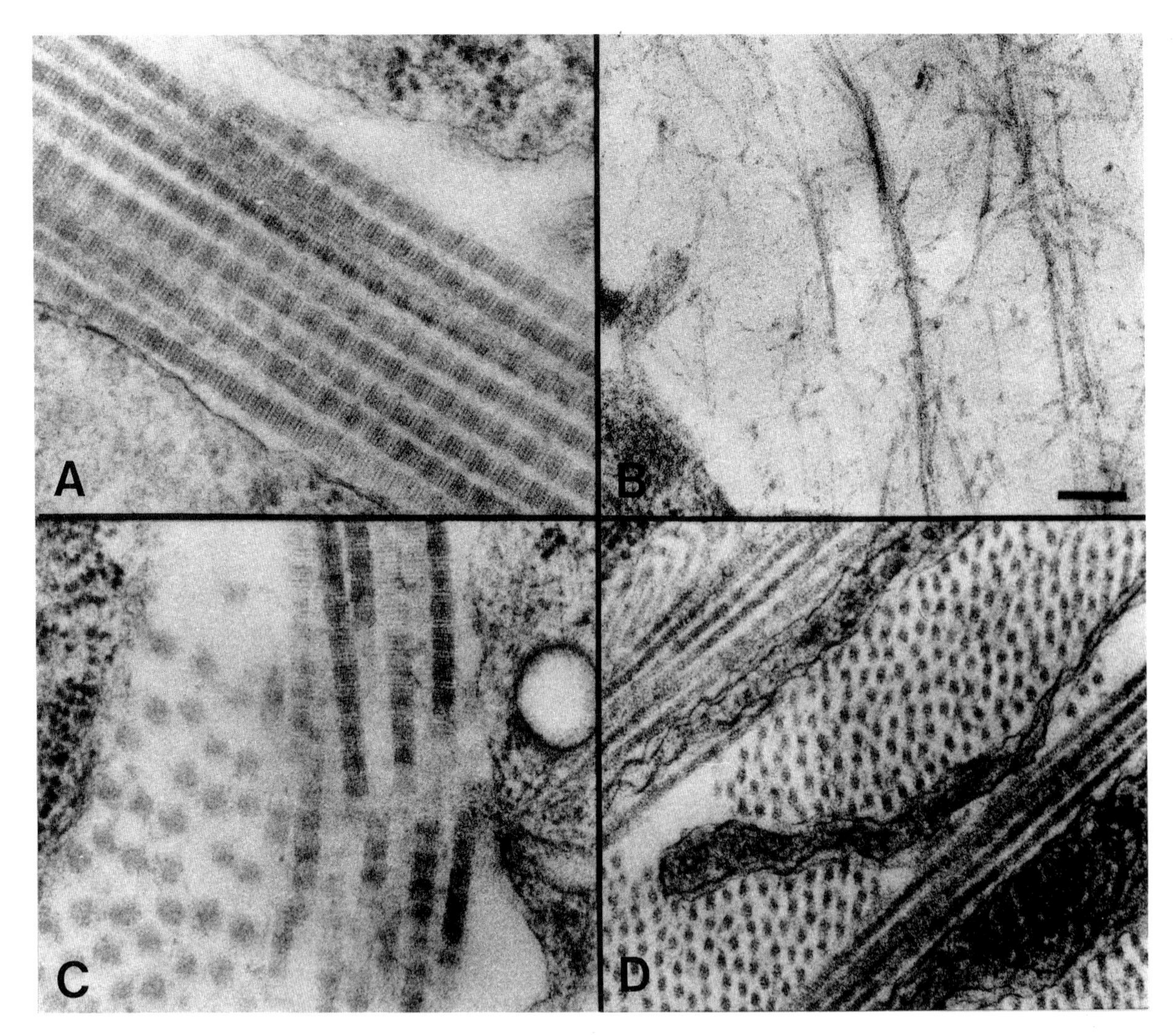

Figure 2. Collagen fibrils in 14-day chick embryo tendon (A), sternal cartilage (B), dermis (C), and corneal stroma (D). (Courtesy of Dr. David Birk).

along the collagen molecules[7]. Cell differentiation and migration during development are influenced by fibrillar collagens, and collagens interact with cells through receptors on cell surfaces[9,24-27].

■ ANTIBODIES

Polyclonal and monoclonal antibodies against all fibrillar collagens from a number of animal species are available. Some of these antibodies are directed against epitopes in the propeptide domains of the procollagens[21]; others are directed against epitopes in the triple-helical domain[28,29]. Several monoclonal antibodies have been used for epitope mapping in conjunction with rotary shadowing and electron microscopy[29,30]. A variety of antibodies are available from commercial sources, including Chemicon, Southern Biotechnology Associates, Inc., Upstate Biotechnology, Inc., DMI/Pasteur Institute in Lyon, and Developmental Studies Hybridoma Bank.

■ GENES

The chicken $\alpha2$(I) collagen gene was the first fibrillar collagen gene to be isolated and completely characterized[31,32]. Since then, cDNAs and genomic clones have been isolated for almost all fibrillar collagen genes, from a number of species[2,10,33,34]. The number and the size of exons are similar in all of these collagen genes. However, the sizes of the introns are more variable[10]. Since the number of entries into GenBank/EMBL Data Bank is very large, the readers are referred to original publications for accession numbers of specific sequences.

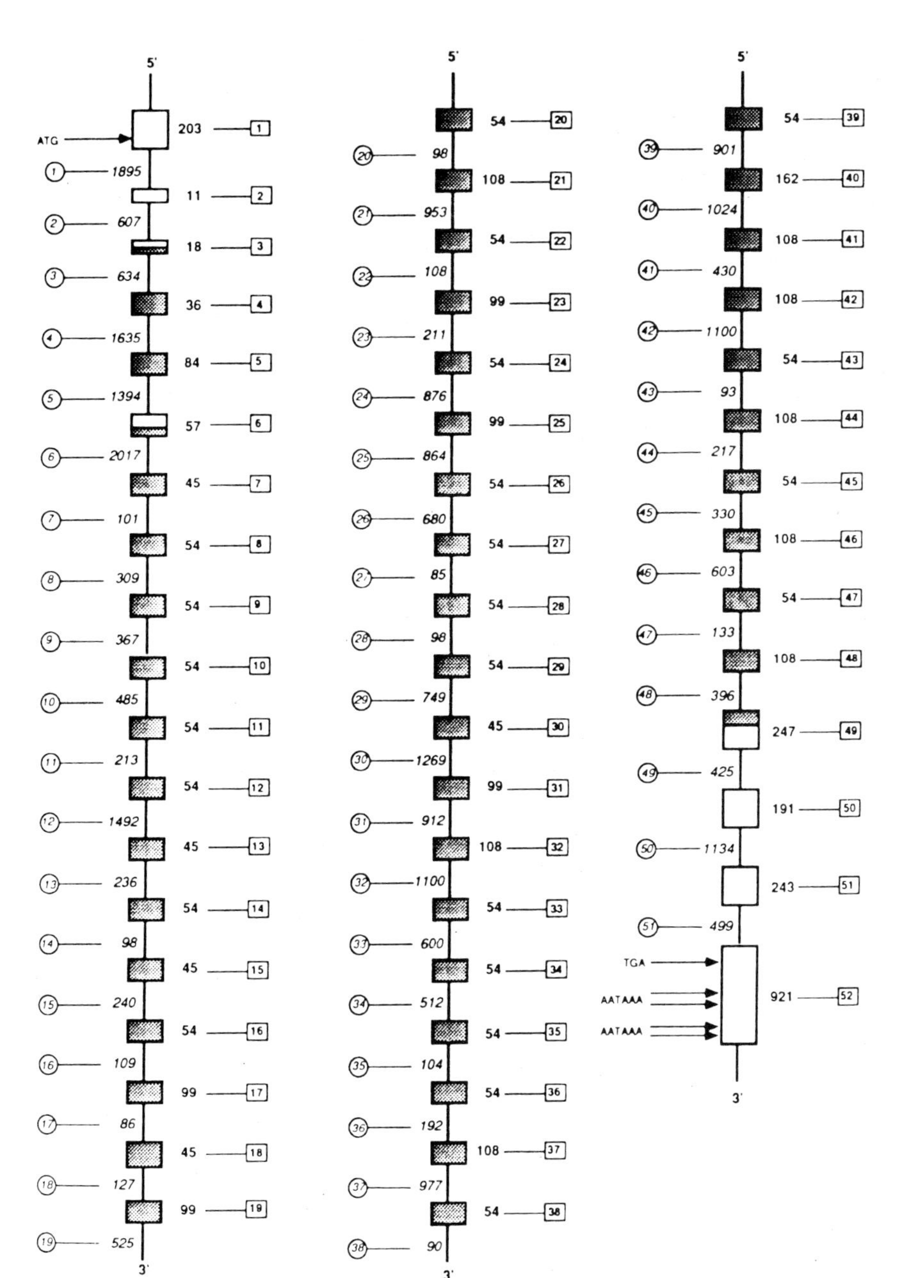

Figure 3. Diagram showing the exon/intron structure of the chicken $\alpha 2(I)$ collagen gene. (From Sandell and Boyd (1990)[10]).

REFERENCES

1. Jacenko, O., Olsen, B.R. and LuValle, P. (1991) In: Crit. Rev. Eukar. Gene Expression (G.S. Stein, J.L. Stein and J.B. Lian, eds.) CRC Press, pp. 327-353.
2. Vuorio, E. and deCrombrugghe, B. (1990) Annu. Rev. Biochem. 59, 837-872.
3. Kivirikko, K.I., Helaakoski, T., Tasanen, K., Vuori, K., Myllylä, R., Parkkonen, T. and Pihlajaniemi, T. (1990) Ann. N.Y. Acad. Sci. 580, 132-142.
4. Kivirikko, K.I., Myllylä, R. and Pihlajaniemni, T. (1989) FASEB J. 3, 1609-1617.
5. Kivirikko, K.I. and Myllylä, R. (1985) Ann. N.Y. Acad. Sci 460, 187-201.
6. Olsen, B.R. (1991) In: Cell Biology of Extracellular Matrix (E.D. Hay, ed.) Plenum Publishing Corp. in press.
7. Kühn, K. (1987) In: Structure and Function of Collagen Types (R. Mayne and R.E. Burgeson, eds.) Academic Press, Inc., pp. 1-42.
8. Linsenmayer, T.F., Fitch, J.M. and Birk, D.E. (1990) Ann. N.Y. Acad. Sci. 580, 143-160.
9. Hemler, M.E. (1990) Annu. Rev. Immunol. 8, 365-400.
10. Sandell, L.J. and Boyd, C.D. (1990) In: Extracellular Matrix Genes (L.J. Sandell and C.D. Boyd, eds.) Academic Press, Inc., pp. 1-56.
11. Byers, P.H. and Bonadio, J.F. (1989) In: Collagen - Molecular Biology, Vol. IV, (B.R. Olsen and M.E. Nimni, eds.) CRC Press, pp. 125-139.

12.Prockop, D.J., Olsen, A., Kontusari, S., Hyland, J., Ala-Kokko, L., Vasan, N.S., Barton, E., Buck, D., Harrison, K. and Brent, R.L. (1990) Ann. N.Y. Acad. Sci. 580, 330-339.
13.Lee, B., Vissing, H., Ramirez, F., Rogers, D. and Rimoin, D. (1989) Science 244, 978-980.
14.Ala-Kokko, L., Baldwin, C.T., Moskowitz, R.W. and Prockop, D.J. (1990) Proc. Natl. Acad. Sci. (USA) 87, 6565-6569.
15.Superti-Furga, A., Steinmann, B., Ramirez, F. and Byers, P. (1989) Hum. Genet. 82, 104-108.
16.Trelstad, R.L. (1982) In: Immunochemistry of the Extracellular Matrix, Vol. I (H. Furthmayr, ed.) CRC Press, pp. 31-41.
17.Mayne, R. and van der Rest, M. (1991) In Extracellular Matrix Molecules - A Practical Approach (M.A. Haralson and J.R. Hassell, eds.) IRL Press, in press.
18.Haralson, M.A., Furuto, D. and Miller, E. (1991) In: Extracellular Matrix Macromolecules - A Practical Approach (M.A. Haralson and J.R. Hassell, eds.) IRL Press, in press.
19.Fessler, J.H. and Fessler, L.I. (1987) In: Structure and Function of Collagen Types (R. Mayne and R.E. Burgeson, eds.) Academic Press, Inc., pp. 81-103.
20.Eyre, D. and Wu, J.-J. (1987) In: Structure and Function of Collagen Types (R. Mayne and R.E. Burgeson, eds.) Academic Press, Inc., pp. 261-281.
21.Pesciotta, D.M., Curran, S.F. and Olsen, B.R. (1982) In: Immunochemistry of the Extracellular Matrix, Vol. I (H. Furthmayr, ed.) CRC Press, pp. 91-109.
22.Kadler, K.E., Hulmes, D.J.S., Hojima, Y. and Prockop, D.J. (1990) Ann. N.Y. Acad. Sci. 580, 214-224.
23.Birk, D.E., Zycband, E.I., Winkelmann, D.A. and Trelstad, R.L. (1990) Ann. N.Y. Acad. Sci. 580, 176-194.
24.Santoro, S.A. (1986) Cell 46, 913-920.
25.Staatz, W.D. (1989) J. Cell Biol. 108, 1917-1924.
26.Elices, M.J. and Hemler, M.E. (1989) Proc. Natl. Acad. Sci. (USA) 86, 9906-9910.
27.Kirchhofer, D., Languino, L.R., Ruoslahti, E. and Pierschbacher, M.D. (1990) J. Biol. Chem. 265, 615-618.
28.Linsenmayer, T.F. (1991) In: Cell Biology of Extracellular Matrix (E.D. Hay, ed.) Plenum Press, in press.
29.Linsenmayer, T.F., Fitch, J.M., Schmid, T.M., Birk, D.E., Bruns, R.R. and Mayne, R. (1989) In: Collagen - Molecular Biology, Vol IV (B.R. Olsen and M.E. Nimni, eds.) CRC Press, pp. 141-170.
30.Mayne, R., Wiedemann, H., Irwin, M.H., Sanderson, R.D., Fitch, J.M., Linsenmayer, T.F. and Kühn, K. (1984) J. Cell Biol. 98, 1637-1644.
31.Ohkubo, H., Vogeli, G., Mudryj, M., Avvedimento, E., Sullivan, M., Pastan, I. and de Crombrugghe, B. (1980) Proc. Natl. Acad. Sci. (USA) 77, 7059-7063.
32.Boedtker, H., Finer, M. and Aho, S. (1985) Ann. N.Y. Acad. Sci. (USA) 460, 85-116.
33.Myers, J. and Dion, A.S. (1990) In: Extracellular Matrix Genes (L.J. Sandell and C.D. Boyd, eds.) Academic Press, Inc., pp. 54-78.
34.Kimura, T., Cheah, K.S.E., Chan, D.H., Lui, V.C.H., Mattei, M.-G., van der Rest, M., Ono, K., Solomon, E., Ninomiya, Y. and Olsen, B.R. (1989) J. Biol. Chem. 264, 13910-13916.

■ *Bjorn Reino Olsen:*
Department of Anatomy and Cellular Biology, Harvard Medical School, Boston, MA, USA

■ *Yoshifumi Ninomiya:*
Department of Molecular Biology and Biochemistry, Okayama University Medical School, Okayama, Japan

Other Collagens (Types VI, VII, XIII)

This is a heterogeneous group of proteins that on a genetic basis does not belong to one of the defined collagen families. They are discussed here as a group only for practical reasons; as new collagen genes are identified and characterized, it is quite likely that the collagens discussed below will be classified as members of their own distinct families.

Type VI collagen is ubiquitously expressed in different tissues as the major component of beaded microfibrils[1]. Each type VI molecule appears in the electron microscope as a 105 nm long triple-helical rod flanked by two globular domains, and contains three different polypeptide subunits α1(VI), α2(VI), and α3(VI). The three chains have apparent molecular masses of about 140, 140 and 250 kDa, respectively[2].

These heterotrimeric type VI molecules form disulphide bonded dimers and tetramers. The tetramers associate end-to-end and generate the microfibrils, which have a characteristic periodicity of 100 nm. The complete primary structures of the α1(VI) and α2(VI) chains and most of the α3(VI) chain have been determined from amino acid and cDNA sequencing[3-6]. The chains contain a central, relatively short triple-helical domain of 335-336 amino acid residues[7]. A cysteinyl residue in this domain probably participates in intramolecular disulphide bonding. All three chains contain a C-terminal nontriple-helical domain composed of two repeats of a 200 residue long segment that is homologous to the A-domains of **von Willebrand factor**[6,8]. In the N-terminal region of the α1(VI) and α2(VI) chains there is a single 200 residue long von Willebrand factor A homology domain, while the α3(VI) chain contains at least nine such repeats in this region[6]. Binding sites for type I collagen have been ascribed to the von Willebrand factor A region[9], and it is possible that the homologous domains in type VI collagen have collagen binding properties as well[3,6]. It is also possible that type VI collagen has a cell adhesion function; several Arg-Gly-Asp sequences are found in the primary sequence of the type VI collagen subunits[3,6,7,10].

■ PURIFICATION

The triple-helical portion of type VI collagen can be obtained by differential precipitation with NaCl from pepsin digests of various tissues in acetic acid or formic acid. Further purification can be accomplished by repre-

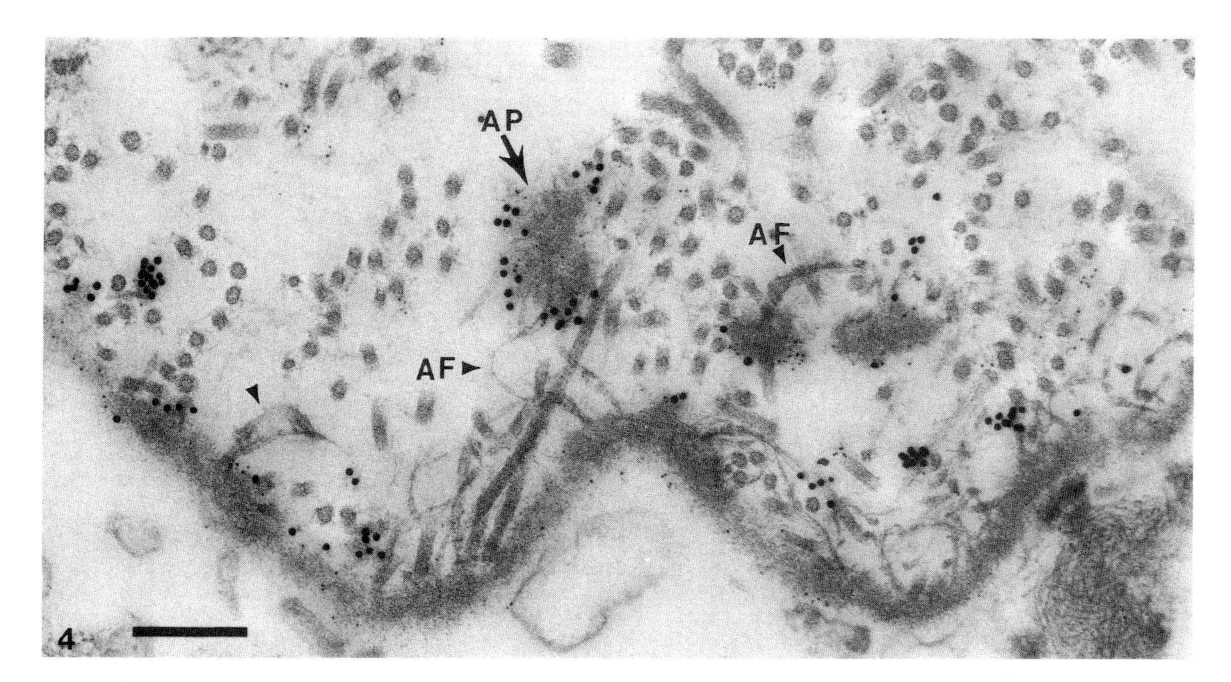

Figure. Ultrastructural immunolocalization of type VII collagen within the dermal-epidermal junction of neonatal human foreskin with gold conjugated antibodies. AF: anchoring fibrils; AP: anchoring plaques. (From Keene et al. (1987)[20]).

cipitation through dialysis against 0.02M Na_2HPO_4, followed by ion exchange or molecular sieve chromatography[1,11]. Intact type VI collagen can be purified by ion exchange and molecular sieve chromatography of guanidine or urea extracts of tissues or cell cultures[1].

■ ACTIVITIES

Type VI collagen molecules assemble into disulphide bonded polymers that form beaded microfibrils[12,13]. The microfibrils frequently aggregate further laterally into crossbanded fibres, referred to as Luse bodies, fusiform bodies or zebra collagen[1,14,15].

■ ANTIBODIES

Several polyclonal and monoclonal antibodies are available against type VI collagen[1]. They have been used for detecting type VI chains or degradation products by immunoblotting, immunoprecipitation, and immunohistochemistry[1]. Monoclonal antibodies have been used for epitope mapping by rotary shadowing electron microscopy[1,16]. Anti-type VI collagen antibodies are available from several commercial sources, including Chemicon, Telios Pharmaceuticals, Inc., Developmental Studies Hybridoma Bank, and Upstate Biotechnology, Inc.

■ GENES

cDNAs encoding all three chains of type VI collagen in both humans and chicken have been isolated and sequenced[3-8,17]. The α2(VI) gene generates transcripts that are alternatively spliced at the 3′ end, giving rise to several mRNA variants[3]. The significance of this is at present unclear. Partial genomic sequences are also available[18].

Type VII collagen is the major collagenous component of anchoring fibrils associated with the basement membranes under stratified squamous epithelia[19]. The fibrils originate from the lamina densa and extend into the upper papillary dermis of skin where they insert into so-called anchoring plaques[20] (Figure). Anchoring fibrils also connect anchoring plaques. Type VII collagen molecules are homotrimers containing a triple-helical domain that is about 50% longer than the triple-helix of fibrillar collagens. This domain is flanked by relatively large nontriple-helical domains, of molecular masses 150 and 30 kDa[19]. The 30 kDa domain appears to be proteolytically cleaved extracellularly, and the processed molecules form anti-parallel dimers, through a C-terminal overlap region. Lateral aggregation of such dimers leads to the formation of the centro-symmetrically banded anchoring fibrils[19,21].

Dystrophic forms of epidermolysis bullosa show ultrastructural abnormalities in anchoring fibrils[22,23]; some cases of severe recessive forms of the disease show no staining with type VII collagen antibodies in the basement membrane zone[24]. Also, analysis of a large pedigree of dominant dystrophic epidermolysis bullosa has demonstrated a strong genetic linkage between the disease and an α1(VII) collagen gene marker[25]. The α1(VII) collagen gene is therefore a strong candidate gene for some forms of this blistering disease.

PURIFICATION

The triple-helical domain of type VII collagen can be solubilized by pepsin extraction of human skin or amnion. Purification is by differential salt precipitation with NaCl, followed by ion exchange chromatography and HPLC[19]. The intact, biosynthetic form of type VII collagen has been partially purified from the media of KB cells (derived from a human oral basal cell carcinoma) and WISH cells (derived from amniotic epithelial cells)[19].

ACTIVITIES

Type VII collagen molecules form the major scaffold in anchoring fibrils in skin, chorioamnion, oral mucosa, cornea, and the uterine cervix[19].

ANTIBODIES

Monoclonal antibodies against human type VII collagen are available[19]. Polyclonal serum from a patient with epidermolysis bullosa aquisita has been used to isolate an α1(VII) cDNA clone from a λgt11 expression library[26].

GENES

cDNAs encoding part of the human α1(VII) chain have been isolated, sequenced, and used to locate the α1(VII) collagen gene to the p21.1-p21.3 region of chromosome 3[26].

Type XIII collagen was discovered by isolation and sequencing of human cDNA and genomic clones[27-29]. Although a polyclonal antipeptide antibody recognize 76 and 62 kDa chains in immunoblots[28], the chain composition and supramolecular structure of this collagen are not known.

From the cDNA sequence the α1(XIII) chain is predicted to contain three short triple-helical domains, separated by nontriple-helical sequences and flanked by nontriple-helical domains at the N- and C-termini[27]. Alternative splicing gives rise to at least five different mRNAs[30]. These RNAs encode polypeptides that are different in their N- and C-terminal thirds. Thus, the length of the α1(XIII) chain varies between 654 and 566 amino acid residues, depending on the mRNA splice variant used for coding. The biological significance of this is not clear.

PURIFICATION

Type XIII collagen has not been isolated from tissues in significant quantities.

ACTIVITIES

Northern and *in situ* hybridization analyses show a wide distribution of α1(XIII) mRNA[31]; however, the polymeric form or function of this collagen is unknown.

ANTIBODIES

A polyclonal antipeptide antibody is available[28].

GENES

Human cDNA and genomic clones have been isolated[27-29].

REFERENCES

1. Timpl, R. and Engel, J. (1987) In: Structure and Function of Collagen Types (R. Mayne and R.E. Burgeson, eds.) Academic Press Inc. pp. 105-143.
2. Trüeb, B. and Winterhalter, K.H. (1986) EMBO J. 5, 2815-2819.
3. Chu, M.-L., Pan, T.-C., Conway, D., Saitta, B., Stokes, D., Kuo, H.-J., Glanville, R.W., Timpl, R., Mann, K. and Deutzmann, R. (1990) Ann. N.Y. Acad. Sci. 580, 55-63.
4. Bonaldo, P., Russo, V., Bucciotti, F., Bressan, G.M. and Colombatti, A. (1989) J. Biol. Chem. 264, 5575-5580.
5. Koller, E., Winterhalter, K.H. and Trueb, B. (1989) EMBO J. 8, 1073-1077.
6. Bonaldo, P., Russo, V., Bucciotti, F., Doliana, R. and Colombatti, A. (1990) Biochemistry 29, 1245-1254.
7. Chu, M.-L., Conway, D., Pan, T., Baldwin, C., Mann, K., Deutzmann, R. and Timpl, R. (1988) J. Biol. Chem. 263, 18601-18606.
8. Chu, M.-L., Pan, T., Conway, D., Kuo, H.-J., Glanville, R.W., Timpl, R., Mann, K. and Deutzmann, R. (1989) EMBO J. 8, 1939-1946.
9. Pareti, F.I., Niiya, K., McPherson, J.M. and Ruggeri, Z.M. (1987) J. Biol. Chem. 262, 13835-13841.
10. Aumailley, M., Mann, K., von der Mark, H. and Timpl, R. (1989) Exp. Cell Res. 181, 463-474.
11. Haralson, M.A., Furuto, D. and Miller, E. (1991) In: Extracellular Matrix Macromolecules - A Practical Approach (M.A. Haralson and J.R. Hassell, eds.) IRL Press, in press.
12. Bruns, R.R. (1984) J. Ultrastruct. Res. 89, 136-146.
13. Bruns, R.R., Press, W., Engvall, E., Timpl, R. and Gross, J. (1986) J. Cell Biol. 103, 393-404.
14. Garron, L.K., Feeney, M.L., Hogan, M.J. and McEwen, W.K. (1958) Am. J. Ophthalmol. 46, 27-35.
15. Luse, S.A. (1960) Neurology, 10, 881-905.
16. Linsenmayer, T.F., Mentzer, A., Irwin, M.H., Waldrep, N.K. and Mayne, R. (1986) Exp. Cell Res. 165, 518-529.
17. Chu, M.-L., Mann, K., Deutzmann, R., Pribula-Conway, D., Hsu-Chen, M., Bernard, M. and Timpl, R. (1987) Eur. J. Biochem. 168, 309-317.
18. Hayman, A.R., Köppel, J., Winterhalter, H. and Trueb,B. (1990) J. Biol. Chem. 265, 9864-9868.
19. Burgeson, R.E. (1987) In: Structure and Function of Collagen Types (R. Mayne and R.E. Burgeson, eds.) Academic Press Inc. pp. 145-172.
20. Keene, D.R., Sakai, L.Y., Lunsturm, G.P., Morris, N.P. and Burgeson, R.E. (1987) J. Cell Biol. 104, 611-621.
21. Burgeson, R.E., Lunstrum, G.P, Rokosova, B., Rimberg, C.S., Rosenbaum, L.M. and Keene, D.R. (1990) Ann. N.Y. Acad. Sci. 580, 32-43.
22. Briggaman, R.A. (1985) J. Invest. Dermatol. 84, 371-373.
23. Bruckner-Tuderman, L., Niemi, K., Kero, M., Schnyder, U.W. and Reunala, T. (1990) Br. J. Dermatol. 122, 383-390.
24. Bruckner-Tuderman, L., Mitsuhashi, Y., Schnyder, U.W. and Bruckner, P. (1989) J. Invest. Dermatol. 93, 3-9.
25. Uitto, J., Ryynänen, M., Parente, M.G., Chung, L., Chu, M.-L. and Knowlton, R. (1992) Clin. Res. in press.

26.Parente, M.G., Chung, L.C., Woodley, D.T., Wynn, K.C., Bauer, E.A., Mattei, M.-G., Chu, M.-L. and Uitto, J. (1991) Proc. Natl. Acad. Sci. (USA) 88, 6931-6935.
27.Pihlajaniemi, T., Tamminen, M., Sandberg, M., Hirvonen, H. and Vuorio, E. (1990) Ann. N.Y. Acad. Sci. 580, 440-443.
28.Pihlajaniemi, T., Myllylä, R., Seyer, J., Kurkinen, M. and Prockop, D.J. (1987) Proc. Natl. Acad. Sci. (USA) 84, 940-944.
29.Tikka, L., Pihlajaniemi, T., Hinttu, P., Prockop, D.J. and Tryggvason, K. (1988) Proc. Natl. Acad. Sci. (USA) 85, 7491-7495.
30.Pihlajaniemi, T. and Tamminen, M. (1990) J. Biol. Chem. 265, 16922-16928.
31.Sandberg, M., Tamminen, M., Hirvonen, H., Vuorio, E. and Pihlajaniemi, T. (1989) J. Cell Biol. 109, 1371-1379.

■ *Bjorn Reino Olsen:*
Department of Anatomy and Cellular Biology, Harvard Medical School, Boston, MA, USA
■ *Yoshifumi Ninomiya:*
Department of Molecular Biology and Biochemistry, Okayama University Medical School, Okayama, Japan

Short Chain Collagens (Types VIII, X)

Types VIII and X, composed of the three chains α1(VIII), α2(VIII), and α1(X), form the subgroup of short chain collagens, so named because their subunits are short (only about 60 kDa) as compared with fibrillar collagen chains. Despite similarities in domain structure, amino acid sequences, and genomic exon configurations, the two types show very different temporal and spatial expression. Given the similarity in exon structure it is likely that the three genes evolved from a common precursor gene to provide a similar function in different tissues.

Type VIII collagen was originally identified as a product of bovine aortic and rabbit corneal endothelial cells but is also synthesized by nonvascular and some tumor cells as well[1]. The molecule is probably a heterotrimer composed of 60 kDa α1(VIII) and α2(VIII) chains in a ratio of two to one[2], but the existence of homotrimeric molecules composed entirely of α1(VIII) or α2(VIII) chains cannot be ruled out[3]. Type X collagen is a product of hypertrophic chondrocytes[4]. The molecule is a homotrimer of 59 kDa α1(X) chains. Both types of collagen share a similar domain structure; a central triple-helical (COL1) domain of 50 kDa is flanked by N-terminal (NC2) and C-terminal (NC1) non-triple-helical domains[5]. Both type VIII and type X molecules appear as 130 nm-long rods with knobs at both ends by electron microscopy after rotary shadowing[4,6]. The COL1 and NC1 domains of both types are encoded by one large exon, whereas the NC2 domain is encoded by a small additional exon[6-8]. Additional exons [one for α1(X) or two for α1(VIII)] encode the 5′ untranslated portion of the mRNAs (Figure). This exon configuration is in stark contrast to the multiexon structure of fibrillar and FACIT collagen genes.

Despite these similarities, a distinct tissue distribution has been found for these two molecules: type X is restricted to hypertrophic cartilage[4], whereas type VIII is distributed in various tissues including Descemet's membrane, vascular subendothelial matrices, heart, liver, kidney, perichondrium, and lung, as well as several malignant tumors including astrocytoma, Ewing's sarcoma and hepatocellular carcinoma[1,9,10].

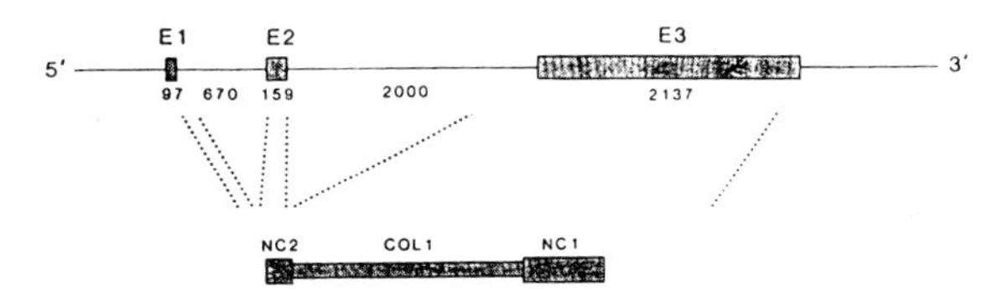

Figure. Diagram of the α1(X) collagen gene and the domain structure of the α1(X) collagen chain.

In Descemet's membrane, type VIII collagen molecules represent major components of a hexagonal lattice structure[11], with type VIII molecules most likely linked together by interactions involving the nontriple-helical end regions (Figure). Type X collagen molecules may form the same kind of polymers in hypertrophic cartilage[12,13]. The expression of type X collagen *in vivo* by hypertrophic chondrocytes is regulated primarily at a transcriptional level [14].

■ PURIFICATION

The triple-helical domain of type VIII collagen can be purified from Descemet's membrane by pepsin extraction[1]. The digested material can be precipitated with NaCl at neutral pH, and purified further by chromatography through agarose and by reverse phase HPLC[15]. Intact type VIII collagen can be recovered from the medium of cultured endothelial cells[1]. Type X collagen can be isolated intact from the medium of chicken hypertrophic chondrocytes kept in long-term culture or as a triple-helical fragment by pepsin extraction of hypertrophic cartilage[4,15].

■ ACTIVITIES

Type VIII collagen is the major constituent of the hexagonal lattice observed in Descemet's membrane, as demonstrated by immunoelectron microscopy[11]. It is possible that the general function of type VIII collagen is to provide an open, porous structure that can withstand compressive force. Type X collagen in hypertrophic cartilage may provide a scaffold to prevent local collapse as the cartilage matrix is removed during endochondral ossification[12,13,16].

ANTIBODIES

Polyclonal antibodies against bovine type VIII have been used for immunofluorescence studies and immunoblots[10]. Monoclonal antibodies against the bovine α1(VIII) chain have been used for immuno-electronmicroscopy to demonstrate that the backbone unit within Descemet's membrane is composed of type VIII collagen[11]. Polyclonal antibodies against sheep type VIII have also been produced[9]. A monoclonal antibody, X-AC9, has been produced against chicken type X and used extensively for investigations on the location of type X, its ultrastructure and thermal stability[4]. However, AC9 reacts only weakly with α1(X) by immunoblotting. Two other monoclonal antibodies, 1A6 and 6F6, both recognize an epitope located in the N-terminal, nontriple-helical domain of type X collagen; they are both useful for immunoblotting analyses[17].

GENES

The primary structures of the α1(VIII) and α2(VIII) chains are strikingly similar to the corresponding structure of α1(X) collagen. cDNA and genomic DNAs encoding rabbit, human and mouse α1(VIII) and α2(VIII)[3,5,6,18] chains have been isolated and characterized. Chromosomal locations for the human α1(VIII) and α2(VIII) genes have been determined on chromosomes 3 and 1, respectively[3,18]. The chicken type X gene was the first to be isolated among the short chain collagen genes[7,8]. The bovine[19], mouse, and human[20] type X genes are currently being characterized. The human α1(X) gene has been localized to the q21-q22 region of chromosome 6[20].

REFERENCES

1. Sage, H. and Bornstein, P. (1987) In: Structure and Function of Collagen Types (R. Mayne and R.E. Burgeson, eds.) Academic Press Inc. pp. 173-194.
2. Mann, K., Jander, R., Korsching, E., Kühn, K. and Rauterberg, J. (1990) FEBS Lett. 273, 168-172.
3. Muragaki, Y., Jacenko, O., Apte, S., Mattei, M.-G., Ninomiya, Y. and Olsen, B.R. (1991) J. Biol. Chem. 266, 7721-7727.
4. Schmid, T.M. and Linsenmayer, T.F. (1987) In: Structure and Function of Collagen Types (R. Mayne and R.E. Burgeson, eds.) Academic Press Inc. pp. 223-259.
5. Yamaguchi, N., Benya, P.D., van der Rest, M. and Ninomiya, Y. (1989) J. Biol. Chem. 264, 16022-16029.
6. Yamaguchi, N., Mayne, R. and Ninomiya, Y. (1991) J. Biol. Chem. 266, 4508-4513.
7. Ninomiya, Y., Gordon, M., van der Rest, M., Schmid, T., Linsenmayer, T. and Olsen, B.R. (1986) J. Biol. Chem. 261, 5041-5050.
8. LuValle, P., Ninomiya, Y., Rosenblum, N.D. and Olsen, B.R. (1988) J. Biol. Chem 263, 18378-18385.
9. Kittelberger, R., Davis, P.F., Flyn, D.W. and Greenhill, N.S. (1990) Connect. Tiss. Res. 24, 303-318.
10. Sage, H. and Iruela-Arispe, M.-L. (1990) Ann. N.Y. Acad. Sci. 580, 17-31.
11. Sawada, H., Konomi, H. and Hirosawa, K. (1990) J. Cell Biol. 110, 219-227.
12. Gordon, M.K. and Olsen, B.R. (1990) Curr. Opinion Cell Biol. 2, 833-838.
13. Jacenko, O., Olsen, B.R. and LuValle, P. (1991) In: Crit. Rev. Eukar. Gene Expression (G.S. Stein, J.L. Stein and J.B. Lian, eds.) CRC Press, pp. 327-353.
14. LuValle, P., Hayashi, M. and Olsen, B.R. (1989) Dev. Biol. 133, 613-616.
15. Mayne, R. and van der Rest, M. (1991) In: Extracellular Matrix Molecules - A Practical Approach (M.A. Haralson and J.R. Hassell, eds.) IRL Press, in press.
16. Olsen, B.R. (1991) In: Articular Cartilage and Osteoarthritis (K. Kuettner and R. Schleyerbach, eds.) Raven Press, in press.
17. Summers, T.A., Irwin, M.H., Mayne, R. and Balian, G. (1988) J. Biol. Chem. 263, 581-587.
18. Muragaki, Y., Mattei, M.-G, Yamaguchi, N., Olsen, B.R. and Ninomiya, Y. (1991) Eur. J. Biochem. 197, 615-622.
19. Thomas, J.T., Kwan, A.P.L., Grant, M.E. and Boot-Handford, R.P. (1991) Biochem. J. 273, 141-148.
20. Apte, S., Mattei, M.-G. and Olsen, B.R. (1991) FEBS Lett. 282, 393-396.

Bjorn Reino Olsen:
Department of Anatomy and Cellular Biology, Harvard Medical School, Boston, MA, USA

Yoshifumi Ninomiya:
Department of Molecular Biology and Biochemistry, Okayama University Medical School, Okayama, Japan

Decorin (DCN)

Decorin (DCN), also known as PG-40, PG-II, PG-2, PG-S2 and CS/DS-PGII, is a small proteoglycan with a single chondroitin or dermatan sulphate chain attached to the fourth amino acid of the secreted 38 kDa protein[1]. Decorin has been found associated with collagen fibrils in virtually all connective tissues[2], perhaps near the d and e bands in the D period[3]. In vitro experiments have shown that decorin can change the kinetics of collagen fibril formation[4], affect the morphology of forming collagen fibrils[5] and bind to TGF-β[6].

Decorin containing a chondroitin-4-sulphate chain is isolated from developing bone[7] while DCN containing a dermatan sulphate chain is generally isolated from articular cartilage[8] or tendon[9]. The DCN is heterogenous with respect to glycosaminoglycan chain size and the average size of the chains differ with tissue and developmental age, however, the Alcian blue or StainsAll bands generally are centered from 100 to 250 kDa. While decorin size may differ in different tissues, it is almost always smaller than the **biglycan** (BGN) proteoglycan in the same tissue. Both the human[10] and bovine[11] cDNA have been published and show a ~38 kDa core protein. DCN changes the kinetics of the generation of **collagen** fibrils *in vitro*[4] and affects the final morphology of the resulting fibrils[5]. Monoclonal antisera have been used in electron microscopic immunolocalization, and have shown DCN to be in the D space, perhaps near the d or e bands[3]. While these results suggest that DCN may play an important role in collagen fibril formation, it is still unknown why the small proteoglycan is maintained on the fibrils throughout life, well after the fibrils are crosslinked and stabilized. Interestingly, DCN has been shown *in vitro* to bind to transforming growth factor-β (TGF-β)[6]. When the expression of high levels of DCN is induced in Chinese hamster ovary (CHO) cells, their morphology and growth properties are dramatically changed[12].

Decorin, which has been localized to human chromosome 12[13], shows obvious homology to two other small proteoglycans, biglycan (BGN)[14] and **fibromodulin**[15]. All three proteoglycans are predominantly composed of 10-12 tandem repeats with each nominal 24 amino acid repeat having a pattern of hydrophobic amino acids[14,15]. These repeat sequences have been used many times in evolution when protein-protein, protein-cell or cell-cell interactions are required. While decorin, like fibromodulin, is found associated with collagen fibrils[15], biglycan appears to be associated on or very near cell surfaces and not collagen bundles[2].

PURIFICATION

Chondroitin sulphate-containing decorin is isolated from developing bone by a series of extraction procedures and routine chromatography[7]. Bone is milled to a fine powder, extracted with denaturing solvents to remove blood and cellular proteins, and the residue extracted with demineralizing buffers. Standard ion exchange and molecular sieve chromatography is performed in denaturing solvents to suppress endogenous protease activity associated with the bone matrix. DCN is too hydrophobic to be chromatographed on standard reverse phase columns using methanol or acetonitrile gradients. Dermatan sulphate containing DCN can be isolated in good yield from cartilage[8], skin[8] and tendon[9] using a similar method as above with the possible use of an octyl-Sepharose and a detergent gradient[7]. The isolation of undenatured DCN in good yield has not been reported.

ACTIVITIES

DCN has been shown to change the kinetics of collagen fibril formation *in vitro*[4] and, because of its ability to bind TGF-β it neutralizes the effect of TGF-β on CHO cells[6].

ANTISERA

Two monoclonal against different parts of the bovine DCN, 6D6 and 7B1 has been reported[3]. A synthetic peptide antiserum made against a region of human DCN, LF-30, has been reported to be useful in immunolocalization[2] and immunoblot analysis[7] in human and monkey. Other antisera in the literature are unclear about the monospecificity, particularly concerning the closely related and often copurifying small proteoglycan, biglycan[14].

GENES

A full length cDNA for human DCN (GenBank M14219) has been reported[10]. A second source has reported a cDNA that is slightly shorter at the 5′ end but longer at the 3′ end in Bluescript plasmid (P2) for human DCN and full length bovine DCN (plasmid PG-28)[11,14]. The human gene has been localized to chromosome 12.

REFERENCES

1. Chopra, R.K., Pearson, C.H., Pringle, G.A., Fackre, D.S. and Scott, P.G. (1985) Biochem. J. 232, 277-279.
2. Bianco, P., Fisher, L.W., Young, M.F., Termine, J.D. and Robey, P.G. (1990) J. Histochem. Cytochem. 38, 1549-1563.
3. Pringle, G.A. and Dodd, C.M. (1990) J. Histochem. Cytochem. 38, 1405-1411.
4. Vogel, K.G., Paulsson, M. and Heinegård, D. (1984) Biochem. J. 223, 587-597.
5. Vogel, K.G. and Trotter, J.A. (1987) Collagen Rel. Res. 7, 105-114.
6. Yamaguchi, Y., Mann, D.M. and Ruoslahti, E. (1990) Nature 346, 281-284.
7. Fisher, L.W., Hawkins, G.R., Tuross, N. and Termine, J.D. (1987) J. Biol. Chem. 262, 9702-9708.
8. Choi, H.U., Johnson, T.L., Subhash, P., Tang, L.-H., Rosenberg, L.C. and Neame, P.J. (1989) J. Biol. Chem. 264, 2876-2884.
9. Vogel, K.G. and Heinegård, D. (1985) J. Biol. Chem. 260, 9298-9306.
10. Krusius, T. and Ruoslahti, E. (1986) Proc. Natl. Acad. Sci. (USA) 83, 7683-7687.
11. Day, A.A., McQuillan, C.I., Termine, J.D. and Young, M.F. (1987) Biochem. J. 248, 801-805.
12. Yamaguchi, Y. and Ruoslahti, E. (1988) Nature 336, 244-246.
13. McBride, O.W., Fisher, L.W. and Young, M.F. (1990) Genomics 6, 219-225.
14. Fisher, L.W., Termine, J.D. and Young, M.F. (1989) J. Biol. Chem. 264, 4571-4576.
15. Oldberg, A., Antonsson, P., Lindblom, K. and Heinegård, D. (1989) EMBO J. 8, 2601-2604.

Larry W. Fisher:
Bone Research Branch, NIDR, NIH,
Bethesda, MD, USA

Elastin

Elastin[1] is the elastic protein that is found in abundance in tissues that undergo repeated stretching, such as major blood vessels and lung. Elastin has an unusual amino acid composition characterized by a low content of acidic and basic amino acids and is correspondingly rich in hydrophobic residues, particularly valine. One third of the residue are glycine and one ninth are proline. In addition, elastin contains several lysine derivatives that serve as covalent crosslinkages between protein monomers. Insoluble elastin is stable under normal physiological conditions, perhaps lasting for the life of the organism[2].

The elastic fibre is a complex structure that contains elastin, microfibrillar proteins, lysyl oxidase, and, perhaps, proteoglycans. Elastin is a predominant protein of mature elastic fibres and endows the fibre with the characteristic property of elastic recoil. Elastic fibres first appear in fetal development as aggregates of 10-12 nm microfibrils arranged in parallel array, often occupying infoldings of the cell membrane. Definitive elastin then is deposited as small clumps of amorphous material within these bundles of microfibrils, which subsequently coalesce to form true elastic fibres. The relative proportion of microfibrils to elastin declines with increasing age with adult elastic fibres having only a sparse peripheral mantle of microfibrillar material[3,4]. It is likely that microfibrils serve to align tropoelastin molecules, the secreted form of elastin, in precise register, so that crosslinking regions are juxtaposed prior to oxidation by lysyl oxidase.

In the extracellular space, tropoelastin appears to be directly and rapidly accreted to the surface of elastic fibres in a monomeric form; there is no evidence that the mole-

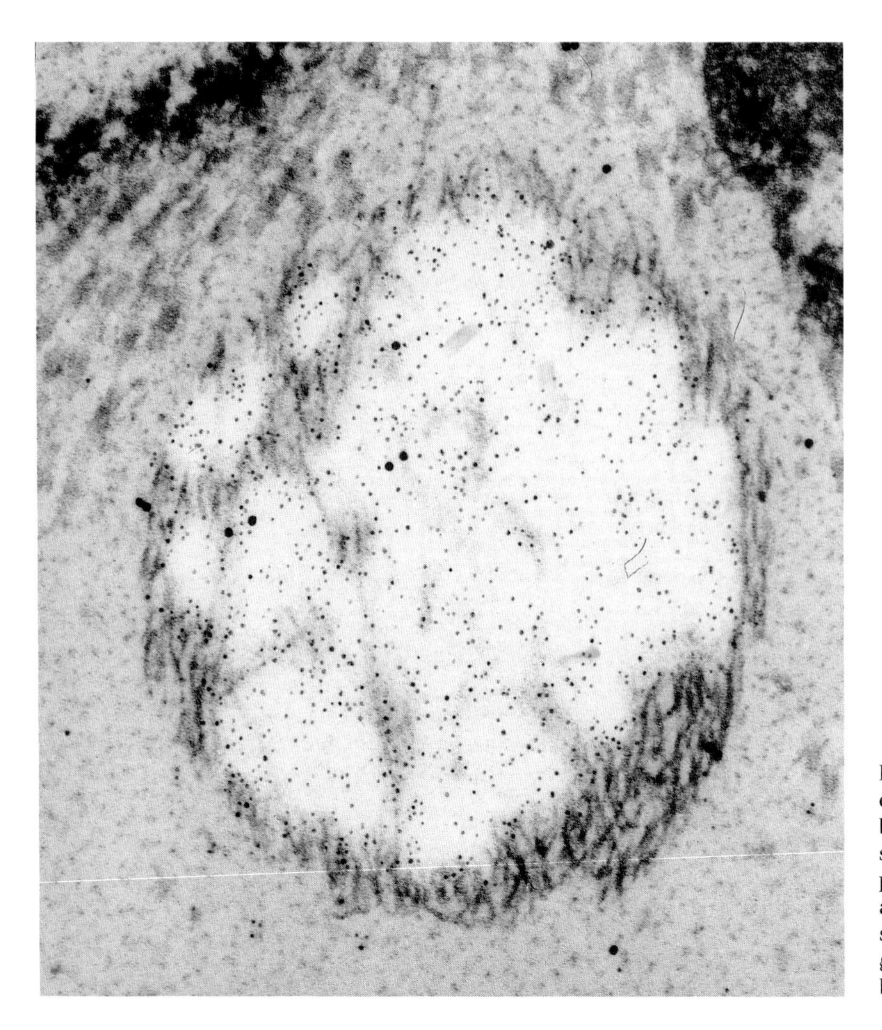

Figure. Electron micrograph of elastic fibre in developing bovine *ligamentum nuchae* showing lightly staining amorphous elastin and tubular-appearing microfibrils. The section was treated with 5 nm gold particles coated with antibodies to elastin. 80,500x.

cule undergoes cleavage prior to crosslinking into the developing fibre. Crosslinking of elastin is inititated by the action of lysyl oxidase which catalyzes the oxidative deamination of lysine to allysine. This appears to be the only enzymatic step involved in elastin crosslinking. The subsequent formation of elastin crosslinks (including the tetrafunctional amino acid isomers desmosine and isodesmosine) probably occurs as a series of spontaneous condensation reactions[5,6].

Elastic fibres can be recognized in the light microscope by characteristic staining reactions, and in the electron microscope by typical ultrastructural appearances (see Figure) Histochemical stains that selectively bind elastic fibres have enabled the visualization of elastic structures in many different tissues. It is clear from these studies that elastin is distributed as interconnected fibres in three morphologically distinct forms. In elastic ligaments, lung and skin, the fibres are small, rope like and variable in length. In major arteries, such as the aorta, elastic fibres form concentric sheets or lamellae while in elastic cartilage a three dimensional honeycomb arrangement of very large anastomosing fibres is apparent. These differing and complex structures are thought to arise as a consequence of the strength and direction of forces put upon the tissue.

PURIFICATION

Elastin is generally isolated from elastic tissues by removing all other connective tissue components by denaturation or degradation. The most successful purification procedures include subjecting the tissue to successive one hour periods of autoclaving until no further protein appears in the supernatant[7], or exposure to 0.1N NaOH at 95°C for 45 min[8]. The insoluble residue left behind is elastin. Several alternative purification methods include treatment of tissues with a combination of proteases, chaotropic and reducing agents, and cyanogen bromide[9]. Generally speaking, however, hot alkali and autoclave techniques give the cleanest elastin. Isolation of tropoelastin is much more difficult since the protein can only be extracted from tissues where the crosslinking pathway is interrupted[9,10].

ACTIVITIES

Elastin is characterized by a high degree of elasticity, including the ability to deform to large extensions with small forces. The precise physicochemical properties that account for elastin's rubber like characteristics, however, have not been fully characterized.

ANTIBODIES

Purified, mature elastin is a weak antigen and anitsera of relatively low titer are obtained when the insoluble protein or elastin derived peptides are used as antigens. Antisera to insoluble elastin or to solubilized peptides contain populations of antibodies that map generally to two immunodominant epitopes on the elastin molecule[11]: antibodies with broad species crossreactivity that map to the alanine-rich crosslinking region and antibodies with little of no species crossreactivity that map to noncrosslinked domains. Antibodies to tropoelastin have higher titers than antibodies to insoluble elastin and tend to show a broader range of crossspecies reactivity[10]. Monoclonal antibodies to elastin have also been described[12].

GENES

Full length cDNAs for human (GenBank M24782)[13], bovine (GenBank J02717)[14,15], and chicken (GenBank M15889)[16,17] elastins have been published. The mRNA for elastin is 3.5 kB of which 2.5 kB encodes the protein. The 3' end of the mRNA has been found to contain a large (1.0-1.2 kB) untranslated region which is highly conserved between the sheep, cow and human[18]. It is interesting that a 258 bp segment in the 3' untranslated region of lysyl oxidase is 93% identical with the 3' untranslated region of elastin[19]. Most evidence currently favours the presence of a single elastin gene[20] of 45 kB with a remarkably high intron to coding exon ration of about 20:1[21,22].

REFERENCES

1. Partridge, S.M. (1962) Adv. Prot. Chem. 17, 227-302.
2. Lefevre, M. and Rucker, R.B. (1980) Biochim. Biophys. Acta 630, 519-529.
3. Greenlee, T.K.J., Ross, R. and Hartman, J.L. (1966) J. Cell Biol. 30, 59-71.
4. Fahrenbach, W.H., Sandberg, L.B. and Cleary, E.G. (1966) Anat. Rec. 155, 563-576.
5. Franzbau, C. (1971) Comprehensive Biochem. 26c, 659-712.
6. Paz, M.A., Keith, D.A. and Gallop, P.M. (1982) Methods Enzymol. 82, 571-587.
7. Partridge, S.M., Davis, H.F. and Adair, G.S. (1955) Biochem. J. 61, 11-21.
8. Lansing, A.I., Roberts, E., Ramasarma, G.B., Rosenthal, T.B. and Alex, M. (1951) Proc. Soc. Exp. Biol. Med. 76, 714-717.
9. Soskel, N.T., Wolt, T.B. and Sandberg, L.B. (1987) Methods Enzymol. 144, 196-214.
10. Prosser, I.W., Whitehouse, L.A., Parks, W.C., Stahle-Bäckdahl, M., Hinek, A., Park, P.W. and Mecham, R.P. (1991) Connect. Tiss. Res. 25, 265-279.
11. Wrenn, D.S. and Mecham, R.P. (1987) Methods Enzymol. 144, 246-259.
12. Wrenn, D.S., Griffin, G.L., Senior, R.M. and Mecham, R.P. (1986) Biochemistry 25, 5172-5176.
13. Indik, Z., Yoon, K., Morrow, S.D., Cicila, B., Rosenbloom, J.C., Rosenbloom, J. and Ornstein-Goldstein, N. (1987) Connect. Tiss. Res. 16, 197-211.
14. Raju, K. and Anwar, R.A. (1987) J. Biol. Chem. 262, 5755-5762.
15. Yeh, H., Ornstein-Goldstein, N., Indik, Z., Sheppard, P., Anderson, N., Rosenbloom, J.C., Cicila, G., Yoon, K. and Rosenbloom, J. (1987) Collagen Rel. Res. 7, 235-247.
16. Bressan, G.M., Argos, P. and Stanley, K.K. (1987) Biochemistry 26, 1497-1503.
17. Tokimitsu, I., Tajima, S., Nishikawa, T., Tajima, M. and Fukasawa, T. (1987) Arch. Biochem. Biophys. 256, 455-461.
18. Rosenbloom, J. (1987) Methods Enzymol. 144(D), 259-288.
19. Trackman, P.C., Pratt, A.M., Wolanski, A., Tang, S.S., Offner, G.D., Troxler, R.F. and Kagan, H.M. (1990) Biochemistry 29, 4863-4870.

20. Fazio, M.J., Mattei, M.G., Passage, E., Chu, M.-L., Black, D., Solomon, E., Davidson, J.M. and Uitto, J. (1991) Am. J. Hum. Genet. 48, 696-703.
21. Kähäri, V.-M., Fazio, M.J., Chen, Y.Q., Bushir, M.M., Rosenblom, J. and Uitto, U. (1990) J. Biol. Chem. 265, 9485-9490.
22. Fazio, M.J., Kähäri, V.-M., Bashir, M.M., Saitta, B., Rosenbloom, J. and Uitto, J. (1990) J. Invest. Dermatol. 94, 191-196.

Robert P. Mecham:
Departments of Cell Biology & Medicine,
Washington University,
St. Louis, MO, USA

Fibrinogen/Fibrin

Fibrinogen is a soluble plasma protein that, after thrombin cleavage, is converted to fibrin monomers; these, in turn, self associate to form an insoluble homopolymeric structure, the fibrin clot. Moreover, fibrinogen binds to the platelet receptor, glycoprotein IIb-IIIa complex (integrin αIIbβ3), and contributes to the formation of the platelet thrombus. The congenital deficiency of fibrinogen results in a bleeding disorder, while increased plasma levels are associated with heightened arterial thrombotic risk.

Fibrinogen is synthesized in hepatocytes[1]. It was thought that platelet fibrinogen derives from synthesis in the megakaryocyte[2], but this concept has been recently disputed[3]. The molecule appears as a trinodular rod, 7x48 nm; it has an apparent molecular mass of 340 kDa and contains three pairs of nonidentical chains Aα (67 kDa), Bβ (56 kDa) and γ (47 kDa)[4] (Figure). The latter exists in two different forms, γ and γ', resulting from alternative mRNA splicing[5]. Both forms are present in plasma fibrinogen (10% contains γ'), but only γ chain is present in platelet fibrinogen[6]. The γ' variant has a 16-residue extension at the C-terminus and, while as effective as γ in clotting, has decreased affinity for platelets[7]. The three genes encoding the fibrinogen subunits are located in close proximity within a 50 kB span on chromosome 4. Each coding sequence is preceded by homologous 5' untranslated regions that may control coordinate expression[8]; however, translation of the corresponding mRNAs is uncoordinated[9]. The rate limiting step for fibrinogen assembly appears to be the synthesis of the Bβ chain, while both Aα and γ chains accumulate in intracellular pools.

Assembly of fibrinogen occurs through complex interchain disulphide bonding involving the N-terminal regions of the three polypeptides; together, these form the central nodular E domain. The globular ends of the molecule represent the D domain, containing the C-termini of Bβ and γ chains. D and E domains are connected by the coiled-coil regions of the three chains, while the Aα C-terminus appears as a flexible appendix of the oblong molecule. One N-linked carbohydrate chain exists on each of the Bβ and γ chains[10].

The active serine protease, thrombin, cleaves an Arg-Gly bond in the N-terminus of the Aα chain, releasing the 16-residue fibrinopeptide A (FPA)[11]. The exposed end of the α chain (so designated after release of FPA) interacts with sites in the γ chain C-termini of neighbouring molecules, initiating the noncovalent end-to-end and lateral assembly of the two molecule thick fibrin protofibrils[12]. Subsequently, a second thrombin cleavage releases a 14-residue peptide (FPB) from the N-terminus of the Bβ chain, and the exposed new β N-terminus reinforces lateral aggregation resulting in the formation of thick fibrin fibres. Ca^{2+}-binding sites in both the E and D domains are important in polymerization. Each fibrinogen molecule contains three high affinity and ten or more low affinity Ca^{2+}-binding sites[13]; the latter correspond to sialic acid residues.

After assembly of the fibrin polymer, the transglutaminase, factor XIII, stabilizes the structure by crosslinking covalently the α and γ chains[14]. Plasmin, a serine protease, degrades proteolytically fibrin, as well as fibrinogen, releasing degradation products (FDPs); this is an essential function for the regulation of normal hemostasis[15]. FDPs have a variety of possible pathophysiological activities[16].

Quantitative defects of the fibrinogen molecule (afibrinogenemia or hypofibrinogenemia) result in an abnormal bleeding tendency[17], while qualitative defects (dysfibrinogenemias) may cause either bleeding or hypercoagulability[18]. Dysfibrinogenemias may result in abnormalities of fibrinopeptide release, fibrin polymerization, crosslinking, or degradation by plasmin[18].

PURIFICATION

Fibrinogen is purified from plasma using differential precipitation methods with either glycine[19] or polyethylene glycol[20]. During the purification procedure, care must be taken to avoid proteolytic degradation, particularly cleavage of the C-terminus of the Aα chain.

ACTIVITIES

In addition to the role in clotting, fibrinogen binds to a specific platelet receptor, the glycoprotein IIb-IIIa complex, and mediates platelet adhesion and aggregation onto thrombogenic surfaces[21]. Fibrinogen also interacts with other distinct **integrin** receptors on endothelial cells[22] and leukocytes[23], but the pathophysiological significance of these interactions remains to be elucidated fully.

A

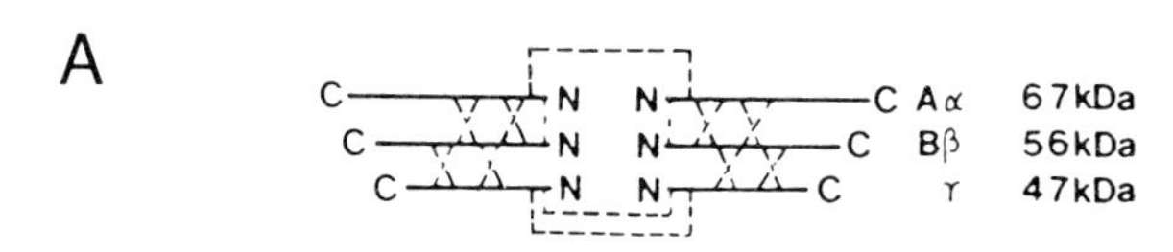

B

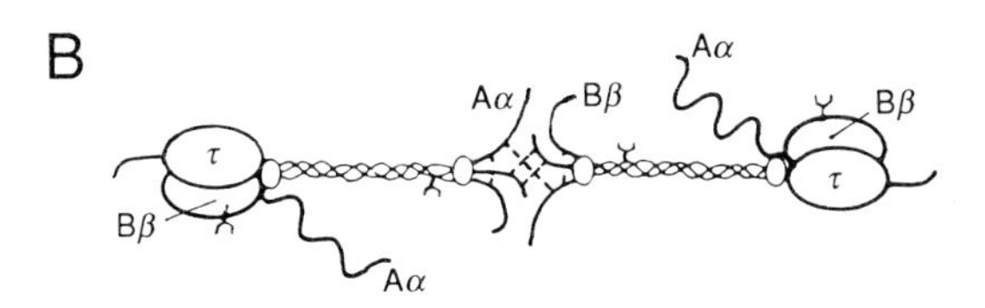

C

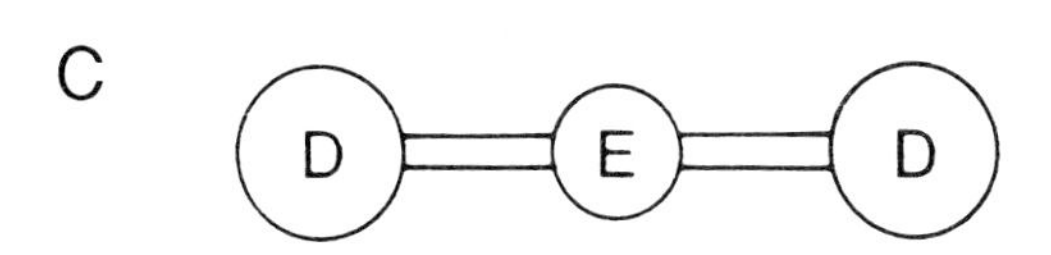

Figure. Schematic representation of the fibrinogen structure (reprinted with permission from Reference 23). (A) Pairs of Aα, Bβ and γ chains are linked by interchain disulphide bonds (represented by broken lines) to form a fibrinogen molecule. (B) The N-terminal domain containing the disulphide bridges is connected to the carboxyl terminal domains by the coiled-coil regions. Note the Aα C-terminus appearing as an appendix extending away from the globular C-termini of the Bβ and γ chains. The forked symbols represent carbohydrate side-chains. (C) Schematic representation of the trinodular structure of fibrinogen. The E domain corresponds to the N-termini of the three pairs of chains, while each D domain contains the C-terminus of one Bβ and γ chain and the random coil region near the C-terminus of the Aα chain. The latter has a flexible C-terminus extending away from the D domain region (shown in b).

ANTIBODIES

Numerous polyclonal and monoclonal antibodies against fibrinogen, its plasmic degradation products, and both FPA and FPB are available commercially (Sigma, St. Louis MO; Accurate Chemical and Scientific Westbury, N.Y.). Monoclonal anti-peptide antibodies have been characterized that react specifically with the putative platelet adhesion sites[22]. Certain monoclonal antibodies react only with surface bound, not soluble fibrinogen, for example fibrinogen bound to its platelet receptor[24,25]. Other antibodies bind to neoantigens on fibrin and not to fibrinogen[26,27].

GENES

Each of the fibrinogen chains is encoded by single copies of three distinct genes located on the long arm of chromosome 4, region q23-q32 (GenBank/EMBL J00128 for α chain; J00129/30/31/32/33 for β chain; M10014 for γ and γ' chains).

REFERENCES

1. Chung, D.W., Que, B.G., Rixon, M.W., Mace Jr., M. and Davie, E.W. (1983) Biochemistry 22, 3244-3250.
2. Uzan, G., Courtois, G., Stanckovic, Z., Crabtree, G.R. and Marguerie, G. (1986) Biochem. Biophys. Res. Commun. 140, 543-549.
3. Louache, F., Debili, N., Cramer, E., Breton-Gorius, J. and Vainchenker, W. (1991) Blood 77, 311-316.
4. Weisel, J.W., Stauffacher, C.V., Bullitt, E. and Cohen, C. (1985) Science 230, 1388-1391.
5. Chung, D.W. and Davie, E.W. (1984) Biochemistry 23, 4232-4236.
6. Francis, C.W., Nachman, R.L. and Marder, V.J. (1984) Thromb. Haemostas. 51, 84-88.
7. Harfenist, E.J., Packham, M.A. and Mustard, J.F. (1984) Blood 64, 1163-1186.
8. Fowlkes, D.M., Mullis, N.T., Comeau, C.M. and Crabtree, G.R. (1984) Proc. Natl. Acad. Sci. (USA) 81, 2313-2316.
9. Yu, S., Sher, B., Kudryk, B. and Redman, C.M. (1983) J. Biol. Chem. 258, 13407-13410.
10. Townsend, R.R., Hilliker, E., Li, Y.T., Laine, R.A., Bell, W.R. and Lee, Y.G. (1982) J. Biol. Chem. 257, 9704-9710.
11. Lewis, S.D., Shields, P.P. and Shafer, J.A. (1985) J. Biol. Chem. 260, 10192-10199.
12. Laudano, A.P. and Doolittle, R.F. (1981) Science 212, 457-459.
13. Nieuwenhuizen, W. and Haverkate, F. (1985) Ann. N.Y. Acad. Sci. 408, 92-96.
14. Doolittle, R.F. (1984) Annu. Rev. Biochem. 53, 195-229.
15. Mihalyi, E. (1983) Ann. N.Y. Acad. Sci. 408, 60-70.
16. Francis, C.W. and Marder, V.J. et al. (1987) In Hemostasis and Thrombosis, Lippincott, pp. 371-373.
17. Mammen, E.F. (1983) Semin. Thromb. Hemostas 9, 1-9.
18. Dang, C.V., Bell, W.R. and Shuman, M. (1989) Am. J. Med. 87, 567-576.
19. Kazal, L.A., Amsel, S., Miller, O.P. and Tocantins, L.M. (1963) Proc. Soc. Exp. Biol. Med. 113, 989-994.
20. Masri, M.A., Masri, S.A. and Boyd, N.D. (1983) Thromb. Haemostas. 49, 116-119.
21. Ikeda, Y., Handa, M., Kawano, K., Kamata, T., Murata, M., Araki, Y., Anbo, H., Kawai, Y., Watanabe, K., Itagaki, I., Sakai, K. and Ruggeri, Z.M. (1991) J. Clin. Invest. 87, 1234-1240.
22. Cheresh, D.A., Berliner, A.S., Vicente, V. and Ruggeri, Z.M. (1989) Cell 58, 945-953.
23. Altieri, D.C., Agbanyo, F.R., Plescia, J., Ginsberg, M.H., Edgington, T.S. and Plow, E.F. (1990) J. Biol. Chem. 265, 12119-12122.
24. Abrams, C.S., Ellison, N., Budzynski, A.Z., Shattil, S.J. (1990) Blood 75, 128-138.
25. Zamarron, C., Ginsberg, M.H. and Plow, E.F. (1989) Blood 74 Suppl. 1, 208a.
26. Scheefers-Borchel, U., Muller-Berghaus, G., Fuhge, P., Eberle, R. and Heimburger, N. (1985) Proc. Natl. Acad. Sci. (USA) 82, 7091-7095.
27. Schielen, W.J.G., Voskuilen, M., Tesser, G.I. and Nieuwenhuizen, W. (1989) Proc. Natl. Acad. Sci. (USA) 86, 8951-8954.

Zaverio M. Ruggeri:
Roon Research Center for Arteriosclerosis and Thrombosis,
Department of Molecular and Experimental Medicine,
Committee on Vascular Biology,
Research Institute of Scripps Clinic,
La Jolla, California, USA

Fibroglycan

Fibroglycan[1] is a major cell surface proteoglycan in human lung fibroblasts. It is an integral type I membrane protein with distinct cytoplasmic, transmembrane and extracellular domains. The extracellular domain is substituted with heparan sulphate chains that mediate high affinity binding interactions with fibronectin and type I collagen fibres.

Fibroglycan is one of the cell surface heparan sulphate proteoglycans expressed by cultured human lung fibroblasts[2,3]. cDNA clones isolated from a human lung fibroblast library[4] and the analysis of the N-terminal sequence of the proteoglycan indicate that human fibroglycan is encoded as a polypeptide of 201 amino acids with a molecular weight of 22,173 which is further processed into a mature protein of 183 amino acids with a molecular weight of 20,218. The cDNA sequence predicts that fibroglycan is a unique protein with discrete cytoplasmic, transmembrane and N-terminal extracellular domains. The extracellular domain contains three serine-glycine sequences which occur in contexts that resemble known glycosaminoglycan attachment sites[5], and has one potential site for N-glycosylation. All are clustered near the N-terminus of the mature protein. Except for some similarities at these glycosylation sites, the sequence of the extracellular domain of fibroglycan is distinct from that of **syndecan**[6,7] and from that of **glypican**[8], two other cell surface proteoglycans that are substituted with heparan sulphate chains. The transmembrane and the cytoplasmic domains of fibroglycan, in contrast, are nearly 60 percent homologous to the corresponding domains of syndecan[6,7]. This structural similarity includes the presence of four tyrosine residues[9]. These findings imply that fibroglycan and syndecan may have distinct extracellular, but similar intracellular interactions. Fibroglycan has a somewhat restricted distribution. It is expressed at high levels by cultured human fibroblasts, but is hardly detectable in several cultured epithelial cell lines.

PURIFICATION

Fibroglycan can be purified from nonionic detergent extracts of cultured human fibroblasts by ion exchange chromatography, gel filtration and affinity chromatography on anti-core protein monoclonal antibodies[3]. The heparitinase treated proteoglycan migrates as a ~48 kDa protein during electrophoresis in SDS-PAGE. It has a strong tendency to self-aggregate, which may lead to the formation of "ladder" patterns.

ACTIVITIES

Fibroglycan from human lung fibroblasts binds with high affinity to type I **collagen** fibres, and to **fibronectin**. This binding is mediated by the heparan sulphate chains.

ANTIBODIES

Two monoclonal antibodies have been isolated that react with human fibroglycan[3]. Both antibodies can be used in immunoblotting and as matrices for affinity purifications.

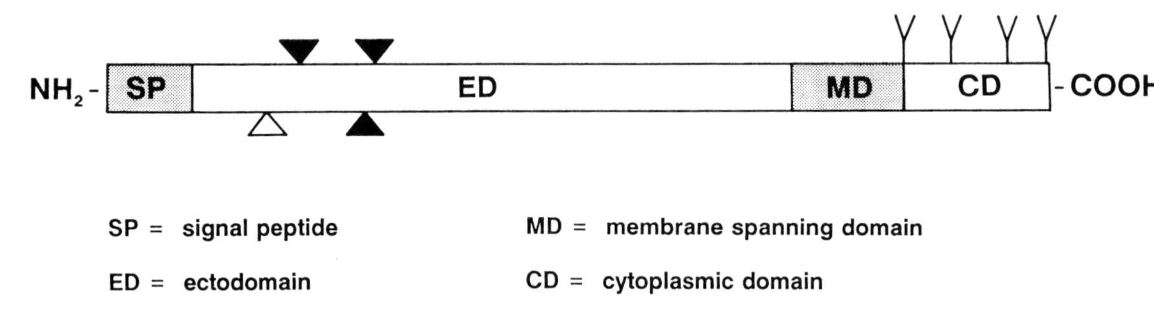

Figure. Model structure of human fibroglycan deduced from its cDNA sequence[4]. The peptide is 202 amino acids long and features a C-terminal cytoplasmic domain (CD), a hydrophobic transmembrane domain (MD), an extracellular domain (ED) with one potential N-glycosylation (△) and three potential glycosaminoglycan attachment (▲) sites, and a typical signal peptide (SP). The mature protein contains no cysteines. The transmembrane and cytoplasmic domains contain four tyrosines (Y), a feature which is also observed in the structurally similar transmembrane and cytoplasmic domains of syndecan[7].

Neither of them works satisfactorily in immunohistochemistry.

GENES

A cDNA sequence for human fibroglycan has been reported[4] (GenBank accession number J04621). Protein sequencing data gathered since the submission of this cDNA sequence suggest that translation is initiated at the first ATG codon of clone 48 K5 (residue 589)[4]. Two major messages can be detected in human lung fibroblasts, and appear to be generated by the alternative use of polyadenylation signals[4]. The gene for fibroglycan has been mapped to chromosome 8q23 by *in situ* hybridisation and the analysis of somatic cell hybrids[4]. Murine fibroglycan cDNA clones have been isolated, revealing 85% identity between human and murine fibroglycan in the extracellular domains and complete sequence identity in the transmembrane and cytoplasmic domains, suggesting that these domains are functionally very important.

REFERENCES

1. David, G. (1990) J. Cell Biol. 111, 9a.
2. Lories, V., De Boeck, H., David, G., Cassiman, J.J. and Van den Berghe, H. (1987) J. Biol. Chem. 262, 854-859.
3. Lories, V., Cassiman, J.J., Van den Berghe, H. and David, G. (1989) J. Biol. Chem. 264, 7009-7016.
4. Marynen, P., Zhang, J., Cassiman, J.J., Van den Berghe, H. and David, G. (1989) J. Biol. Chem. 264, 7017-7024.
5. Bourdon, M.A., Krusius, T., Campbell, S., Schwartz, N.B. and Ruoslahti, E. (1987) Proc. Natl. Acad. Sci. (USA). 84, 3194-3198.
6. Saunders, S., Jalkanen, M., O'Farrell, S. and Bernfield, M. (1989) J. Cell Biol. 108, 1547-1556.
7. Mali, M., Jaakkola, P., Arvilommi, A.M. and Jalkanen, M. (1990) J. Biol. Chem. 265, 6884-6889.
8. David, G., Lories, V., Decock, B., Marynen, P., Cassiman, J.J. and Van den Berghe, H. (1990) J. Cell Biol. 111, 3165-3176.
9. David, G. (1990) Biochem. Soc. Trans. 18, 805-807.

Guido David:
Center for Human Genetics, University of Leuven,
Campus Gasthuisberg 0&N,
Herestraat 49, B-3000
Leuven, Belgium

Fibromodulin

Fibromodulin (or 59 kDa cartilage protein) is a keratansulphate proteoglycan present in many types of connective tissues, e.g. cartilage, tendon and skin. Fibromodulin is structurally related to the dermatan sulphate/chondroitin sulphate proteoglycans decorin and biglycan. Fibromodulin binds to collagen and affects the collagen fibrillogenesis in vitro[1].

The fibromodulin protein backbone consists of 357 amino acid residues (42 kDa) which can be divided into three structural domains[2]. The N-terminal domain has four cysteine residues of which two are involved in an intrachain disulphide bond. This region of the protein also contains five to seven closely spaced tyrosine sulphate residues (Antonsson, Heinegård and Oldberg; unpublished). The central domain, which constitutes 60% of the protein, consists of ten repeats of 25 amino acid residues. This central repeat domain, with preferentially leucine residues in conserved positions, is homologous to similar repeats in a number of proteins including the interstitial proteoglycans **decorin** and **biglycan**[2]. The C-terminal domain contains two cystein residues which form an intrachain disulphide bond.

Fibromodulin from cartilage, tendon and sclera contains asparagine-linked keratan sulphate chains[2,3] (Antonsson, Heinegård and Oldberg; unpublished). Four of the five potential N-glycosylation sites in fibromodulin from bovine articular cartilage is substituted with keratan sulphate chains[3].

PURIFICATION

Fibromodulin can be purified from cartilage using a combination of CsCl gradient centrifugation and ion exchange chromatography[4].

ACTIVITIES

Fibromodulin binds to type I and II **collagen** with a K_d of 35 nM. The protein also delays the collagen fibrillation *in vitro* and causes the formation of thinner fibrills. This collagen binding property is shared by decorin but not by the structurally related biglycan[5,6].

ANTIBODIES

Antibodies raised against the bovine protein crossreacts with human fibromodulin[4].

GENES

The human fibromodulin gene is composed of three exons

and is present as a single copy in the genome (Antonsson, Heinegård and Oldberg; unpublished). The bovine fibromodulin cDNA has the EMBL Data Bank number X16483.

REFERENCES

1. Heinegård, D. and Oldberg, Å. (1989) FASEB J. 3, 2042-2051.
2. Oldberg, Å., Antonsson, P., Lindblom, K. and Heinegård, D. (1989) EMBO J. 8, 2601-2606.
3. Plaas, A.H.K., Neame, P.G., Nivens, C.M. and Reiss, L. (1990) J. Biol. Chem. 265, 20634-20640.
4. Heinegård, D., Larsson, T., Sommarin, Y., Franzen, A., Paulsson, M. and Hedbom, E. (1986) J. Biol. Chem. 261, 13866-13872.
5. Hedbom, E. and Heinegård, D. (1989) J. Biol. Chem. 264, 6898-6905.
6. Brown, D.C. and Vogel, K.G. (1989) Matrix 9, 468-478.

Åke Oldberg:
Department of Physiological Chemistry,
University of Lund,
P.O.Box 94, 22100
Lund, Sweden

Fibronectins

Fibronectins[1,2] are high molecular weight glycoproteins found in many extracellular matrices and in blood plasma. They promote cell adhesion and affect cell morphology, migration and differentiation and cytoskeletal organization. Each subunit is made up of a series of repeating units which in turn form structural and functional domains specialized for binding to cell surface integrin receptors or other extracellular matrix molecules. Different fibronectin isoforms arise by alternative splicing of the transcript of a single gene.

The subunits of fibronectins vary in size between approximately 235 and 270 kDa plus carbohydrate (see below). Each subunit is made up largely of repeating modules of three types (I, II, III). There are 12 type I repeats, each around 45 amino acids long, clustered in three groups (Figure 1), two adjacent type II repeats, each 60 amino acids long, and 15-17 type III repeats, each about 90 amino acids long. Type I and II repeats each contain two disulphide bonds whereas type III repeats lack disulphide bonds. There are two free cysteines per subunit. All three types of fibronectin repeats have been identified in other proteins. Each repeat is encoded by one or two exons with

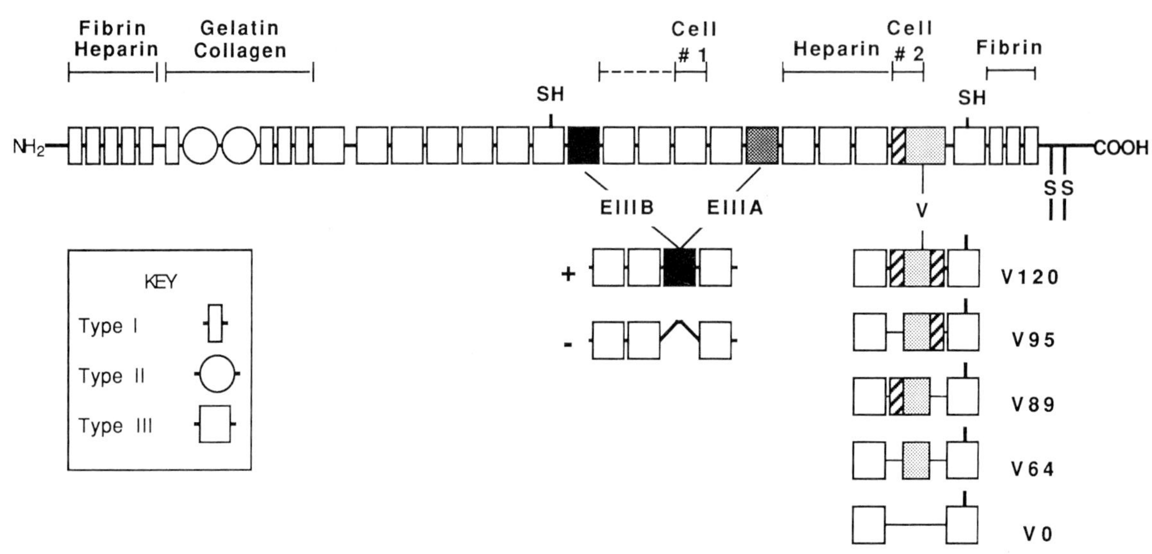

Figure 1. Structure and variants of fibronectins. Diagram shows modular structure of one fibronectin subunit composed of three types of repeat. At three positions (EIIIB, EIIIA and V) alternative splicing produces variations in structure. The splicing of the V region is that seen in humans where the V region varies from 0 to 120 amino acids by inclusion or omission of three segments. Binding sites for other molecules and for cells are marked. The two cell binding sites are recognized by different integrin receptors; #1 which comprises three residues, RGD, is recognized by $\alpha_5\beta_1$ integrin while #2 which comprises EILDV is recognized by $\alpha_4\beta_1$ integrin[9,10]. The efficacy of cell binding site #1 is enhanced by sequences N-terminal to it[8].

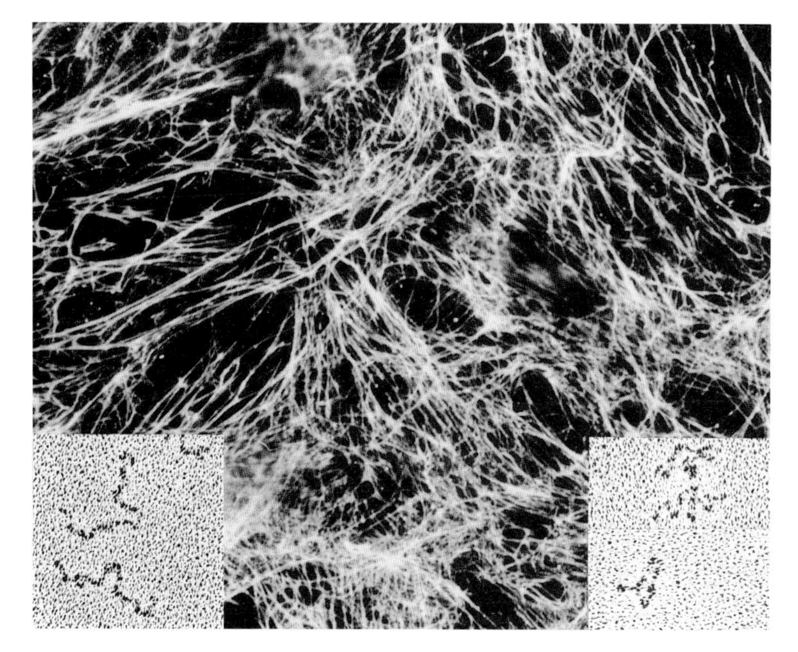

Figure 2. (Main panel) Fibrillar matrix of fibronectin surrounding cells of a confluent layer of fibroblasts as revealed by immunofluorescence. (Insets). Show fibronectin molecules in extended (high salt) form (left) and compact (low salt) form (right) as revealed by rotary shadowing electron microscopy[21].

introns precisely separating repeats. The initial secreted form of fibronectin is a dimer of two subunits held together by a pair of disulphide bonds near their C-termini. Dimeric fibronectins are soluble molecules but, in extracellular matrix fibrils (Figure 2), fibronectins are further disulphide bonded into high molecular weight polymers[1,2].

The variations in subunit size arise from alternative splicing of three segments; two type III repeats (EIIIA or ED.A and EIIIB or ED.B) and a third nonhomologous segment known as V or IIICS. These three alternative splices occur in mammals, birds and amphibians although the precise details of splicing of the V region vary among species. A wide variety of cell types and tissues express fibronectins. The splicing patterns are cell type specific and regulated during development and physiological processes[1,2].

Each subunit carries 5-7 asparagine linked complex carbohydrate side chains and one or two O-linked chains. Fibronectins are typically about 5% carbohydrate but, in some tissues, higher levels of glycosylation occur, usually through further elaboration of the N-linked side chains. Other posttranslational modifications include tyrosine sulphation of the V region and serine/threonine phosphorylation of several sites near the C-terminus. The functions of these posttranslational modifications are unknown except that glycosylation protects the protein from proteolysis[1,2].

The repeating modules of fibronectin subunits fold independently with 20-35% β-structure and no α-helix. The molecule as a whole is extended and is best characterized as a tightly packed string of beads which is flexible. In solution these strings of beads are partially folded, giving a Stokes radius of 10-11 nm and sedimentation coefficient of 13-15 $S_{20,w}$. At extremes of salt or pH, or when assembled in extracellular matrix fibrils the subunits unfold into elongated forms 2-3 nm in diameter; each subunit is 60-70 nm long (Figure 2).

Extended polypeptide segments in certain parts of the molecule are highly susceptible to proteolysis, which generates a series of protease resistant domains, each comprising several of the repeating modules (Figure 1). These domains contain a variety of binding sites for other molecules, including **collagens**, **fibrin**, heparin/heparan sulphate, and cell surface receptors (**integrins**)[1,2].

Fibronectins are widely expressed in embryos and adults especially in regions of active morphogenesis, cell migration and inflammation. Tumour cells show reduced levels of fibronectin and levels in plasma fall in various forms of trauma. In contrast, fibronectin levels are elevated during wound healing and fibrosis[1,2].

PURIFICATION

Fibronectins are most commonly purified by their affinity for denatured collagen (gelatin)[1-4]. Fibronectins bind firmly and specifically to gelatin-Sepharose and can be eluted by chaotropes or low pH.

ACTIVITIES

Fibronectins promote the adhesion and spreading of many cell types by binding to several different integrin receptors[5-10]. Fibronectins also promote cell migration and assembly of ***actin*** microfilament bundles[1,2].

ANTIBODIES

Although relatively conserved in sequence among species, fibronectins are highly immunogenic. Polyclonal and monoclonal antibodies are readily raised and many are available commercially. The polyclonal sera tend to be widely species crossreactive, while the monoclonals tend to be species specific.

GENES

There is a single fibronectin gene in species where this question has been studied. The human gene is at 2q32-36 and the mouse gene is in a syntenic segment of chromosome 1[11]. Complete cDNA sequences have been published for human (GenBank X02761)[12], and rat (GenBank X15906)[13,14] and partial sequences for mouse (GenBank M18194)[15], chicken (GenBank M21554/5, X06533)[16,17] and *Xenopus laevis* (DeSimone, Norton and Hynes; unpublished). A complete amino acid sequence is available for bovine plasma fibronectin[18]. Genomic clones for human (X07717/8)[19], rat (GenBank X05831/2/3/4)[13,14] and chicken[17,20] are available.

REFERENCES

1. Hynes, R.O. (1990) Fibronectins. Springer-Verlag, New York.
2. Mosher, D.F. ed. (1989) Fibronectin. Academic Press, New York.
3. Ruoslahti, E., Hayman, E.G., Pierschbacher, M. and Engvall, E. (1982) Methods Enzymol. 82A, 803-883.
4. Yamada, K.M. (1982) In Immunochemistry of the Extracellular Matrix, vol 1: ed. H. Furthmayer. CRC Press.
5. Hynes, R.O. (1987) Cell 48, 549-554.
6. Ruoslahti, E. and Pierschbacher, M.D. (1987) Science 238, 491-497.
7. Ruoslahti, E. (1988) Ann. Rev. Biochem. 57, 375-413.
8. Obara, M., Kang, M.S. and Yamada, K.M. (1988) Cell 53, 649-657.
9. Guan, J.-L. and Hynes, R.O. (1990) Cell 60, 51-63.
10. Mould, A.P., Wheldon, L.A., Komoriya, A., Wayner, E.A., Yamada, K.M. and Humphries, M.J. (1990) J. Biol. Chem. 265, 4020-4024.
11. Schurr, E., Skamene, E., Forget, A. and Gros, P. (1989) J. Immun. 142, 4507-4513.
12. Kornblihtt, A.R., Umezawa, K., Vibe-Pedersen, K. and Baralle, F.E. (1985) EMBO J. 4, 1755-1759.
13. Patel, R.S., Odermatt, E., Schwarzbauer, J.E. and Hynes, R.O. (1987) EMBO J. 6, 2565-2572.
14. Schwarzbauer, J.E., Patel, R.S., Fonda, D. and Hynes, R.O. (1987) EMBO J. 6, 2673-2580.
15. Blatti, S.P., Foster, D.N., Ranganthan, G., Moses, H.L. and Getz, M.J. (1988) Proc. Natl. Acad. Sci. (USA) 85, 1119-1123.
16. Norton, P.A. and Hynes, R.O. (1987) Mol. Cell Biol. 7, 4297-4307.
17. Kubomura, S., Obara, M., Karasaki, Y., Taniguchi, H., Gotoh, S., Tsuda, T., Higashi, K., Ohsato, K. and Hirano, H. (1987) Biochim. Biophys. Acta 910, 171-181.
18. Skorstengaard, K., Jensen, M.S., Sahl, P., Petersen, T.E. and Magnusson, S. (1986) Eur. J. Biochem. 161, 441-453.
19. Paolella, G., Henchcliffe, C., Sebastio, G. and Baralle, F.E. (1988) Nucl. Acids Res. 16, 3547-3557.
20. Hirano, H., Yamada, Y., Sullivan M., deCrombugghe, B., Pastan, I. and Yamada, K.M. (1983) Proc. Natl. Acad. Sci. (USA) 80, 46-50.
21. Erickson, H.P. and Carrell, N.A. (1983) J. Biol. Chem. 258, 14539-14544.

Richard Hynes:
Howard Hughes Medical Institute and Center for Cancer Research, Department of Biology, Massachusetts Institute of Technology, Cambridge, MA, USA

Fibulin

Fibulin, is an extracellular matrix and plasma glycoprotein[1,2]. It has a multidomain structure containing repeated EGF like elements and a repeated cysteine motif having a potential disulphide loop structure resembling that of the complement component anaphylatoxins C3a, C4a and C5a, as well as proteins of the albumin gene family. Alternative splicing of a prefibulin RNA results in the expression of at least three transcripts encoding forms of fibulin differing in their C-terminal regions.

Fibulin is a cysteine-rich, Ca^{2+}-binding protein first isolated by virtue of its affinity for a synthetic peptide representing the cytoplasmic domain of the **integrin** β1 subunit. The significance of this interaction remains unclear as fibulin was later found to be a secreted glycoprotein. In SDS-PAGE fibulin has a reduced electrophoretic mobility of 100 kDa. In the absence of reducing agents it has a higher electrophoretic mobility and an apparent molecular mass of 80 kDa.

Immunofluorescence staining of cultured fibroblasts with anti-fibulin antibodies has shown that fibulin initially accumulates at cell-substratum attachment sites and eventually forms a web-like extracellular matrix. Purified fibulin, when added to cultured fibroblast monolayers, will also become assembled into a fibrillar matrix (Figure 1). The fibrillar pattern of fibulin accumulation closely resembles that of **fibronectin**. In addition to fibulin being an extracellular matrix molecule it is also present in plasma at a level of 33 μg/ml. Kluge et al.[3] have reported on a 90 kDa extracellular matrix and blood protein from mouse which they term BM-90. Comparison of the sequences of peptide fragments of BM-90 with sequences of human fibulin indicates that the two proteins are likely interspecies-homologues, in that there was 85% identity over the 199 residues compared. BM-90 is a constituent of several basement membranes including Riechart's membrane. These observations agree with immunohistological localization studies of fibulin in the early chicken embryo that showed

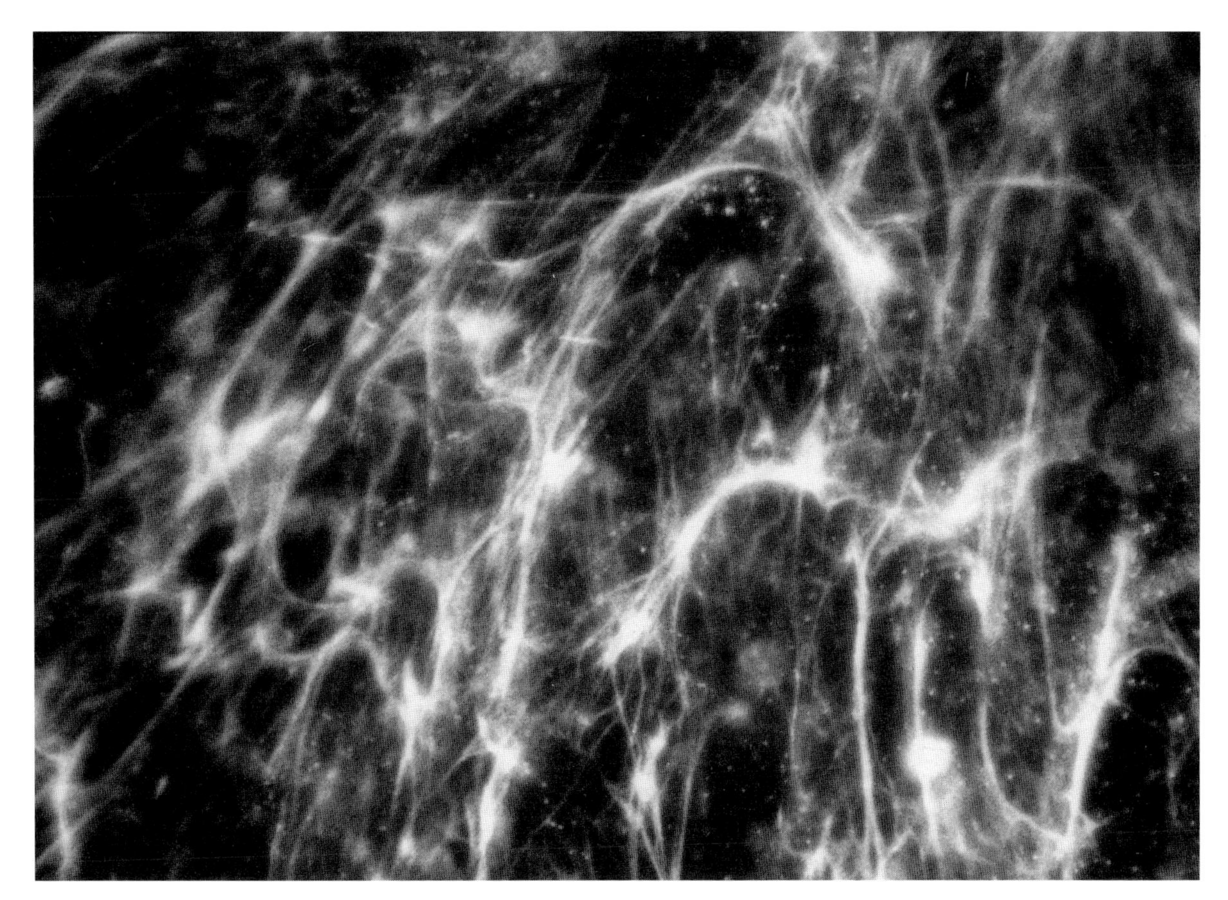

Figure 1. Fluorescent image of a monolayer of human gingival fibroblast after incubation with biotinylated fibulin followed by fluorescein labelled avidin. Exogenously added fibulin is assembled into fibrils in much the same pattern as endogenously expressed fibulin.

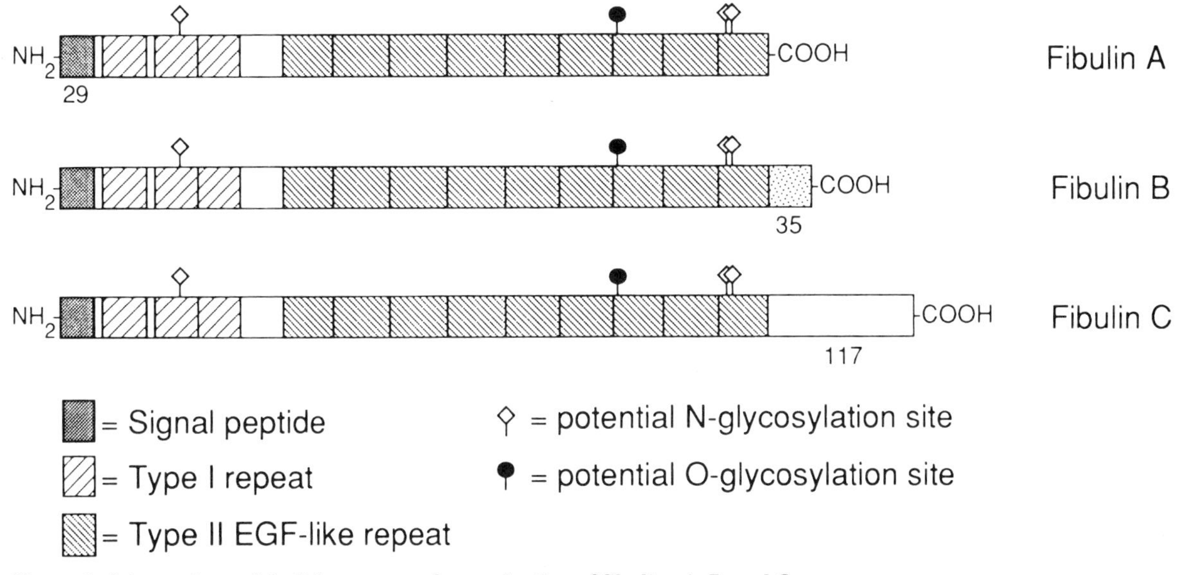

Figure 2. Schematic model of the structural organization of fibulins A, B, and C.

it to be present in many basement membranes and at sites of epithelial-mesenchymal transitions[4].

The complete amino acid sequence of human fibulin has been deduced from cDNA sequences[2]. Alternative splicing of prefibulin mRNA leads to the expression of three fibulin transcripts, 2.2, 2.4 and 2.6 kB in length. The three transcripts encode fibulin polypeptides of 566, 601 and 683 amino acids. The polypeptides are identical in residues 1 through 566. Fibulins B and C differ from A in that they have additional 35 and 117 amino acid C-terminal segments, respectively (Figure).

Analysis of the primary sequence of fibulin reveals that it is a modular protein having several repeated motifs including nine EGF like elements and an element, repeated three times, whose cysteine pattern has homology to that of the complement component polypeptides C3a, C4a and C5a as well as proteins of the albumin gene family. Embodied within four of the EGF like repeats is a consensus sequence for asparagine/aspartic acid hydroxylation[3]. Such hydroxylated residues have been found within the EGF like elements of the vitamin K-dependent blood coagulation proteins and have been implicated in Ca^{2+}-binding. Amino acid analysis of fibulin confirms that it contains ß-hydroxylated asparagine/aspartic acid[5]. Whether such residues play a role in the Ca^{2+}-binding activity of fibulin remains to be determined.

PURIFICATION

Fibulin is purified from extracts of human placenta by affinity chromatography on a synthetic peptide corresponding to the cytoplasmic domain of the integrin β_1 subunit[1] or by immunoadsorbtion on mouse monoclonal anti-fibulin IgG Sepharose[2]

ACTIVITIES

Fibulin incorporates into a fibrillar extracellular matrix when expressed by cultured fibroblasts or added exogenously to cell monolayers[2]. Fibulin is a high affinity Ca^{2+}-binding protein[1]. Recent findings have revealed that fibulin binds to fibronectin and **fibrinogen**.

ANTIBODIES

Rabbit polyclonal and mouse monoclonal antibodies to human placental fibulin have been prepared[1,2]. The polyclonal serum has been shown to crossreact with mouse and rat fibulin.

GENES

The sequences for three alternatively spliced human placental fibulin cDNAs have been published[2] (EMBL/GenBank accession numbers X53741, X53742 and X53743; PIR accession numbers A36346, B36346, C36346).

REFERENCES

1. Argraves W. S., Dickerson, K., Burgess, W.H. and Rouslahti, E. (1989) Cell 58, 623-629.
2. Argraves W. S., Tran, H., Burgess, W.H. and Dickerson, K. (1990) J. Cell Biol. 111, 3155-3164.
3. Kluge, M., Mann, K., Dziadek, M. and Timpl, R. (1990) Eur. J. Biochem. 193, 651-659.
4. Spence, S.G., Argraves, W.S., Walters, L., Hungerford, J.E. and Little, C.D. (1992) Dev. Biol. 151, 473-484.
5. Stenflo J., Ohlin, A.-K., Owen, W.G. and Schneider, W.J. (1988) J. Biol. Chem. 263, 21-24.

W. Scott Argraves:
Biochemistry Laboratory,
American Red Cross,
Rockville, MD, USA

Glypican

Glypican[1] is a glycosyl phosphatidylinositol linked heparan sulphate proteoglycan which is expressed at the surface of cultured human lung fibroblasts and of several other cell types. It binds to fibronectin, type I collagen fibers and to antithrombin III through the intermediate of its heparan sulphate chains. This glypiated proteoglycan may therefore modulate the attachment and shape of the cells, the deposition and assembly of the extracellular matrix, and the activity of extracellular proteinases.

Glypican is one of the cell surface heparan sulphate proteoglycans expressed by cultured human lung fibroblasts[2,3]. It is specifically and selectively depleted from the liposome intercalatable proteoglycan fraction of these cells after exposure to phosphatidylinositol-specific phospholipase C[1]. cDNA clones isolated from a human lung fibroblast library indicate that the glypican core protein is encoded as a polypeptide of 558 amino acids, with a predicted M_r of 61,649[1]. It has a unique sequence which distinguishes glypican from other known membrane associated heparan sulphate proteoglycans[4,5]. The C-terminal end of the encoded polypeptide features a short hydrophobic signal peptide like sequence which may function as a marker for the attachment of the glycosyl

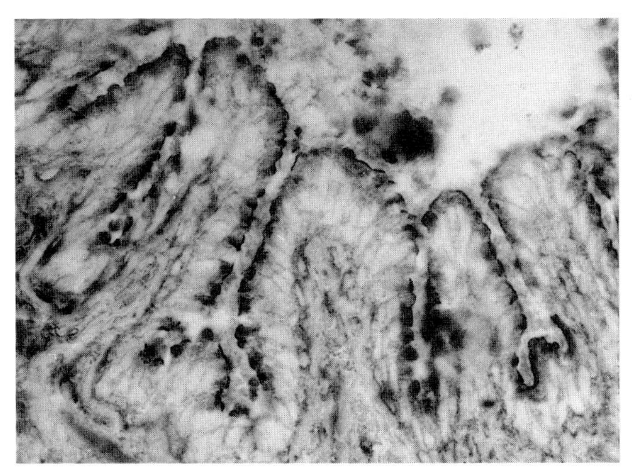

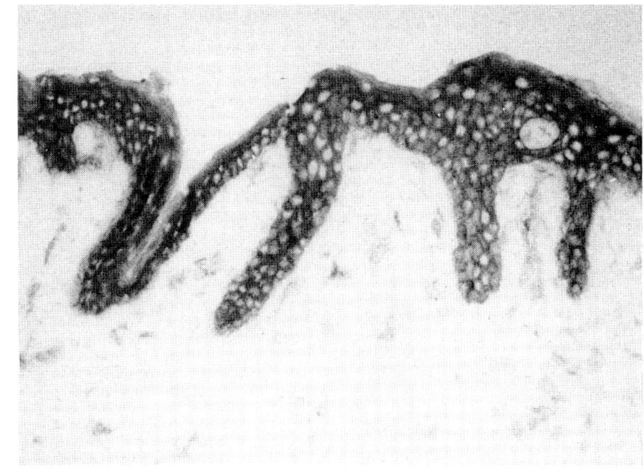

Figure. Indirect immunoperoxidase staining of glypican in a human bronchiole (left) and in human skin (right) using monoclonal antibody S1. The apical pole of the respiratory epithelial cells is heavily stained. In skin the stain surrounds the epidermal keratinocytes and dermal fibroblasts.

phosphatidylinositol membrane anchor. The mature glypiated core carries N-linked oligosaccharide and up to four heparan sulphate chains, and migrates as a peptide with an apparent molecular mass of ~56 kDa after complete reduction and deglycosylation[3].

Glypican is rapidly and quantitatively shed to the culture medium by confluent fibroblasts. The shed and membrane associated proteoglycans have similar N-termini, but the core protein of the shed proteoglycan is slightly smaller (3 kDa) than the core protein of the membrane associated form and lacks the hydrophobic properties of the latter[1]. Glypican is also expressed by human endothelial, smooth muscle and several epithelial cells. Its possible relationship to glypiated heparan sulphate proteoglycans that have been described in liver[6], ovarian granulosa[7] and Schwann cells[8] remains to be determined. Histological studies suggest that in certain epithelia glypican is mainly accumulating at the apical pole of the cells (Figure).

■ PURIFICATION

The shed form of glypican can be purified from conditioned culture media by CsCl density gradient ultracentrifugation, ion exchange chromatography and gel filtration[9] or by immunoaffinity chromatography using anti-glypican monoclonal antibodies[1]. The membrane-associated form has been purified from nonionic detergent extracts of the cells, by a combination of ion exchange chromatography, intercalation into liposomes and immunoaffinity chromatography[1,3].

■ ACTIVITIES

As isolated from human lung fibroblasts, glypican binds with high affinity to type I **collagen** fibers and to **fibronectin**. Glypican is also specifically enriched in the cell surface heparan sulphate proteoglycan fraction from human umbilical vein endothelial cells that binds with high affinity to antithrombin III. The physiological relevance of these binding interactions remains to be established. They imply, however, that glypican may play a role in processes of cell attachment and in controlling the anticoagulant properties of the vascular wall.

■ ANTIBODIES

Two monoclonal antibodies have been isolated that react with the core protein of human glypican[1,10]. Both can be used for immunoblotting, immunoprecipitation and immunohistochemistry. Monoclonal antibody S1 reacts only with the unreduced core protein[10]. Monoclonal antibody F81-1G11 reacts with both the reduced and unreduced forms of the core protein and with a recombinant β galactosidase- glypican fusion protein[1].

■ GENES

A cDNA sequence is available for human glypican (EMBL Data Base access number X54232). A single ~3.8 kB message was detected in human lung fibroblasts1.

■ REFERENCES

1. David, G., Lories, V., Decock, B., Marynen, P., Cassiman, J.J. and Van den Berghe, H. (1990) J. Cell Biol. 111, 3165-3176.
2. Lories, V., De Boeck, H., David, G., Cassiman, J.J. and Van den Berghe, H. (1987) J. Biol. Chem. 262, 854-859.
3. Lories, V., Cassiman, J.J., Van den Berghe, H. and David, G. (1989) J. Biol. Chem. 264, 7009-7016.
4. Marynen, P., Zhang, J., Cassiman, J.J., Van den Berghe, H. and David, G. (1989) J. Biol. Chem. 264, 7017-7024.
5. Saunders, S., Jalkanen, M., O'Farrell, S. and Bernfield, M. (1989) J. Cell Biol. 108, 1547-1556.
6. Ishihara, M., Fedarko, N.S. and Conrad, H.E. (1987) J. Biol. Chem. 262, 4708-4716.

7.Yanagashita, M. and McQuillan, D.J. (1989) J. Biol. Chem. 264, 17551-17558.
8.Carey, D.J. and Evans, D.M. (1989) J. Cell Biol. 108, 1891-1897.
9.Lories, V., David, G., Cassiman, J.J. and Van den Berghe, H. (1986) Eur. J. Biochem. 158, 351-360.
10.Deboeck, H., Lories, V., David, G., Cassiman, J.J. and Van den Berghe, H. (1987) Biochem. J. 247, 765-771.

Guido David :
Center for Human Genetics,
University of Leuven, Campus Gasthuisberg 0&N,
Herestraat 49, B-3000
Leuven, Belgium

HB-GAM

HB-GAM (heparin binding growth associated molecule) is a developmentally regulated secretory protein that is exceptionally rich in cationic amino acids (24%) and in cysteine (7%). It is strongly expressed in brain during the rapid growth phase of axons and dendrites, and it promotes the outgrowth of neurites in vitro. HB-GAM is suggested to function as a differentiation factor of neural cells.

HB-GAM (p18) was isolated from early postnatal rat brain[1,2] and from adult bovine brain[3] as a protein involved in neuronal differentiation, and from bovine uterus as a protein with both mitogenic and neurite outgrowth promoting activities (pleiotrophin)[4,5]. HB-GAM binds strongly to heparin; it is eluted from heparin-Sepharose at 1.1-1.3 M NaCl in salt gradients. The apparent molecular mass in SDS-PAGE is about 18 kDa. These properties are strikingly similar to those of the basic fibroblast growth factor (bFGF), but HB-GAM is immunochemically distinct from bFGF[1].

The HB-GAM sequence consists of 136 amino acids preceded by a cleavable secretion signal[2,5]. The sequence is highly conserved across human, rat and bovine species[5]. The high content of cationic amino acids (24%) is mainly due to lysine residues, which are clustered to the N- and C-terminal parts of the molecule. The cysteine residues have been found to form five intrachain disulphide bonds[3]. The sequence contains several potential O-glycosylation sites

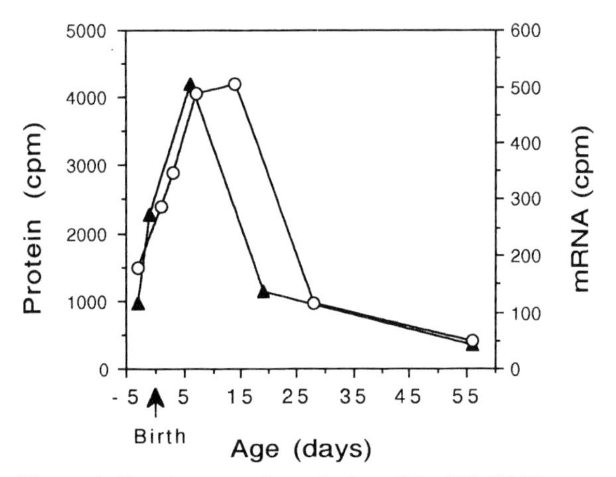

Figure 1. Developmental regulation of the HB-GAM mRNA (o-o) and its protein product (▲-▲) during the development of rat brain.

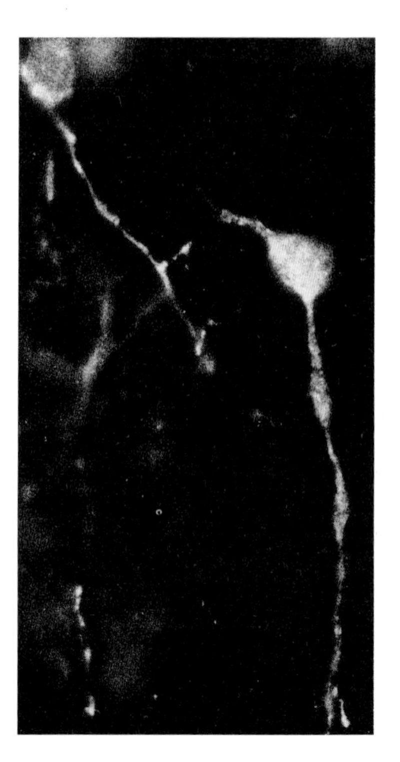

Figure 2. Immunofluorescence staining of HB-GAM in rat brain neurons with the aid of anti-synthetic peptide antibodies that bind to the N-terminal sequence of the protein[1].

but no N-glycosylation sites. The protein from adult bovine brain was reported to lack carbohydrate[3].

The HB-GAM sequence is about 50% identical as compared to the retinoic acid induced gene MK (retinoic acid induced differentiation factor)[6,7]. The HB-GAM and MK genes thus appear to define a novel gene family that plays a role in cell differentiation. A 19 kDa heparin binding protein from chicken basement membranes may belong to the same protein family since the reported N-terminal sequence[8] is clearly homologous with that of HB-GAM.

HB-GAM is strongly expressed in brain, where both the mRNA and the protein peak in a characteristic manner, closely corresponding to the rapid postnatal spurt of growth of rat brain (Figure 1). In dissociated rat brain cells HB-GAM is mainly detected on neurons by immunofluorescence microscopy[1], where the protein is found in developing axonal processes (Figure 2). HB-GAM is also found in the cell body in a granular pattern suggesting that it is a secretory component of neuronal cells[1]. A developmentally regulated expression occurs in some other tissues[2,5,9], for example in uterus, intestine, kidney, muscle, lung and skin. The cell types in these tissues that express HB-GAM have not been identified. No expression or a very low expression was found in liver and spleen[2].

PURIFICATION

HB-GAM can be purified by heparin-Sepharose chromatography using a gradient of 0.1-2 M NaCl for elution[1]. Further purification can be achieved by recycling through heparin-Sepharose or on Affi-Gel blue[1], or by using FPLC[4] or HPLC[3,5].

ACTIVITIES

The biological role of HB-GAM as a neural differentiation factor is suggested by the following findings: (1) The developmental expression follows the rapid phase of axonal growth, and HB-GAM is found as associated to developing axonal processes[1,2]. (2) HB-GAM enhances neurite outgrowth *in vitro* in rat brain cells[1,5] and in PC 12 cells[3]. (3) The sequence of HB-GAM is strongly homologous with the MK gene implicated in the differentiation of teratocarcinoma cells[6,7]. Controversial results have been reported on the mitogenic role of HB-GAM[3-5].

ANTIBODIES

Rabbit antibodies to N-terminal peptides[1,5] and to the protein isolated from rat brain[2] have been published.

GENES

cDNA's that encode the whole amino acid sequence of rat brain HB-GAM (GenBank™/EMBL Data Bank J05657)[2] and a partial N-terminal sequence[10] have been published. The same sequence is reported for rat brain pleiotrophin (PTN)[5]. Human and bovine sequences (EMBL nucleotide sequence data bases X52945 and X52946) that exhibit 98% identity with the rat sequence have been published[5].

REFERENCES

1. Rauvala, H. (1989) EMBO J. 8, 2933-2941.
2. Merenmies, J. and Rauvala, H. (1990) J. Biol. Chem. 265, 16721-16724.
3. Kuo, M.-D., Oda, Y., Huang, J.S. and Huang, S.S. (1990) J. Biol. Chem. 265, 18749-18752.
4. Milner, P. G., Li, Y.-S., Hoffman, R.M., Kodner, C.M., Siegel, N.R. and Deuel, T.F. (1989) Biochem. Biophys. Res. Commun. 165, 1096-1103.
5. Li, Y.-S., Milner, P.G., Chauhan, A.K., Watson, M.A., Hoffman, R.M., Kodner, C.M., Milbrandt, J. and Deuel, T.F. (1990) Science 250, 1690-1694.
6. Kadomatsu, K., Tomomura, M. and Muramatsu, T. (1988) Biochem. Biophys. Res. Commun. 151, 1312-1318.
7. Matsubara, S., Tomomura, M., Kadomatsu, K. and Muramatsu, T. (1990) J. Biol. Chem. 265, 9441-9443.
8. Vigny, M., Raulais, D., Puzenat, N., Hartmann, M.P., Jeanny, J.C. and Courtois, Y. (1989) Eur. J. Biochem.186, 733-740.
9. Tezuka, K., Takeshita, S., Hakeda, Y., Kumegawa, M., Kikuno, R. and Hashimoto-Kotoh, T. (1990) Biochem. Biophys. Res. Commun. 173, 246-251.
10. Kovesdi, I., Fairhurst, J.L., Kretschmer, P.J. and Böhlen, P. (1990) Biochem. Biophys. Res. Commun. 172, 850-854.

Heikki Rauvala and Jussi Merenmies:
Departmment of Medical Chemistry and the
Institute of Biotechnology,
University of Helsinki,
Helsinki, Finland

Hyaluronan and Hyaluronan Binding Proteins (Hyaladherins)

Hyaluronan (HA) is a high molecular weight, highly anionic polysaccharide composed of repeating disaccharides of β-1,4-glucuronate-β-1,3-N-acetylglucosamine. HA is found in extracellular matrices, where it plays a structural role, and at cell surfaces, where it influences cell behaviour. Important to both types of function are its unique physicochemical properties and its interactions with hyaluronan binding proteins (HABPs). HA binding proteoglycans and link protein contribute to the structure of extracellular matrices via interactions with HA; cell surface associated HABPs are thought to act as HA receptors which mediate the effects of HA on cell behavior. These two groups of HABPs have homologous HA binding domains and are termed the hyaladherins[1,2].

Hyaluronan varies in molecular mass between ~100-5,000 kDa. In aqueous solution HA adopts a voluminous configuration due to intramolecular charge repulsion and, at concentrations found in several extracellular matrices, it forms an entangled, continuous network via molecular overlapping. Consequent on these properties are its high viscosity and its dramatic effects on molecular exclusion, flow resistance, tissue osmosis, etc.[3,4], all of which are likely to be important in HA rich matrices such as in synovial fluid, umbilical cord, dermis, and subcutaneous tissue. Reduction of HA concentrations in these tissues would result in osmotic imbalances, tissue adhesions, or decreased matrix hydration. In cartilage, the major function of HA is its contribution to matrix integrity via ternary and multivalent interaction with proteoglycan and **link protein**; this type of interaction contributes to the structure of other tissues also[5].

Cell surface HA binding proteins that may act as HA receptors have been partially characterized. Interactions of HA with these HABPs mediate or influence aspects of cell behaviour, especially cell adhesion and migration[1,2]. The best characterized of the HABPs is the 85 kDa HA receptor of BHK cells[6]. Recently this protein was found to be the same as, or closely related to **CD44**, a lymphocyte surface protein involved in homing to mucosal lymph nodes and in other immune functions[7]. These and related HABPs are present on the surface of a wide variety of cell types, especially embryonic and transformed cells[1,2].

The HA-binding domains of link protein[8] and of the core proteins from the large proteoglycans of cartilage (**aggrecan**)[9] and fibroblasts (**versican**)[10] have been sequenced and found to contain extensive homologies. Likewise CD44, and thus the 85 kDa HA receptor, has been shown to contain a region of significant homology to these proteoglycans and link protein[11,12]. However the HA binding properties of the matrix and cell surface classes of HABPs are not identical[1,2]. The precise sequence requirements for interaction of these proteins with HA is not yet known.

PURIFICATION

HA can be extracted and purified by a variety of methods, depending on the source. Common methods are extraction by protease digestion or with chaotropic agents, followed by ion exchange chromatography, molecular sieve chromatography or CsCl density gradient centrifugation[13]. Complete purification of cell surface HABPs has not been published; partial methods include hydroxyapatite, lectin affinity, HA affinity, ion exchange and molecular sieve chromatography[15-17]. Antibodies are now available[6,17] and further purification by immunoaffinity chromatography is in progress.

ACTIVITIES

Interaction of HA with proteoglycan is usually measured by increase in size due to aggregation, as assessed by molecular sieve chromatography[13]. Binding of radiolabelled HA to proteins can be measured by ammonium sulphate precipitation, where free HA remains soluble and protein-bound HA precipitates[18]. Binding of HA to cells or membranes is also readily measured by the latter method, or by centrifugation subsequent to incubation[19].

ANTIBODIES

Monoclonal antibodies to the 85 kDa HA receptor (MAb K3)[6] and to a related HABP isolated from chick embryo brain (MAb IVd4)[17] are available. The former appears to recognize only hamster antigen, whereas the latter recognizes chick, mouse, rat, bovine, and human antigen; both antibodies have been used successfully for immunoblotting, immunoprecipitation and immunohistology. A variety of antibodies are available to CD44[7], link protein[20] and the core proteins of HA-binding proteoglycans[20,21].

GENES

Complete cDNA sequences are published for several variants of CD44[11,12,22-24], (GenBank M22452, M61874 and M61875), link protein[8] (GenBank M13212), and the core proteins of HA-binding proteoglycans[9,10] (GenBank M55172 and J02814).

REFERENCES

1. Toole, B.P. (1990) Current Opinion Cell Biol. 2, 839-844.
2. Toole, B.P. (1991) In: "Cell Biology of Extracellular Matrix", ed. E. Hay, 2nd edition, Plenum, New York, in press.
3. Comper, W.D. and Laurent, T.C. (1978) Physiol. Rev. 58, 255-315.
4. Meyer, F. (1983) Biochim. Biophys. Acta 755, 388-399.
5. Gallagher, J.T. (1989) Current Opinion Cell Biol. 1, 1201-1218.
6. Underhill, C.B., Green, S.J., Comoglio, P.M. and Tarone, G. (1987) J. Biol. Chem. 262, 13142-13146.
7. Culty, M., Miyake, K., Kincade, P.W., Sikorski, E., Butcher, E.C. and Underhill, C. (1990) J. Cell Biol. 111, 2765-2774.
8. Deak, F., Kiss, I., Sparks, K.J., Argraves, W.S., Hampikian, G. and Goetinck, P.F. (1986) Proc. Nat. Acad. Sci. (USA) 83, 3766-3770.
9. Doege, K.J., Sasaki, M., Kimura, T. and Yamada, Y. (1991) J. Biol. Chem. 266, 894-902.
10. Zimmermann, D.R. and Ruoslahti, E. (1989) EMBO J. 8, 2975-2981.
11. Stamenkovic, I., Amiot, M., Pesando, J.M. and Seed, B. (1989) Cell 56, 1057-1062.
12. Goldstein, L.A., Zhou, D.F.H., Picker, L.J., Minty, C.N., Bargatze, R.F., Ding, J.F. and Butcher, E.C. (1989) Cell 56, 1063-1072.
13. Heinegård, D. and Sommarin, Y. (1987) Methods Enzymol. 144, 319-372.
14. Baker, J.R. and Neame, P.J. (1987) Methods Enzymol. 144, 401-412.
15. Underhill, C.B., Thurn, A.L. and Lacy, B.E. (1985) J. Biol. Chem. 260, 8128-8133.
16. Turley, E.A., Moore, D. and Hayden, L.J. (1987) Biochemistry 26, 2997-3005.
17. Banerjee, S.D. and Toole, B.P. (1991) Dev. Biol. 146, 186-197.
18. Underhill, C.B., Chi-Rosso, G. and Toole, B.P. (1983) J. Biol. Chem. 258, 8086-8091.
19. Goldberg, R.L., Seidman, J.D., Chi-Rosso, G. and Toole, B.P. (1984) J. Biol. Chem. 259, 9440-9446.
20. Caterson, B., Calabro, T. and Hampton, A. (1987) In: "Biology of Proteoglycans", ed. T. Wight and R. Mecham, Academic, New York, pp. 1-26.
21. Krusius, T., Gehlsen, K.R. and Ruoslahti, E. (1987) J. Biol. Chem. 262, 13120-13125.
22. Stamenkovic, I., Aruffo, A., Amiot, M. and Seed, B. (1991) EMBO J. 10, 343-348.
23. Gunthert, U., Hofmann, M., Rudy, W., Reber, S., Zoller, M., Haussman, I., Matzku, S., Wenzel, A., Ponta, H. and Herrlich, P. (1991) Cell 65, 13-24.
24. Idzerda, R.L., Carter, W.G., Nottenburg, C., Wayner, E.A., Gallatin, W.M. and St. John, T. (1989) Proc. Nat. Acad. Sci. (USA) 86, 4659-4663.

Bryan P. Toole:
Dept. of Anatomy and Cellular Biology,
Tufts University,
Boston, MA 02111, USA

J1 Glycoproteins (Janusin)

The J1 glycoproteins form a family of extracellular matrix glycoproteins that are involved in cell-substrate interactions[1,2]. The high molecular weight group of J1 glycoproteins in the mouse are identical to tenascin in chicken. The lower molecular weight group of the J1 glycoproteins is structurally related to the murine tenascin. It has been designated janusin, since it shows a dual face in cell interactions: it mediates adhesion of astrocytes and repulsion of neuronal cell bodies and growth cones in the nervous system.

Janusin is expressed by differentiating oligodendrocytes in the CNS, but not by Schwann cells in the PNS of mammals[3]. Unlike **tenascin**, janusin has so far not been found in nonneural cells. The major forms are a rod like monomeric structure (previously designated J1-160) and a trimeric molecule consisting of disulphide linked subunits of 180 and 200 kDa (J1-180)[3]. Like tenascin, janusin is a multidomain protein consisting of EGF like and **fibronectin** type III repeats and a **fibrinogen** like globular domain[4,5].

Both tenascin and janusin are expressed in the central nervous system by astrocytes and oligodendrocytes, respectively. They mediate neuron-glia recognition which results in a transient adhesion between the J1 substrates and the cell bodies, but then reverses into an interaction that is repulsive for the neuronal cell bodies[6-8]. This reduced interaction appears to underlie the increased rate of neurite extension on the J1 glycoprotein substrates[7]. On tenascin, the neurite outgrowth promoting properties appear to be mediated by a fibronectin type III domain that is distinct from the cell binding domain for fibroblasts[7]. While janusin induces repulsion of neurons in culture after several hours, it promotes adhesion of astrocytes, suggesting that it exerts differential effects on different cell types[6].

Janusin appears first at the initial stages of myelination at the oligodendrocyte-axon contact and at the myelinating loops of oligodendrocyte processes (Bartsch, Pesheva and Schachner, unpublished). In the adult, janusin remains expressed in association with myelin and accumulated at the nodes of Ranvier[9].

PURIFICATION

Janusin is isolated by immunoaffinity chromatography using monoclonal antibody columns from adult mouse brain[3,6].

ACTIVITIES

The biological activity of janusin is best measured *in vitro* in cell-substrate interaction assays[3,6].

ANTIBODIES

Monoclonal antibodies to mouse janusin have been described[3,6].

GENES

The cDNA sequence for janusin is partially known.

REFERENCES

1. Kruse, J., Mailhammer, R., Wernecke, H., Faissner, A., Sommer, I., Goridis, C. and Schachner, M. (1984) Nature 311, 153-155.
2. Kruse, J., Keilhauer, G., Faissner, A., Timpl, R. and Schachner, M. (1985) Nature 316, 146-148.
3. Pesheva, P., Spiess, E. and Schachner, M. (1989) J. Cell Biol. 109, 1765-1778.
4. Fuss, B. and Schachner, M. (1991) Soc. Neurosci. Abstracts, in press.
5. Fuss, B., Pott, U., Fischer, P., Schwab, M.E. and Schachner, M. (1991) J. Neurosci. Res. 29, 299-307.
6. Morganti, M.C., Taylor, J., Pesheva, P. and Schachner, M. (1990) Exp. Neurol. 109, 98-110.
7. Lochter, A., Lloyd, V., Kaplony, A., Prochiantz, A., Schachner, M. and Faissner, A. (1991) J. Cell Biol. 113, 1159-1171.
8. Faissner, A. and Kruse, J. (1990) Neuron 5, 627-637.
9. French-Constant, C., Miller, R.H., Kruse, J., Schachner, M. and Raff, M.C. (1986) J. Cell Biol. 102, 844-852.

M. Schachner:
Department of Neurobiology,
Swiss Federal Institute of Technology,
Hönggeberg, 8093 Zurich,
Switzerland

Laminin

Laminin[1-3] is the best known member of a family of basement membrane proteins. Other members of this family are merosin[4] and S-laminin[5]. At least one member of the laminin family is present in all basement membranes where they are thought to affect adhesion, migration, growth and differentiation of cells.

Laminin is composed of three subunits, the 400 kDa A-chain and two 200 kDa B-chains. The B1- and B2-chains are homologous to each other and to the N-terminal two-thirds of the A-chain. The three subunits assemble into a crosslike structure with one long and three short arms. The N-termini of the B1-, B2-, and A-subunits separately form the three short arms. The stem of the long arm is composed of all three chains folded together in a coiled-coil structure. The C-terminal one-third of the A-chain forms a large globular domain at the end of the long arm. Merosin is a laminin variant with a different heavy chain, the M-chain, which is homologous to the A-chain. The S-chain is homologous to the B1-chain and can replace B1 in both laminin and merosin. The five known subunits can

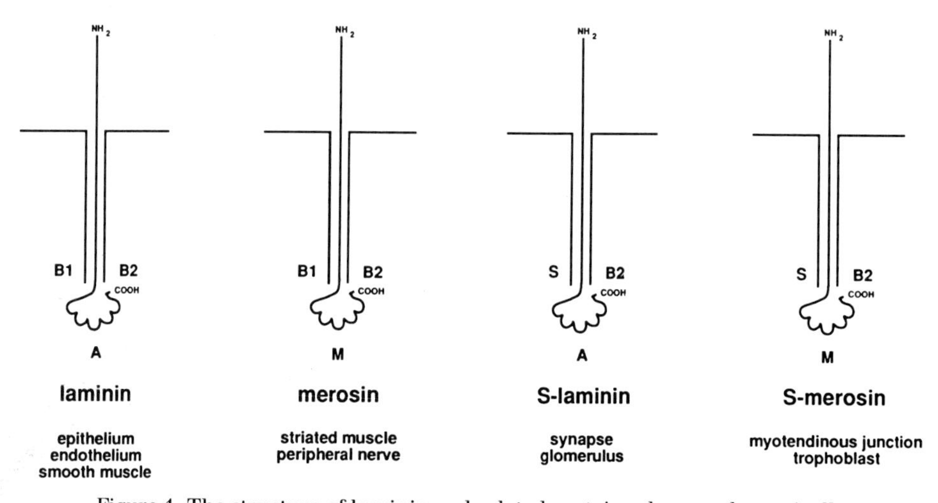

Figure 1. The structure of laminin and related proteins shown schematically.

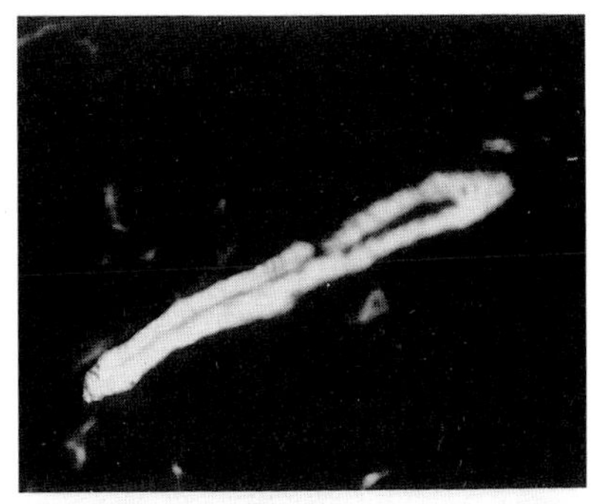

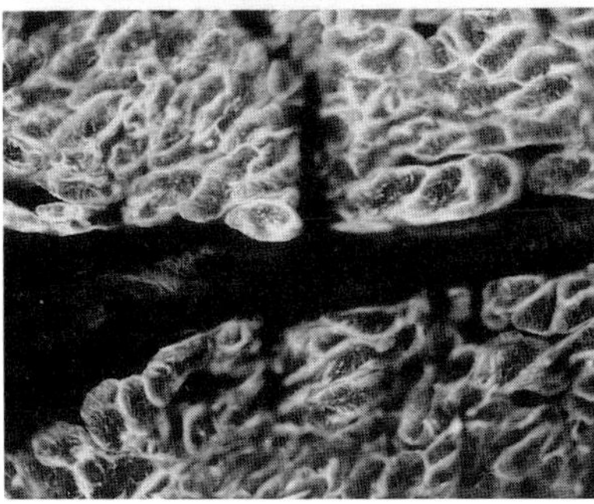

Figure 2. Indirect immunofluorescence showing the distribution of laminin and merosin in heart. The laminin A-chain (top panel) is present in blood vessels, while the merosin M-chain is present in the heart muscle basement membrane.

assemble into four different isoforms with different tissue distributions[6] (Figures 1 and 2).

Laminin has profound effects on cells. *In vitro* it promotes adhesion, spreading, migration, differentiation, and growth. The most studied activities of laminin are its cell adhesive and neurite promoting activities. These are mediated by integrin type cell surface receptors, at least six of which bind to laminin[7]. The major sites of interaction of these receptors with laminin appear to be located at the end of the long arm of laminin and near the center of the laminin cross. A number of synthetic peptides made from sequences in laminin subunits have been found to affect cells in various ways. Whether these peptides mimic the biological function of laminin or express nonreceptor mediated activities has not yet been established. Laminin binds to itself in a Ca^{2+}-dependent manner and also interacts with other basement membrane components including entactin, heparan sulphate proteoglycan, and type IV collagen.

TABLE
Genes for laminin subunits; cDNA sequences and chromosomal locations

Species	subunit	GenBank #	chromosome location	references
Human	A		8p11.3	(18,19)
	M	M32076		(4)
	B1	J02778	7q22	(see 2)
	B2	J03202	1q31	(see 2)
Mouse	AJ	04064		(see 2)
	B11			(see 2; 20)
	B2	J034841		(see 2)
Rat	S			(5)
Drosophila	B1			(see 2)
	B2	X07806		(see 2)
A	M75882			(21)

PURIFICATION

Laminin has been purified from embryonal mouse (EHS) and rat tumours[8], human placenta[9,10], chick heart[11], and *Drosophila* cell cultures[12]. Merosin has been purified from mouse and bovine heart[13,14] and from human placenta[4]. Conventional biochemical methods have been used as well as antibody[10] and heparin[4] affinity chromatography. The isolated S-chain has been purified in a denatured form from renal cortex[5].

ACTIVITIES

The cell binding and neurite outgrowth promoting activities of laminin are tested on plastic substrates coated with laminin[15]. Neurons may require addition of nonspecific adhesive agents such as polylysine or polyornithine[16]. Merosin has similar ativities to laminin in these assays[17].

ANTIBODIES

Polyclonal antibodies against mouse laminin from the EHS tumour or rat laminin from yolk sac tumours (with the chain composition A,B1,B2) and against human placental laminin (a mixture of laminin variants) are commercially available (see Linscott's Directory, Mill Valley, CA). Many monoclonal antibodies against mouse, rat, chicken, and human laminin as well as against laminin variants and subunits are available from the Developmental Hybridoma bank, the American Type Culture Collection, commercially, or from individual investigators[4-6,10,11].

GENES

The genes, location on chromosomes and the available GenBank numbers are listed in the table.

REFERENCES

1. Timpl, R. (1989) Eur. J. Biochem. 180, 487-502.
2. Beck, K., Hunter, I. and Engel, J. (1990) FASEB J. 4, 148-160.
3. Yurchenco, P.D. and Schittny, J.C. (1990) FASEB J. 4, 1577-1590.
4. Ehrig, K., Leivo, I., Argraves, W.S., Ruoslahti, E. and Engvall, E. (1990) Proc. Natl. Acad. Sci. 87, 3264-3268.
5. Hunter, D.D., Shah, V., Merlie, J.P. and Sanes, J.R. (1989) Nature 338, 229-234.

6.Engvall, E., Earwicker, D., Haaparanta, T., Ruoslahti, E. and Sanes, J.R. (1990) Cell Reg. 1, 731-740.
7.Ruoslahti, E. (1991) J. Clin. Invest. 87, 1-5.
8.Timpl, R., Rohde, H., Gehron Robey, P., Rennard, S.I., Foidart, J.-M. and Martin, G.R. (1979) J. Biol. Chem. 254, 9933-9937.
9.Dixit, S.N. (1985) Conn. Tiss. Res. 14, 31-40.
10.Wewer, U., Albrechtsen, R., Manthorpe, M., Varon, S., Engvall, E. and Ruoslahti, E. (1983) J. Biol. Chem. 258, 12654-12660.
11.Brubacher, D., Wehrle-Haller, B. and Chiquet, M. (1991) Exp. Cell Res. 197, 290-299.
12.Fessler, L.I., Campbell, A.G., Duncan, K.G. and Fessler, J.H. (1987) J. Cell Biol. 105, 2383-2391.
13.Paulsson, M. and Saladin, K. (1989) J. Biol. Chem. 264, 18726-18732.
14.Paulsson, M., Saladin, K. and Engvall, E. (1991) J. Biol. Chem. 266, 17545-17551.
15.Dillner, L., Dickerson, K., Manthorpe, M., Ruoslahti, E. and Engvall, E. (1988) Exp. Cell Res. 177, 186-198.
16.Manthorpe, M., Engvall, E., Ruoslahti, E., Longo, F.M., Davis, G.E. and Varon, S. (1983) J. Cell Biol. 97, 1882-1890.
17.Engvall, E., Earwicker, D., Day, A., Muir, D., Manthorpe, M. and Paulsson, M. (1992) Exp. Cell Res. 198, 115-123.
18.Haaparanta, T., Uitto, J., Ruoslahti, E. and Engvall, E. (1991) Matrix 11, 151-160.
19.Nagayoshi, T., Mattei, M.-G., Passage, E., Knowlton, R., Chu, M.-L. and Uitto, J. (1989) Genomics 5, 932-935.
20.Elliott, R.W., Barlow, D. and Hogan, B.L.M. (1985) In Vitro 21, 477-484.
21.Garrison, K., MacKrell, A.J. and Fessler, J.H. (1991) J. Biol. Chem. 266, 22899-22904.

Eva Engvall:
La Jolla Cancer Research Foundation,
La Jolla, CA 92037, USA

Laminin Binding Proteins (LBP, CBP 35)

Cellular proteins that bind the basement membrane glycoprotein laminin include several glycoprotein heterodimers of the integrin family that function as cell surface laminin receptors. In addition, a number of nonintegrin proteins that range in size from 32 kDa to 110 kDa bind laminin with relatively high affinity. Foremost in this group is a 67 kDa laminin binding protein (LBP) originally thought to be the major laminin receptor, although its role and that of other nonintegrin LBPs in mediating cellular interactions with laminin has remained controversial. The most prominent nonintegrin laminin binding protein expressed on the surface of several cell types is carbohydrate binding protein 35 (CBP 35). Since the structure and carbohydrate binding properties of this protein are well defined, it represents the first nonintegrin LBP whose surface function can be examined rigorously.

CBP 35 is a soluble or S-type lectin[1] that binds to the carbohydrate chains on laminin[2] (Figure 1). CBP 35 is identical to the murine macrophage differentiation antigen Mac-2[3], the rat neuronal antigen RL-29[4], and the human/murine metastasis lectin L-34[5,6], and it exhibits high homology with IgE binding protein[7]. The major form is a monomeric molecule of 29-35 kDa that contains one carbohydrate binding domain. Multimeric forms provide multivalency for carbohydrate binding[8]. CBP 35 is present in the nucleus, cytoplasm, and on the cell surface[9] where it behaves as a peripheral membrane protein. Phosphorylated isoforms of CBP 35 exist and may affect the protein's localization among different cellular compartments[10]. The affinity of CBP 35 for galactose containing oligosaccharides is poly-N-acetyllactosamine > lactosamine > lactose > galactose[11]. This specificity explains its high affinity for laminin because laminin is one of the few known glycoproteins to express an abundance of poly-lactosamine carbohydrates. cDNA cloning and sequencing revealed that CBP 35 is comprised of two distinct domains: an N-terminal domain that has a high content of proline and glycine residues and a C-terminal domain that contains a highly conserved carbohydrate (galactose) binding sequence[12].

CBP 35 is expressed by many cell types including macrophages, epithelial cells, fibroblasts, trophoblasts, and carcinoma cells. Expression may correlate with processes (differentiation, inflammation, invasion) that involve cell migration on laminin containing substrata (Figure 2). However, the involvement of this LBP in mediating cellular interactions with laminin, though likely, has not been demonstrated. The putative transmembrane cell surface ligand for CBP 35 is also not yet identified.

PURIFICATION

CBP 35 is purified routinely by affinity chromatography of cell extracts using laminin[2]-, asialofetuin[13]-, or lactose-Sepharose. High salt (0.5 M NaCl) removes most proteins that bind to these affinity matrices with lower affinity with the exception of a 14 kDa galactose lectin expressed by some cell types. CBP 35 can be eluted from affinity columns with 0.4 M lactose. CBP 35 is removed from the surface of intact cells by washing with a 50 mM lactose containing physiological buffer. Highly purified protein can be obtained by immuoaffinity chromatography.

ACTIVITIES

CBP 35 binds glycoconjugates that contain poly-N-acetyllactosamine type carbohydrates $(3Ga1\beta1, 4G1cNAc\beta1)_n$[11].

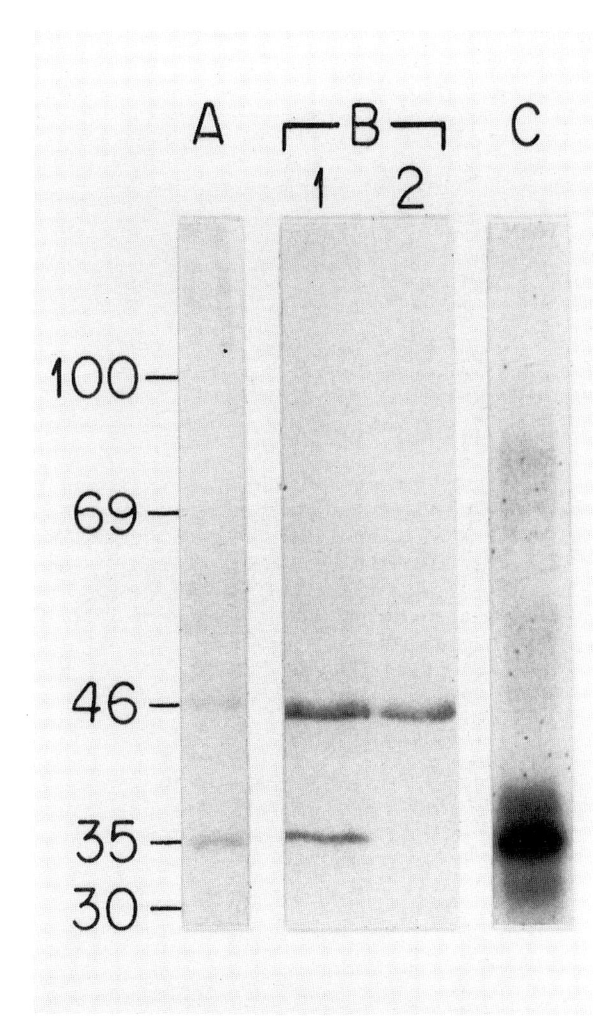

Figure 1. Laminin-Sepharose chromatography of macrophage extracts. Detergent extracts obtained from mouse macrophages were incubated in the presence of laminin-Sepharose, washed with a Tris-Cl buffer (pH8.0) containing 0.5 M NaCl and analyzed by gel electrophoresis. (A) Coomassie stain of gel obtained from unlabelled cells. (B) Fluorogram of gel from ^{35}S-methionine labelled cells. Lane 1, laminin-Sepharose; lane 2, unconjugated Sepharose. (C) Autoradiogram of gel from ^{125}I-surface radiolabelled cells. Note that a 35 kDa protein is the predominant LBP expressed by these cells. Similar results have been obtained with other cell types.

■ ANTIBODIES

The M3/38 hybridoma (Mac-2 antigen)[14] available through ATCC (TIB 166) recognizes CBP 35 in both rodent and human cells. This monoclonal antibody is useful for immunoblotting, immunoprecipitation, and immunostaining. A number of other monoclonal antibodies[15], and polyclonal antisera[9] have been described.

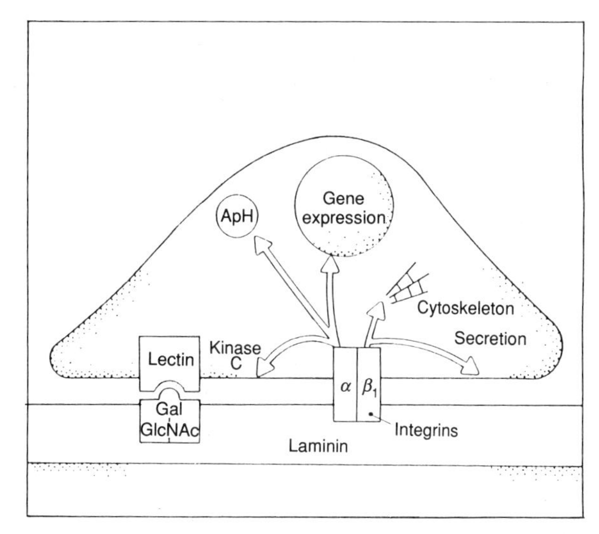

Figure 2. Hypothetical model of cell adhesion to laminin. Putative signalling functions regulated by integrin binding are noted. The putative role of the CBP 35 lectin in stabilizing adhesion or migration is depicted.

■ GENES

Full length cDNAs for mouse (GenBank J03723[12]; X16074[6]) and human CBP 35 (GenBank M35368[16]) are published.

■ REFERENCES

1. Drickamer, K. (1988) J. Biol. Chem. 263, 9557-9560.
2. Woo, H.-J., Shaw, L.M., Messier, J.M. and Mercurio A.M. (1990) J. Biol. Chem. 265, 7097-7099.
3. Cherayil, B.J., Weiner, S.J. and Pillai, S. (1989) J. Exp. Med. 170, 1959-1972.
4. Hynes, M.A., Gitt, M., Barondes, S.H., Jessell, T.M. and Buck, L.B. (1990) J. Neurosci. 10, 1004-1013.
5. Lotan, R., Lotan, D. and Carralero, D.M. (1989) Cancer Lett. 48, 115-122.
6. Raz, A., Pazerini, G. and Carmi, P. (1989) Cancer Res. 49, 3489-3493.
7. Laing, J.G., Robertson, M.W., Gritzmacher, C.A., Wang, J.L. and Liu, F.-T. (1989) J. Biol. Chem. 264, 1907-1910.
8. Woo, H.-J., Letz, M.M., Jung, J.U. and Mercurio, A.M. (1991) J. Biol. Chem. 266, 18419-18422.
9. Moutsatsos, I.K., Davis, J.M. and Wang, J.L. (1986) J. Cell Biol. 102, 477-483.
10. Cowles, E.A., Agrwal, N., Anderson, R.L. and Wang, J.L. (1990) J. Biol. Chem. 265, 17706-17712.
11. Zhou, Q. and Cummings, R.D. (1990) Arch. Biochem. Biophys. 281, 27-35.
12. Jia, S. and Wang, J.L. (1988) J. Biol. Chem. 263, 6009-6011.
13. Agrwal, N., Wang, J.L. and Voss, P.G. (1989) J. Biol. Chem. 264, 17236-17242.
14. Ho, M.-K. and Springer, T.A. (1982) J. Immunol. 128, 1221-1228.

15.Raz, A., Meromsky, L., Carmi, P., Karakashi, R., Lotan, D. and Lotan, R. (1984) EMBO J. 3, 2979-2983.
16.Cherayil, B.J., Chaitovitz, S., Wong, C. and Pillai, S. (1990) Proc. Natl. Acad. Sci. (USA) 87, 7324-7328.

Arthur M. Mercurio and Hee-Jong Woo:
Lab of Cancer Biology,
Deaconess Hospital,
Harvard Medical School, Boston, MA, USA

Link Protein

Link protein is an extracellular glycoprotein found predominantly in cartilage where it forms a ternary complex with the cartilage proteoglycan, aggrecan, and hyaluronic acid. The function of link protein is to stabilize the interaction between aggrecan and hyaluronic acid. As many as 100 aggrecan and link protein molecules can interact with a single hyaluronic acid polymer.

The complete sequence has been determined for link protein of chicken[1], rat[2,3] and human[4,5] either from protein or cDNA data. The primary structure derived from protein sequencing[2] has allowed the precise determination of the disulphide bonds. Although there are multiple forms of link protein (i.e. 41, 46 and 51 kDa for rat) these result from posttranslational modifications involving differential glycosylation or proteolysis. Link protein has also been shown to be able to form oligomers in a variety of concentrated salt solutions[6].

The gene for link protein is large and is present in a single copy in the chicken[7] and rat[4] genome. The relation-

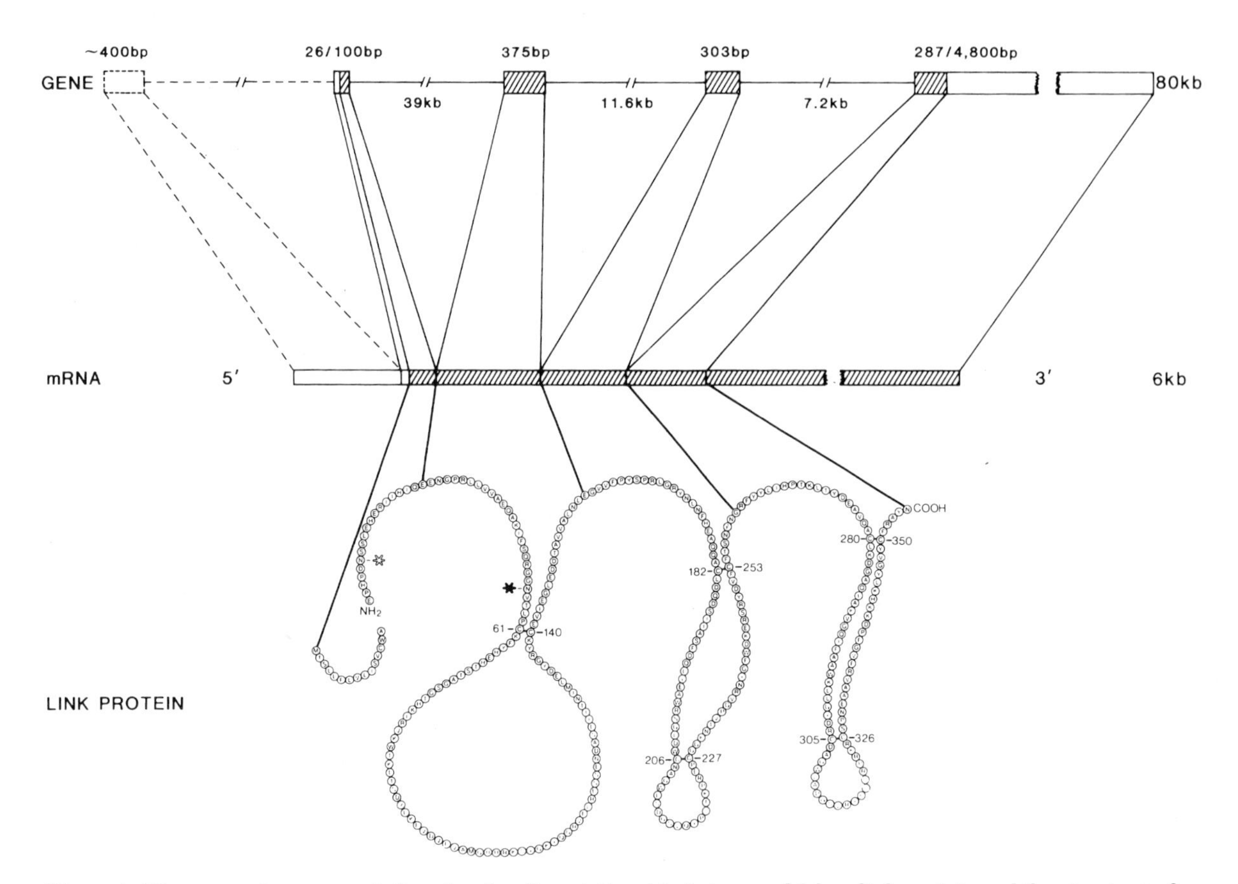

Figure 1. Diagrammatic representation showing the relationship between chicken link protein and the structure of the chicken link protein gene. Each domain of link protein is encoded by a separate exon. The disulphide bonds are based on the structure of link protein determined from amino acid sequence[2]. The stars on the link protein diagram indicate glycosylation sites.

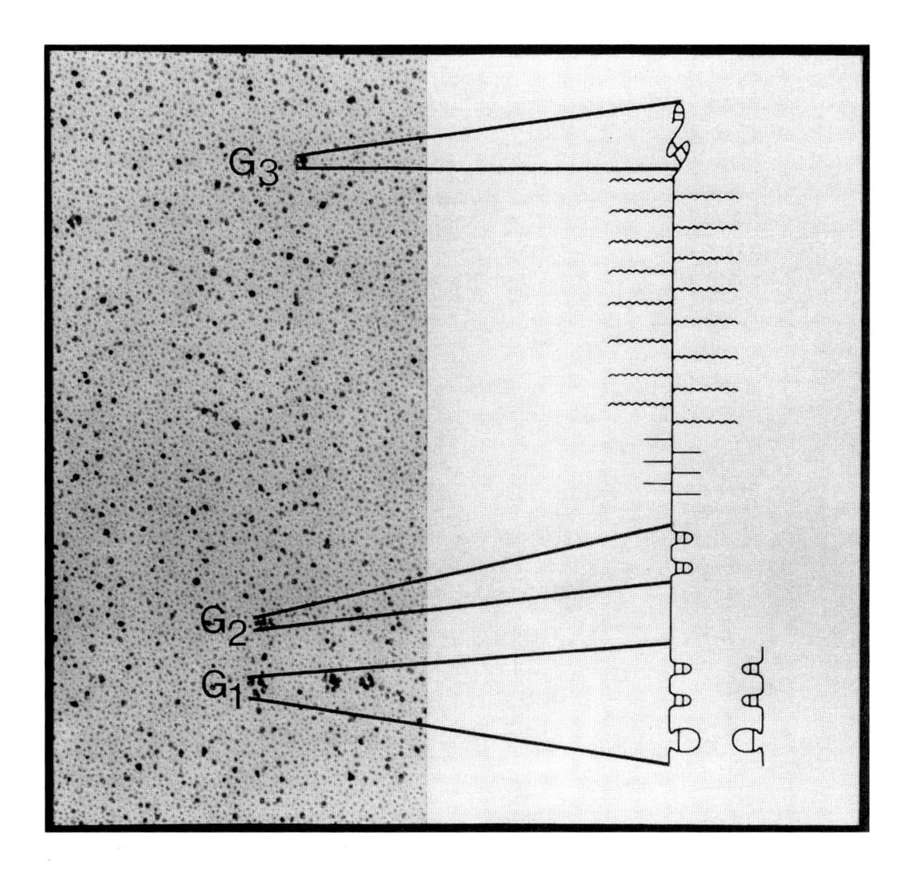

Figure 2. Relationship between the structure of link protein and the globular domains of aggrecan. Rotary shadowing prepartion of chick embryonic aggrecan. Three globular domains (G_1, G_2, G_3) are identified. The diagrammatic representation of aggrecan is based on the rat sequence. The diagram to the extreme right is the one for link protein.

ship between chicken link protein and the structure of the chicken link protein gene is given in Figure 1. A 15 amino acid signal peptide together with approximately 19 amino acids of the N-terminus of the mature protein is encoded by a single exon. Each subsequent domain is also encoded by a single exon. A similar situation exists in the organization of the rat gene except that the introns in the rat gene are smaller than in the chicken. In addition, transcripts resulting from alternative splicing of RNA have been reported in the rat[3]. The domain structure of link protein has homology with that of two proteoglycans. This is illustrated for aggrecan in Figure 2. The rotary shadowing preparation of chick embryonic aggrecan reveals three globular domains (G_1,G_2,G_3). The diagrammatic representation of the proteoglycan to the right of the electron micrograph is based on the sequence of rat aggrecan[8]. The diagram to the extreme right is the one for link protein. The structure of the entire link protein is homologous to the structure of the G_1 domain. The tandemly repeated domains of link protein are also homologous to the G_2 domain of aggrecan. The structural motif of the N-terminus the fibroblast proteoglycan, versican[9], is also identical to the entire structure of link protein. A structural motif homologous to one of the tandemly repeated domains of link protein is present in CD44[10,11] (hyaluronic acid receptor, Hermes antigen, Pgp-1).

The tandemly repeated domains of link protein are involved in its interaction with hyaluronic acid (hyaluronan)[12]. Evidence for this is derived from inhibitions of the interaction with the monoclonal antibody 8A4[13] and with synthetic peptides covering sequences in that region. The two epitopes recognized by 8A4 are in the tandem repeats of link protein. In addition, the tandemly repeated domains are protected from proteolysis in mixtures of link protein with hyaluronic acid[14] whereas the immunoglobulin like domain of link protein is protected when aggrecan and link protein are mixed. The latter observation suggests that the immunoglobulin like domain of link protein is involved in its interaction with aggrecan.

Although the major source of link protein is cartilage, the protein has also been found in noncartilaginous tissues such as the aorta[15] and the chick embryonic mesonephros[16]. These tissues do not have any demonstrable quantities of aggrecan and, therefore, the role of link protein in these tissues is unknown. Either it stabilizes the interaction between hyaluronic acid and a proteoglycan, different from aggrecan, or link protein may have additional functions besides stabilizing proteoglycan aggregates.

PURIFICATION

The purification of link protein from cartilage is based on extractions in 4 M guanidine-HCl solutions in the presence of protease inhibitors[6]. Under these conditions the link protein-aggrecan-hyaluronic acid complex is dissociated. Dialysis of the extracts to 0.4 M or less result in the re-establishment of the aggregate. Centrifugation of these extracts in CsCl gradients with an initial density of 1.6 g/ml results in the sedimentation of link protein stabilized proteoglycan aggregates to the lower densities of the gradient (A1 fraction). When the A1 fraction is returned to dissociative conditions (4 M guanidine-HCl) and subjected to CsCl gradient centrifugation the aggregate dissociates and aggrecan sediments to the bottom of the gradient (A1D1 fraction) and link protein to the top (A1D6 fraction). Hyaluronic acid sediments in the middle region of the gradient. Link protein can be further purified by DEAE chromatography to remove remaining traces of low buoyant density proteoglycans. A method for isolating link protein using wheat germ agglutenin has also been reported[17].

ACTIVITIES

The biological function of link protein is to stabilize the interaction of aggrecan with hyaluronic acid. These link protein stabilized proteoglycan aggregates contribute to the growth of embryonic cartilaginous rudiments and to the compressibility of adult cartilage.

ANTIBODIES

Polyclonal antisera against link protein from bovine[18,19] and chicken[20] have been described. Monoclonal antibody 8A4[13] is well characterized and can be purchased from the "Developmental Studies Hybridoma Bank" (The Johns Hopkins University School of Medicine, Baltimore, MD). A polyclonal antiserum against a synthetic peptide covering a sequence of the immunoglobulin domain of chicken link protein has been reported[12].

GENES

Complete cDNA sequences are available for chicken (GenBank M13212)[1], rat[3] (GenBank M13191) and human[4,5] (EMBL X17405). The human link protein gene has been mapped to chromosome 5q13-q14.1[21].

REFERENCES

1. Deák, F., Kiss, I., Sparks, K., Argraves,W.S., Hampikian, G. and Goetinck, P.F. (1986) Proc. Natl. Acad. Sci. (USA) 83, 3766-3770.
2. Neame, P.J., Christner, J.E. and Baker, J.R. (1986) J. Biol. Chem. 261, 3519-3535.
3. Rhodes, C., Doege, K., Sasaki, M. and Yamada, Y. (1988) J. Biol. Chem. 263, 6063-6067.
4. Doege, K., Rhodes, C., Sasaki, M., Hassell, J.R. and Yamada, Y. (1990) In: Extracellular Matrix Genes 137-155.
5. Dudhia, J. and Hardingham, T.E. (1990) Nucleic Acids Res. 18, 1292.
6. Baker, J.R. and Neame, P.J. (1987) Methods in Enzymol. 144, 401-413.
7. Kiss, I., Deák, F., Mestric, S., Delius, H., Soos, T., Dékány, K., Argraves, W.S., Sparks, K.J. and Goetinck,P.F.. (1987) Proc. Natl. Acad. Sci. (USA) 84, 6399-6403.
8. Doege, K., Sasaki, M., Horigan, E., Hassell, J.R. and Yamada, Y. (1987) J. Biol. Chem. 262, 17757-17767.
9. Zimmerman, D.R. and Ruoslahti, E. (1989) EMBO J. 8, 2975-2981.
10. Stamenkovic, I.M., Amiot, M., Pesando, J.M. and Seed, B. (1989) Cell 56, 1057-1062.
11. Goldstein, L.A., Zhou, D.F.H., Picker, L.J., Minty, C.N., Bargatze, R.F., Ding, J.F. and Butcher, E.C. (1989) Cell 56, 1063-1072.
12. Goetinck, P.F., Stirpe, N.S., Tsonis, P.A. and Carlone, D. (1987) J. Cell Biol. 105, 2403-2408.
13. Caterson, B., Calabro, T. and Hampton, A. (1987) In Biology of Proteoglycans. T. Wight and R. Mecham, Eds.: pp 1-16. Academic Press. New York.
14. Périn, J.-P., Bonnet, F., Thurieau, C. and Jollés, P. (1987) J. Biol. Chem. 262, 13269-13272.
15. Gardell, S., Baker, J.R., Caterson, B., Heinegård, D. and Rodén, L. (1980) Biochem. Biophys. Res. Commun. 95, 1823-1831.
16. Stirpe, N.S., Dickerson, K.T. and Goetinck, P.F. (1990) Dev. Biol. 137, 419-424.
17. Choi, H.U., Tang, L.-H., Johnson, T.L. and Rosenberg, L. (1985) J. Biol. Chem. 260, 13370.
18. Baker, J.R., Caterson, B. and Christner, J.E. (1982) Methods in Enzymol. 83, 216-235.
19. Poole, A.R. and Reiner, A. (1980) J. Biol. Chem. 255, 9295-9305.
20. McKeown-Longo, P.J., Sparks, K.J. and Goetinck, P.F. (1982) Collagen Rel. Res. 2, 232-244.
21. Osborne-Lawrence, S.L., Sinclair, A.K., Hicks, R.C., Lacey, S.W., Eddy Jr., R.L., Byers, M.G., Shows, T.B. and Duby, A.D. (1990) Genomics 8, 562-567.

Paul F. Goetinck:
Cutaneous Biology Research Center,
Massachusetts General Hospital East, Building 149,
13th Street, Charlestown, MA 02129, USA

Mucins

The term mucin is used to describe a diverse group of glycoproteins ranging from the very large, secreted gelling mucins of the gastric and intestinal tract to the simpler, integral membrane mucin produced by many glandular epithelial tissues. Although different mucins show no sequence homology and the genes are located on different chromosomes, they all share some common features. Mucins are large molecular weight glycoproteins containing 50-80% carbohydrate which is O-linked through N-acetylgalactosamine to serine and/or threonine; they contain high levels of serine, threonine, glycine, alanine and proline, and perhaps the most striking feature common to all the epithelial mucins so far cloned, is the presence within the core protein of a large domain of tandemly repeated amino acids. Mucins are thought to act in a lubricative and protective manner, protecting the gastrointestinal, respiratory and genital tracts from physical damage, dehydration and bacterial infection.

The apparent molecular mass of native mucin can range from about 300 kDa for the mammary or polymorphic epithelial mucin (PEM) to about 10,000 kDa for the gastric and tracheal mucins (see Table 1 for mucin nomenclature). However, due to the vast amounts of carbohydrate and the possibility of intra- and intermolecular aggregates, the native molecular weight is difficult to interpret[1]. Moreover, the molecular weight of the core protein of each mucin may be different from individual to individual as many of the mucin genes show a genetic polymorphism due to varying numbers of tandem repeats[2,3]. A translated tandem repeat element is common to all the epithelial mucin genes (Table 1). While all the repeats are rich in proline, serine and/or threonine, the size and amino acid composition of the tandem repeats vary widely among mucin types. However, this element appears to have a common function in that it acts as a scaffold for O-linked carbohydrate, the serines and threonines forming the attachment sites and the prolines preventing α-helix formation and so allowing the necessary conformation of the close packing of the oligosaccharides. The C-termini of mucins vary according to the mucin type. The mammary

Table 1

Epithelial Mucins that have been cloned

Mucin	Gene	Chromosome	Tandem Repeat or Secreted	Integral Membrane	References
Human polymorphic epithelial mucin PEM, PUM, episialin, MAM6, DF3 antigen, H23 antigen, PAS-O, EMA, NPGP, NCRC11 antigen	MUC1	1q21	20 amino acids	Integral Membrane	6, 12, 16, 17, 18
Mouse mammary mucin	MUC1	?	20 & 21 amino acids	Integral Membrane	19
Human acidic intestinal mucin	MUC2	11	23 amino acids	Secreted[1]	13
Human neutral intestinal mucin	MUC3	7	17 amino acids	Secreted[1]	14
Human tracho-bronchial mucin	MUC4	3	16 amino acids	Secreted[2]	15
Porcine sub-maxillary mucin	-	?	81 amino acids	Secreted[1]	20

[1] Only partial cDNA available, however, no transmembrane domain at C-terminus

[2] Partial cDNA only

mucin or PEM contains a membrane spanning domain and cytoplasmic tail, whereas the secreted porcine submaxillary gland mucin and human intestinal mucin have cysteine rich C-termini that may be involved in inter- or intramolecular bonds. Although the vast majority of the carbohydrate is O-linked[4] to the protein core, there are some potential N-glycosylation sites.

Mucins are found on the luminal surface of epithelial cells, e.g. PEM, or can be secreted into the mucus, e.g. the intestinal and tracheal mucins. However, the tissue expression appears to be rather complex; although specific mucins have been isolated from particular tissues, a single tissue can express a number of different mucins[5]. In this context the MUC1 protein (Table 1) appears to be almost ubiquitous, since it is produced by most glandular epithelia. Moreover, the same mucin core protein can be differently glycosylated by individual tissue types resulting in different profiles of antigenic epitopes[6]. In tumours, many mucins appear to be aberrantly glycosylated resulting in the exposure of epitopes which are generally masked in normal cells[7,8].

PURIFICATION

Mucins are traditionally purified by CsCl density gradient centrifugation and gel filtration[9]. Simpler mucins may be purified by affinity chromatography[7].

ACTIVITIES

Mucins are thought to act in a lubricative and protective manner, protecting the gastrointestinal, respiratory and genital tracts from physical damage, dehydration and bacterial infection[4]. However, the membrane anchored mucin may have additional functions as its cytoplasmic tail is highly conserved among species and this mucin has been shown to be involved in ***actin*** binding[10].

ANTIBODIES

Many of the monoclonal antibodies that react with human tumour associated antigens react with mucins. Monoclonal antibodies are available that react with carbohydrate and with peptide epitopes[11]. Many of the antibodies react with formalin fixed, paraffin embedded tissues and can be used for immunoprecipitation and immunoblotting. Most of these antibodies react with elements within the tandem repeat of the particular mucin and are species specific. Polyclonal antisera are also available to a number of mucins and to the chemically deglycosylated protein cores[12,13-15]. Although most appear to be species specific, polyclonal antiserum made to the cytoplasmic tail of the human mammary mucin crossreacts with the mouse homologue.

GENES

Full length cDNA for the human polymorphic epithelial mucin, MUC1 (GenBank J05581, J05582, J05288) and the mouse mammary mucin, MUC1 (GenBank M64928) have been published. Genomic sequence (GenBank 54350 and 54351) for MUC1 is also available. Partial cDNA sequences have been published for human acidic intestinal mucin, MUC2, (GenBank J04638); human neutral intestinal mucin, MUC3; human tracheal-bronchial mucin, MUC4, and porcine submaxillary mucin (GenBank M61883). See Table 1 for references.

REFERENCES

1. Reid, L. and Clamp, J.R. (1978) Brit. Med. Bull. 34, 5-8.
2. Swallow, D.M., Gendler, S., Griffiths, B., Corney, G., Taylor-Papadimitriou, J. and Bramwell, M.E. (1987) Nature 328, 82-84.
3. Griffiths, B. (1990) Ann. Hum. Genet. 54, 277-285.
4. Jentoft, N. (1990) Trends Biochem. Sci 15, 291-294.
5. Byrd, J.C., Ho, J.J., Lamport, D.T., Ho, S.B., Siddiki, B., Huang, J., Yan, P.S. and Kim, Y.S. (1991) Cancer Res. 51, 1026-1033.
6. Lan, M.S., Batra, S.K., Qi, W.-N., Metzgar, R.S. and Hollingsworth, M.A. (1990) J. Biol. Chem. 265, 15294-15299.
7. Burchell, J., Gendler, S., Taylor-Papadimitriou, J., Girling, A., Lewis, A., Millis, R. and Lamport, D. (1987) Cancer Res. 47, 5476-5482.
8. Boland, C.R. and Deshmukh, G.O. (1990) Gastroenterology 98, 1170-1177.
9. Rose, M.C. (1989) Methods Enzymol. 179, 3-17.
10. Parry, G., Beck, J.C., Moss, L., Bartley, J. and Ojakian, G.K. (1990) Exp. Cell Res. 188, 302-311.
11. Taylor-Papadimitriou, J. (1991)Report on the First International Workshop on Carcinoma-associated Mucins, San Francisco, Int. J. Cancer. 49, 1-5.
12. Gendler, S., Lancaster, C., Taylor-Papadimitriou, J., Duhig, T., Peat, N., Burchell, J., Pemberton, L., Lalani, E.-N. and Wilson, D. (1990) J. Biol. Chem. 265, 15286-15293.
13. Gum, J.R., Byrd, J., Hicks, J., Toribara, N., Lamport, D. and Kim, Y. (1989) J. Biol. Chem. 264, 6480-6487.
14. Gum, J.R., Hicks, J.W., Swallow, D., Lagace, R.L., Byrd, J.C., Lamport, D.T.A., Siddiki, B. and Kim, Y.S. (1990) Biochem. Biophys. Res. Commun. 171, 407-415.
15. Porchet, N., van Cong, N., Dufosse, J., Audie, J.P., Guyonnet-Duperat, V., Gross, M.S., Denis, C., Degand, P., Bernheim, A. and Aubert, J.P. (1991) Biochem. Biophys. Res. Commun. 175, 414-422.
16. Lancaster, C., Peat, N., Duhig, T., Wilson, D., Taylor-Papadimitriou, J. and Gendler, S. (1990) Biochem. Biophys. Res. Commun. 173, 1019-1029.
17. Ligtenberg, M., Vos, H., Gennissen, A. and Hilkens, J. (1990) J. Biol. Chem. 265, 5573-5578.
18. Wreschner, D.H., Hareuveni, M., Tsarfaty, I., Smorodinsky, N., Harev, J., Zaretsky, J., Kotkes, P., Weiss, M., Lathe, R., Dion, A. and Keydar, I. (1990) Eur. J. Biochem. 189, 463-473.
19. Spicer, A.P., Parry, G., Patton, S. and Gendler, S.J. (1991) J. Biol. Chem. 266, 15099-15109.
20. Eckhardt, A.E., Timpte, C.S., Abernethy, J.L., Zhao, Y. and Hill, R.L. (1991) J. Biol. Chem. 266, 9678-9686.

Joy Burchell and Joyce Taylor-Papadimitriou:
Epithelial Cell Biology Laboratory,
Imperial Cancer Research Fund,
London, UK

Nidogen/Entactin

The extracellular matrix protein nidogen[1] or entactin[2] is a typical basement membrane component and an early embryonic product. It shows binding activity for laminin, collagen type IV, cells and calcium.

Nidogen also known as entactin consists of a single polypeptide of 148 kDa[3] and is composed of three globular domains connected by a 15 nm long rod or a flexible link (Figure A, B)[4]. It is frequently found in a tight (K_d ~1 nM), noncovalent complex with laminin[5]. The amino acid sequence (1217-1219 residues) is known for mouse[6,7] and human[8] nidogen and shows 85% identity. The data demonstrate a multidomain structure including seven individual EGF-like repeats (Figure A). Posttranslational modifications include several N- and 0-linked glycosylation sites[3] and O-sulphation of one to two tyrosines[9]. Tissue distribution studies by immunofluorescence (Figure C) and immunoelectron microscopy shows the ubiquitous and almost exclusive occurrence of nidogen in basement membranes[1,10-13] and its appearance at the 8-16 cell stage of mouse development[13].

■ PURIFICATION

The most convenient procedure involves production of recombinant protein in stable mammalian cell clones and the isolation from culture medium (10-20 μg/ml) by binding to a cobalt loaded chelating column[4]. Partially denatured forms are obtained by dissociating the laminin-nidogen complex from the mouse EHS tumour with 2-6 M guanidine-HCl followed by one to three chromatographic steps[3,5]. Purification by SDS-PAGE from cell cultures has also been used[2]. Normal tissues have not yet been utilized.

■ ACTIVITIES

Binding of nidogen to a rod-like segment in the short arms of laminin[5] requires its C-terminal globular domain[4,14]. A different nidogen domain binds to collagen IV[4,15]. These interactions allow the binding of laminin to collagen IV indicating a key role for nidogen in supramolecular assembly. Weak cell adhesion and spreading on nidogen[7] may depend on a single RGD sequence (Figure A). Ca^{2+}-binding was predicted[6-8] to occur with one to two

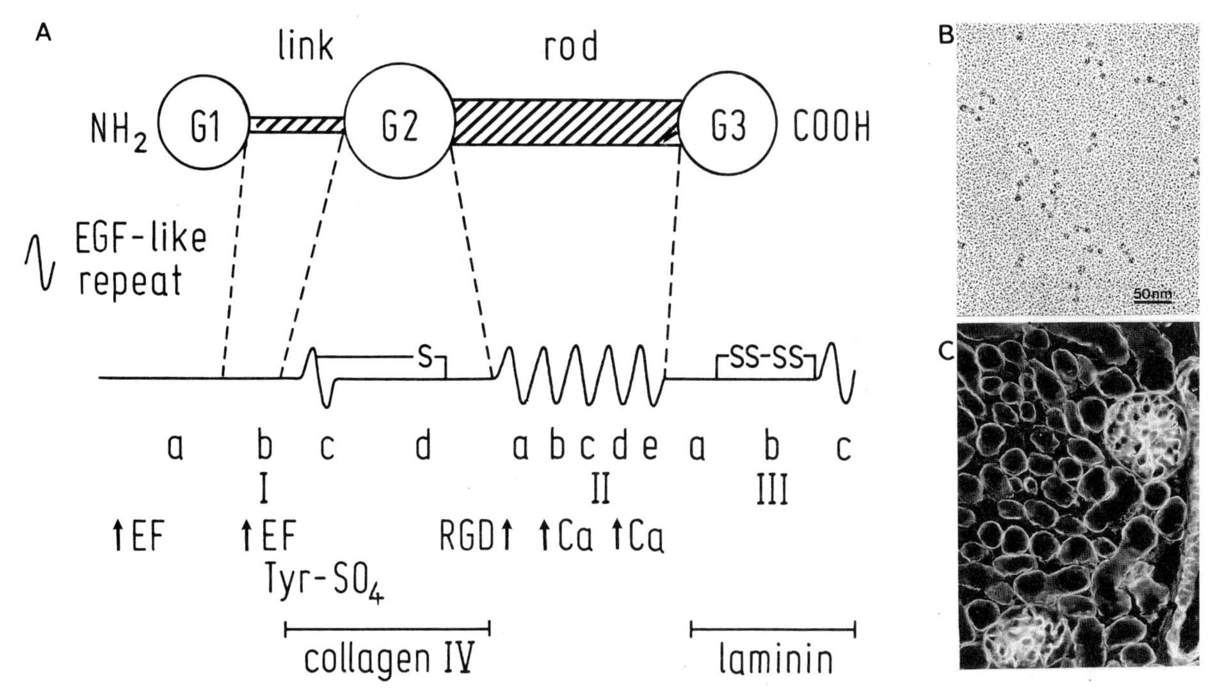

Figure. (A) Correlation between shape and sequence-predicted domain structure of nidogen (modified form[4,7]). Binding sites for laminin, collagen IV, cells (RGD) and calcium within EF hand motifs or EGF-like repeats (Ca) and for tyrosine sulphation (Tyr-SO_4) are indicated. (B) Rotary shadowing images of recombinant nidogen. Bar 50 nm. (C) Immunofluorescence staining for nidogen in mouse kidney by a monoclonal antibody (by courtesy of M. Dziadek).

EF hand motifs in the N-terminal globule and two EGF-like repeats in the rod-like segment (Figure A).

ANTIBODIES

Polyclonal rabbit antisera against mouse nidogen and several of its fragments have been characterized[1,3,14] by immunofluorescence, immunoprecipitation, blotting and radioimmunoassay studies[10-14]. A rat monoclonal antibody to mouse nidogen has been described[16].

GENES

cDNA probes covering the coding and some untranslated regions of mouse and human nidogen[6-8] are available (GenBank Nos. M30269, X14194). The single copy gene was localized to human chromosome 1q43[17].

REFERENCES

1. Timpl, R., Dziadek, M., Fujiwara, S., Nowack, H. and Wick. G. (1983) Eur. J. Biochem. 137, 455-465.
2. Carlin, B., Jaffe, R., Bender, B. and Chung, A.E. (1981) J. Biol. Chem. 256, 5209-5214.
3. Paulsson, M., Deutzmann, R., Dziadek, M., Nowack, H., Timpl, R., Weber, S. and Engel, J. (1986) Eur. J. Biochem. 156, 467-478.
4. Fox, J.W., Mayer, U., Nischt, R., Aumailley, M., Reinhardt, D., Wiedemann, H., Mann, K., Timpl, R., Krieg, T., Engel, J. and Chu, M.-L. (1991) EMBO J. 10, 3137-3146.
5. Paulsson, M., Aumailley, M., Deutzmann, R., Timpl, R., Beck, K. and Engel, J. (1987) Eur. J. Biochem. 166, 11-19.
6. Durkin, M.E., Chakravarti, S., Bartos, B., Liu, S.-H., Friedman, R.L. and Chung, A.E. (1988) J. Cell Biol. 107, 2749-2756.
7. Mann, K., Deutzmann, R., Aumailley, M., Timpl, R., Raimondi, L., Yamada, Y., Pan, T.-C., Conway, D. and Chu, M.-L. (1989) EMBO J. 8, 65-72.
8. Nagayoshi, T., Sanborn, D., Hickock, N.J., Olsen, D.R., Fazio, M.J., Chu, M.-L., Knowlton, R., Mann, K., Deutzmann, R., Timpl, R. and Uitto, J. (1989) DNA 8, 581-594.
9. Paulsson, M., Dziadek, M., Suchanek, C., Huttner, W.B. and Timpl, R. (1985) Biochem. J. 231, 571-579.
10. Hogan, B.L.M., Taylor, A., Kurkinen, M. and Couchman, J.R. (1982) J. Cell Biol. 95, 197-204.
11. Wu, T.C., Wan, Y.-J., Chung, A.E. and Damjanov, I. (1983) Dev. Biol. 100, 496-505.
12. Schittny, J.C., Timpl, R. and Engel, J. (1988) J. Cell Biol. 107, 1599-1610.
13. Dziadek, M. and Timpl, R. (1985) Dev. Biol. 111, 372-382.
14. Mann, K., Deutzmann, R. and Timpl, R. (1988) Eur. J. Biochem. 178, 71-80.
15. Aumailley, M., Wiedemann, H., Mann, K. and Timpl, R. (1989) Eur. J. Biochem. 184, 241-248.
16. Dziadek, M., Clements, R., Mitrangas, K., Reiter, H. and Fowler, K. (1988) Eur. J. Biochem. 172, 219-225.
17. Olsen, D.R., Nagayoshi, T., Fazio, M., Mattei, M.-G., Passage, D., Weil, D., Timpl, R., Chu, M.-L. and Uitto, J. (1989) Am. J. Hum. Genet. 44, 876-885.

Rupert Timpl:
Max-Planck-Institut für Biochemie,
Martinsried, Germany

Osteopontin/Secreted Phosphoprotein (OPN/SPP)

Osteopontin (OPN)[1], an acidic, secreted RGD containing adhesion protein found in milk, bone, kidney, inner ear, uterus, and decidua, and to a lesser extent in plasma, ovary and testis, binds to cells and promotes cell attachment and spreading. Produced at higher levels by activated T cells, OPN, also called ETA, contributes to early resistance to bacterial infections. OPN is induced by calcitriol suggesting an involvement in bone metabolism; its induction by tumour promoters and its constitutive production by many transformed cells may imply that OPN contributes to the oncogenic phenotype.

Osteopontin, also known as 2ar or secreted phosphoprotein, is a 44 kDa bone sialophosphoprotein that mediates cell-matrix and possibly cell-cell interactions, probably via the $\alpha_V\beta_3$ integrin, the vitronectin receptor[2]. It facilitates cell attachment and spreading, and increases the cell's tolerance for heat shock[3]. OPN is induced in mouse skin by the tumour promoting phorbol ester 12-O-tetradecanoylphorbol-13-acetate and is found at high levels in the serum of patients with disseminated carcinomas or gram-negative sepsis. It is also the major phosphoprotein secreted by many transformed (particularly *ras*-transformed) cells[4,5]. OPN maps to the rickettsia-resistance locus Ric[c] on murine chromosome 5 (and human chromosome 4q)[6,7] and its expression is enhanced in activated T lymphocytes in response to certain infections[8]. The protein is a major noncollagenous protein in bone, but is also found in urine (Hoyer, personal communication), breast milk, serum, endometrial glands of secretory phase uterus, decidual cells, several tissues of developing inner ear, and Hertwig's epithelial root sheath of the developing tooth[1,6,9,10]. The mRNA has been detected in developing bone, kidney tubules (Figure 1), uterine epithelium, ovary, the skin of pregnant and lactating mice, sensory epithelium of the inner ear, and granulated metrial gland cells of the decidua[1,11]. Expression in osteoblastic cells is stimulated by 1,25-dihydroxyvitamin D3, TNF, FGF and TGF-β, and is inhibited by parathyroid hormone and dexamethasone[1,12-14].

OPN is an acidic protein of about 300 amino acids[1,15] (Figure 2). The overall amino acid homology between mouse, rat, human and pig is high and several sequence elements are conserved, including a stretch of seven to

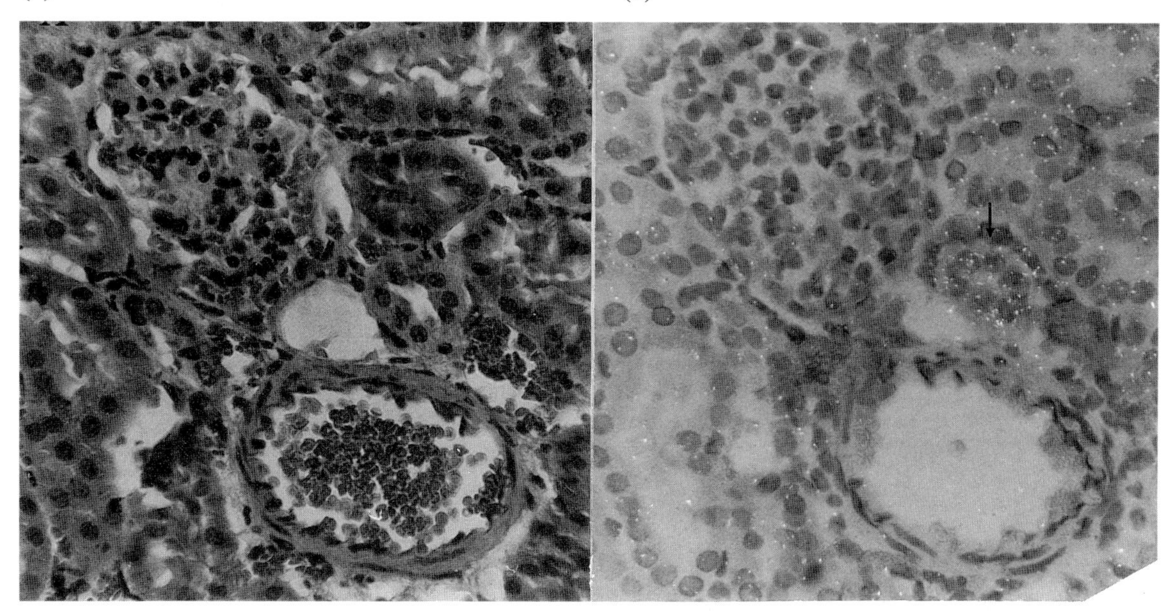

Figure 1. Two adjacent 6 µm sections of fixed, paraffin-embedded mouse kidney were (A) stained with hematoxylin-eosin or (B) hybridized with a ^{35}S-labelled riboprobe specific for the OPN mRNA and then counterstained with hematoxylin-eosin. These and other results indicate that expression is maximal in the thick ascending limb of the long loops of Henle, shown here in crosssection at the tip of the arrow (Courtesy Cecilia Lopez).

nine residues of ASP/GLU, that may represent a calcium or hydroxylapatite binding site, an RGD cell adhesion and a thrombin cleavage site, the latter two which are adjacent on the sequence DGRGDS(L/V)(V/A)YGLRSKS(K/R). SXE sequences, which possibly represent phosphorylation or O-linked glycosylation sites are repeated throughout the protein. OPN from rat bone (Figure 2) possesses some 30 monosaccharides present in one N-linked and up to five to six O-linked oligosaccharides and up to ten residues of sialic acid[1,16]. Mineralized tissue formed by rat bone marrow cells in culture also contains tyrosine-sulphated OPN[17]. OPN is usually phosphorylated (12 phosphoserines and 1 phosphothreonine in rat bone[16]), although a nonphosphorylated form of OPN is produced both by normal rat kidney cells[18] and by calcitriol-treated JB6 cells[19]. What influence these modifications have on the cell adhesion and Ca^{2+}-binding activities of OPN is not known, though phosphorylation or the N-linked oligosaccharides may control OPN binding to fibronectin[18,20]. Multiple electrophoretic species are commonly seen in SDS-PAGE, and significantly different apparent molecular weights can be derived depending upon the specifics of the experiment[12].

cDNA clones of murine mRNAs isolated from macrophages and activated T lymphocytes have a different 5' noncoding exon from that characterized in JB6 epidermal cells and osteoblasts, and comparisons of the chicken, rat, human, porcine and mouse cDNA sequences reveal differences indicative of differential splicing and suggesting that different promoters may be used in different cell types[6,8,11,15,21]. A vitamin D response element has been identified in the murine promoter, which is also responsive to β-estradiol and progesterone[11,13].

PURIFICATION

OPN is isolated from a guanidine hydrochloride-EDTA extract of bone by gel filtration and anion exchange chromatography using HPLC[16,22,23]. It is isolated from plasma, milk and cell culture medium by barium citrate precipitation and reversed phase HPLC[9].

ACTIVITIES

The function(s) of OPN are poorly defined. Its ability to mediate cell attachment suggests that it could determine tissue organization and cell migration. OPN colocalizes with the vitronectin receptor to the clear zone at the site where osteoblasts are anchored to bone[2]. The high negative charge on the protein may help create a permeability barrier. It may act to buffer calcium levels, to signal cells what the external $[Ca^{++}]$ concentration is, and to prevent precipitation of calcium salts in urine.

ANTIBODIES

Polyclonal antisera against intact rat, bovine, and human bone OPN[16,23] and the C-terminal fragment of pig bone OPN[22] have been raised. Also polyclonal antisera have been raised against murine β-galactosidase fusion proteins[24] and against OPN from NRK cells[18] and ts B77-Rat 1 cells[4].

```
: : : : : H H H H H H H H H H H : : : H H H H H H H H H H : :

: : : : : E E E E T T T T : : : H H H H H H H H H : : : T T T

*             = = = = = = = = =                     o o o o o o o o o o
T T T T : : : : : : : : : : : : : : : : : : : : : : : : : : :

o o o o o o o o o o o o o o
: : : : : : : : : : : : : : : : : : : : E E E E E E E E E E E

      ▒▒▒▒▒               «  »
T T T T T T E E E E E : T T T T T T T T T T T T T : : : : H H

                                                          + +
H H H H H H H H H H H H H H H H H H E E E E E E E E E T T T T

+ + + + + + + + + +
T T T T T T T T : : : : : : : : : : H H H H H H H H H H H H H

H H : : : : : : H H H H H H H H H H : H H H H H H H H H H H H

H H H H : : H H H H H H H H H H H H H H H H H H H H : : : : :

H H H H H H H H H H H H : : : : : : : : : :
```

Figure 2 OSTEOPONTIN STRUCTURAL FEATURES:

Each symbol (:) / letter represents one residue in the rat OPN sequence.

H = predicted as alpha helix		E = predicted as beta sheet	
T = predicted as turn		: = no predicted secondary structure	
«» = thrombin cleavage site		▒ = RGD cell adhesion site	

* = N-linked oligosaccharide attachment site
+++ = homologous to 12 residue EF-hand calcium binding loops
=== = 9 aspartic acid residues
ooo = 8 repeats of Sxx (potential sites of PO_4 or O-glycosplation)

■ GENES

The amino acid sequences of rat (GenBank M14656), mouse (GenBank X14882, X13986, X51834), human (X13694, J04765) and pig (X16575) have been deduced from the sequences of the cloned cDNAs.

■ REFERENCES

1. Butler, W.T. (1989) Connect. Tissue Res. 23, 123-136.
2. Reinholt, F.P., Hultenby, K., Oldberg, A. and Heinegård, D. (1990) Proc. Natl. Acad. Sci. (USA) 87, 4473-4475.
3. Sauk, J.J., van Kampen, C.L., Norris, K., Foster, R. and Somerman, M.J. (1990) Biochem. Biophys. Res. Comm. 172, 135-142.
4. Senger, D.R., Perruzzi, C.A. and Papadopoulos, A. (1989) AntiCancer Res. 9, 1291-1300.
5. Craig, A.M., Bowden, G.T., Chambers, A.F., Spearman, M., Greenberg, A.H., Wright, J., McLeod, M. and Denhardt, D.T. (1990) Int. J. Cancer 46, 133-137.
6. Young, M.F., Kerr, J.M, Termine, J.D., Wewer, U.M., Wang, M.G., McBride, O.W. and Fisher, L.W. (1990) Genomics 7, 491-502.
7. Fet, V., Dickinson, M.E. and Hogan, B.L.M. (1990) Genomics 5, 375-377.
8. Patarca, R., Freeman, G.J., Singh, R.P., Wei, F.Y., Durfee, T., Blattner, F., Regnier, D.C., Kozak, C.A., Mock, B.A., Morse III, H.C., Jerrels, T.R. and Cantor, H. (1989) J. Exp. Med. 170, 146-161.
9. Senger, D.R., Perruzzi, C.A., Papadopoulos, A. and Tenen, D.G. (1989) Biochim. Biophys. Acta 996, 43-48.
10. Somerman, M.J., Shroff, B., Agraves, W.S., Morrison, G., Craig, A.M., Denhardt, D.T., Foster, R.A. and Sauk, J.J. (1991) J. Biol. Buccale 18, 207-214.
11. Craig, A.M. and Denhardt, D.T. (1991) Gene, 100, 163-171.
12. Kubota, T., Zhang, Q., Wrana, J.L., Ber, R., Aubin, J.E., Butler, W.T. and Sodek, J. (1989) Biochem. Biophys. Res. Comm. 162, 1453-1459.
13. Noda, M., Vogel, R.L, Craig, A.M., Prahl, J., DeLuca, H.F. and Denhardt, D.T. (1990) Proc. Natl. Acad. Sci. (USA) 87, 9995-9999.
14. Rodan, S.B., Wesolowski, G., Yoon, K. and Rodan, G.A. (1989) J. Biol. Chem. 264, 19934-19941.
15. Miyazaki, Y., Setoguchi, M., Yoshida, S., Higuchi, Y., Akizuki, S. and Yamamoto, S. (1990) J. Biol. Chem. 265, 14432-14438.
16. Prince, C.W., Oosawa, T., Butler, W.T., Tomana, M., Bhown, M. and Schrohenloger, R.E.S. (1987) J. Biol. Chem. 262, 2900-2902.
17. Nagata, T., Todescan, R., Goldberg, H.A., Shang, Q. and Sodek, J. (1989) Biochem. Biophys. Res. Comm. 165, 234-240.
18. Nemir, M., DeVouge, M.W. and Mukherjee, B.B. (1989) J. Biol. Chem. 264, 18202-18206.

19.Chang, P.-L. and Prince, C.W. (1991) Cancer Res. 51, 2144-2150.
20.Singh, K., DeVouge, M.W. and Mukherjee, B.B. (1990) J. Biol. Chem. 265, 18696-18701.
21.Moore, M.A., Gotoh, Y., Rafidi, K. and Gerstenfeld, L.C. (1991) Biochemistry 30, 2501-1508.
22.Zhang, Q., Domenicucci, C., Goldberg, H.A., Wrana, J.L. and Sodek, J. (1990) J. Biol. Chem. 265, 7583-7589.
23.Fisher, L.W., Hawkins, G.R., Tuross, N. and Termine, J.D. (1987) J. Biol. Chem. 262, 9702-9708.
24.Craig, A.M., Smith, J.H. and Denhardt, D.T. (1989) J. Biol. Chem. 264, 9682-9689.

■ *David T. Denhardt:*
Department of Biological Sciences,
Rutgers University,
New Brunswick,
New Jersey, USA
■ *Charles W. Prince:*
Department of Nutrition Sciences,
University of Alabama,
Birmingham, Alabama, USA

Perlecan

Perlecan is the largest and most common proteoglycan of basement membranes, and consists of a multiglobular protein core and two to three terminally attached heparan sulphate side chains. Perlecan appears to have multiple functions including cell attachment, basement membrane assembly (protein core), glomerular ionic filtration, control of serine protease activity, and possible storage of basic fibroblast growth factor (side chains).

Perlecan was initially[1] isolated from the EHS (Englebreth-Holm-Swarm) murine tumour and most of our information regarding the structure of this proteoglycan is based on material purified from this and other tumour lines. The core protein of perlecan is one of the largest of any proteoglycan. Removal of heparan sulphate side chains by heparatinase produces a core protein of 400-450 kDa as determined by SDS-PAGE[2]. Rotary shadowing of the intact proteoglycan (Figure 1) shows the core protein to be 83 nm long and to consist of five to seven continuous globules resembling a string of pearls[3,4]. The name of the proteoglycan, "perlecan", is derived from the middle English word for pearl; *perle*. Limited treatment with trypsin produces a 200 kDa core protein fragment that lacks heparan sulphate side chains[2]. Two to three heparan sulphate side chains are attached to one end of the core protein[3,4]. The size of these side chains varies from 30-60 kDa, with a length of 100 to 170 nm depending upon the cell type.

Although perlecan was originally isolated from a tumour cell basement membrane, several lines of evidence indicate that this gene product is widespread in normal basement membranes. First, antibodies to perlecan's core protein react with most if not all basement membranes[5], particularly in the lamina densa region[6], whereas cationic dyes that bind the side chains predominantly stain the lamina lucida region[7]. Second, anti-perlecan antibodies immunoprecipitate a 400 kDa precursor protein which is produced by a variety of basement membrane secreting cells and cell lines[8]. For example, heparan sulphate proteoglycan produced by endothelial cells[9] is immunologically related and structurally identical to EHS proteoglycan. In the glomerulus the core protein of perlecan may be shortened by proteolytic clipping[10]. Third, antibodies to perlecan have also been used to obtain cDNA clones to two different regions of the core protein[11] with distinct homology to laminin and NCAM. These clones hybridize to a 12-14 kB message which is detectable in a number of different tissues. Additional cDNA clones show that perlecan is most homologous to the N-terminal third of the laminin A chain corresponding to the entire top arm of the cross structure of laminin (Hassell and Noonan; unpublished). This indicates that perlecan belongs to the laminin gene family.

Perlecan is an integral component of the glomerular basement membrane (GBM) and is partially responsible for the charge selective filtration properties of the renal glomerulus (Figure 2). A decrease in GBM associated anionic sites due to the presence of heparan surface[12], as well as a decrease in perlecan core protein within the lamina densa of the GBM[13], and lowered mRNA levels for perlecan[14] are associated with increased glomerular permeability in puromycin aminonucleoside (PAN) nephrosis.

■ PURIFICATION

Perlecan is extracted from the EHS tumour using dissociative solvents and then is sequentially purified using CsC1 density gradient centrifugation, and chromatography on DEAE-Sepharose and Sepharose C1-4B[2]. Similar methodology is used for purification from native sources such as cells in culture or glomeruli.

■ ACTIVITIES

Perlecan self associates in solution[15], promotes the attachment of cells in culture[16] and appears to participate in the assembly of basement membranes in conjunction with other components of basement membranes[17]. It is possible that perlecan may also serve to immobilize bFGF[18] and serine proteases[19].

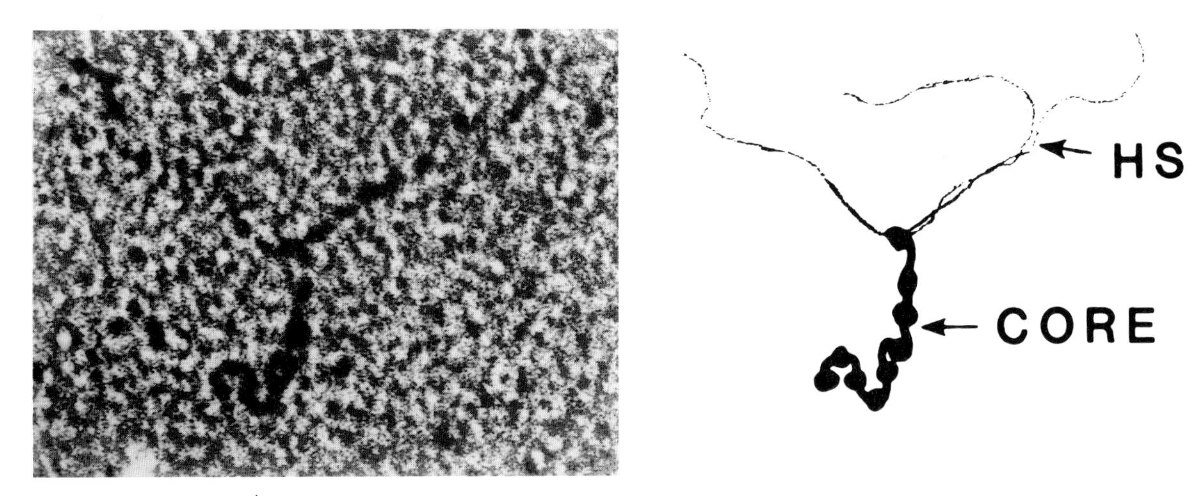

Figure l. Electron micrograph and schematic model of perlecan. Perlecan consists of a globular protein core and two to three long heparan sulphate side chains extending from one terminus.

ANTIBODIES

Polyclonal antibodies have been raised against perlecan isolated from EHS tumour (murine) and glomerular (rat) basement membranes. A monoclonal antibody raised against chick muscle basement membrane reacts with perlecan and this antibody is commercially available[20].

GENES

Mouse[11] and human perlecan[21,22] has been partially cloned and sequenced. Recently, the perlecan gene has been assigned to human chromosome 1, band p 36.1 and to the distal part of mouse chromosome 4[23,24]. GenBank number M-77174 contains the complete sequence of perlecan (mouse).

REFERENCES

1. Hassell, J.R., Robey, P.G., Barrach, H., Wilczek, J., Rennard, S.I. and Martin, G.R. (1980) Proc. Natl. Acad. Sci. 77, 4494-4498.
2. Ledbetter, S.R., Fisher, L.W. and Hassell, J.R. (1987) Biochem. 26, 988-995.
3. Paulsson, M., Yurchenco, P.D., Ruben, G.C., Engel, J. and Timpl, R. (1987) J. Mol. Biol. 197, 297-313.
4. Laurie, G.W., Inoue, S., Bing, J.T. and Hassell, J.R. (l988) Am. J. Anat. 181, 320-326.
5. Horiguchi, Y., Couchman, J.R., Ljubimov, A.V., Yamasaki, H. and Fine, J. (1989) J. Histochem. Cytochem. 37, 961-970.
6. Grant, D.S. and Leblond, C.P. (1988) J. Histochem Cytochem. 36, 271-283.
7. Farquhar, M.G. (1981) In: Cell Biology of the Extracellular Matrix, E.D. Hay, ed. Plenum, N.Y. pp. 335-378.
8. Hassell, J.R., Leyshon, W.C., Ledbetter, S.R., Tyree, B., Suzuki, S., Kato, M., Kimata, K. and Kleinman, H.K. (1985) J. Biol. Chem. 260, 8098-8105.

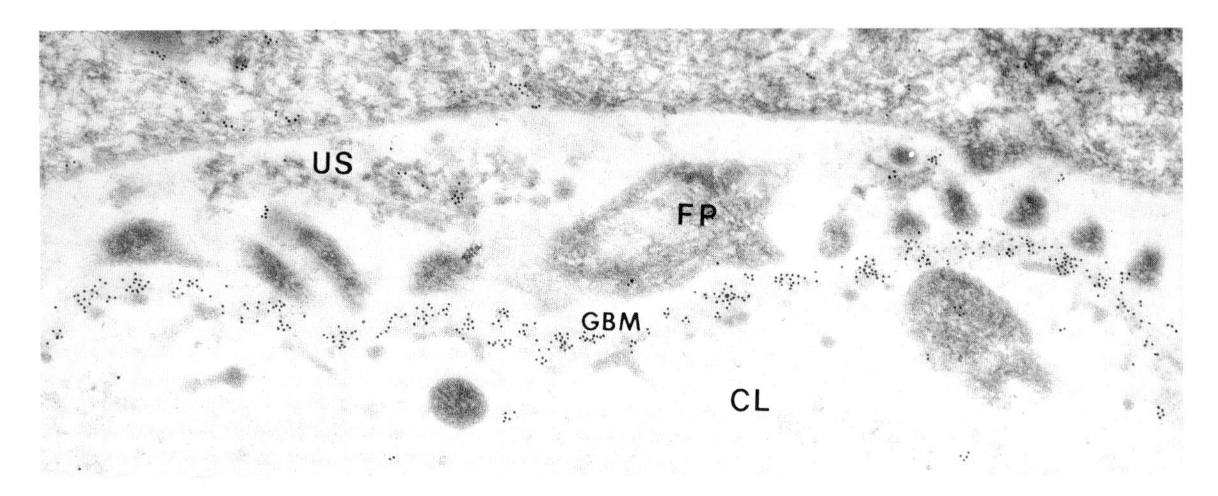

Figure 2. Electron micrograph of a portion of rat renal glomerulus sequentially labelled with rabbit anti-Perlecan antibodies and immunogold. Gold particles are distributed across the entire thickness of the glomerular basement membrane (GBM). FP, foot process; US, urinary space; CL, capillary lumen. Original magnification x 36,000.

9.Saku, T. and Furthmayr, H. (1989) J. Biol. Chem. 264, 3514-3523.
10.Klein, D.J., Brown, D.M., Oegema, T.R., Brenchley, P.E., Anderson, J.C., Dickinson, M.A.J., Horigan, E.A. and Hassell, J.R. (1988) J. Cell Biol. 106, 963-970.
11.Noonan, D.M., Horigan, E.A., Ledbetter, S.R., Vogeli, G., Sasaki, M., Yamada, Y. and Hassell, J.R. (1988) J. Biol. Chem. 263, 16379-16387.
12.Caulfield, J.P. and Farquhar, M.G. (1978) Lab. Invest. 39, 505-512.
13.Mynderse, L., Hassell, J.R., Kleinman, H.K., Martin, G.R. and Martinez-Hernandez, A. (1983) Lab. Invest. 48, 292-302.
14.Nakamura, T., Ebihara, I., Shirato, I., Tomino, Y. and Koide, H. (1991) Lab. Invest. 64, 640-647.
15.Yurchenco, P.D., Cheng, Y. and Ruben, G.C. (1987) J. Biol. Chem. 262, 17668-17676.
16.Clement, B., Segui-Real, B., Hassell, J.R., Martin, G.R. and Yamada, Y. (1989) J. Biol. Chem. 264, 12467-2471.
17.Laurie, G.W., Bing, J.T., Kleinman, H.K., Hassell, J.R., Aumailley, M., Martin, G.R. and Feldman, R.J. (1986) J. Mol. Biol. 189, 205-2l6.
18.Vlodavski, I., Barshavit, R., Ishaimichaeli, R., Bashkin, P., Fuks, Z. and Levi, E. (1991) Trends Biochem. Sci. 16, 268-271.
19.Pejler, G., Bäckström, G. and Lindahl, U. (1987) J. Biol. Chem. 262, 5036-5043.
20.Bayne, E.K., Anderson, M.J. and Fambrough, D.M. (1984) J. Cell Biol. 99, 1486-1501.
21.Dodge, G.R., Kovalszky, I., Chu, M.-L., Hassell, J.R. and Iozzo, R.V. (1990) J. Cell Biol. 111, 152a.
22.Kallunki, P. and Tryggvason, K. (1990) J. Cell Biol. 111, 399a.
23.Wintle, R.F., Kisilevsky, R., Noonan, D. and Duncan, A.M.V. (1990) Cytogenet. Cell Genet. 54, 60-61.
24.Chakravarti, S., Phillips, S.L. and Hassell, J.R. (1991) Mammalian Genome 1, 270-272.

Gordon, W. Laurie[1], Jody, A. Rada, Shukti Chakravarti and John R. Hassell:
The Eye and Ear Institute of Pittsburgh and the Department of Ophthalmology, University of Pittsburgh, Pittsburgh, PA 15213, USA
[1] Department of Anatomy and Cell Biology, University of Virginia, Charlottesville, VA 22908, USA

Plasminogen

Plasminogen is the zymogen of the serine proteinase plasmin, which is produced from plasminogen by highly specific proteolytic cleavage. It is a glycoprotein and has binding interactions with several adhesion proteins of the pericellular matrix, as well as cell surfaces. Plasmin functions in fibrinolysis and in the destruction of extracellular matrix which accompanies cellular migration and invasion (especially tumour invasion), tissue remodelling and inflammation.

Plasminogen (plg) is a single chain glycoprotein produced by hepatocytes and found in plasma at 1.5-2.0 μM, where it has a half-life of ~two days. The major form of 92 kDa consists of a 790 amino acid sequence[1] with 24 disulphide bridges and N-terminal glutamic acid (Glu-plg), but other forms produced by limited plasmin cleavage of the N-terminal sequence are collectively known as Lys-plg[2] (Figure). The glycosylation variants distinguishable by ion exchange chromatography or electrophoresis are known as Glu-plg's I and II. Glu-plg II has higher affinity for fibrin than Glu-plg I, but Glu-plg I is more easily activated to plasmin. Lys-plg differs markedly from Glu-plg in conformation, enabling higher affinity binding interactions and more rapid activation to plasmin[3]. The binding interactions are mediated by the lysine-binding sites contained in five triple-looped internal repeat structures known as kringles, which have extensive homology to similar structures present in the N-termini of prothrombin, tissue plasminogen activator, urokinase, and lipoprotein(a)[4]. Kringle interactions serve to localize and protect plasmin activity against plasma anti-proteases, e.g. alpha-2-antiplasmin[5], but not aprotinin[6]. Interactions involving kringles 1-3 (e.g. with fibrin) and kringle 4 (e.g. with tetranectin) are known, both of which can be disrupted by lysine analogues such as 6-amino-hexanoic acid and trans-4-aminomethylcyclohexanoic acid. Binding proteins include fibrin, tetranectin[7], fibronectin[8], vitronectin[9], thrombospondin[10], laminin[11], histidine-rich glycoprotein (HRGP)[12], platelet glycoproteins IIb/IIIa[13], proteoglycans, immunoglobulin G and very low density lipoprotein (VLDL). Binding of plasminogen to streptokinase does not involve the kringles, but an interaction with the portion corresponding to the catalytic B chain of plasmin[14]. Many types of cells bind plasminogen and plasmin via kringle interactions[13], including endothelial cells, platelets, monocytes and several tumour cell lines derived from carcinoma, fibrosarcoma, rhabdomyosarcoma, and monocytic leukemia, as well as several types of bacteria[15]. However, with the exception of bacteria, the binding affinity in all cases is not higher than with any of the adhesion proteins listed above (i.e. about 1 μM), and a specific membrane receptor protein has proven elusive. Some regulators of plasminogen binding to cells have been found, such as thrombin (endothelial cells and platelets) and dexamethasone (fibrosarcoma HT-1080). Binding of plasminogen to cells enhances considerably the rate of plasmin generation[16], and plasmin formed on the cell surface is able to catalyse the activation of prourokinase while remaining protected from plasma proteinase inhibitors[6].

Plasminogen activation occurs by specific proteolytic

cleavage of Arg-560-Val-561, to produce the plasmin A- and B-chains (65 kDa and 25 kDa respectively) linked by two disulphide bridges (Figure). The catalytic triad of the serine proteinase consists of His-602, Asp-645 and Ser-740, which are all located in the B-chain, while the five kringles are in the A-chain (see Figure). Plasmin behaves as a trypsin-like serine proteinase with a relatively broad specificity, acting on most of the extracellular and basement membrane proteins to which plasminogen binds, namely fibrin, fibronectin, thrombospondin and laminin[17,18]. Plasmin is also able to activate the latent forms of collagenase and growth factors such as TGF-β[18].

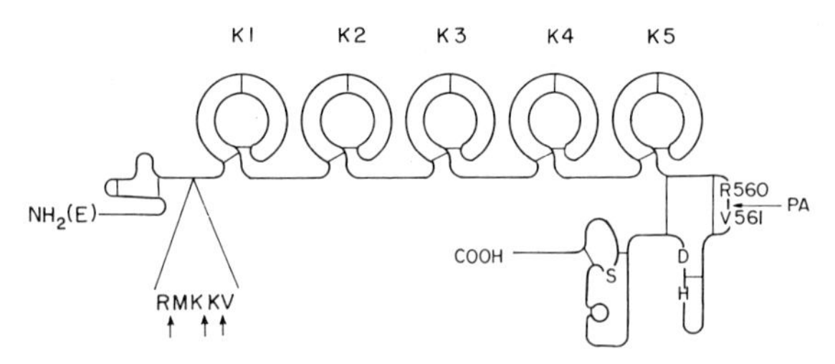

Figure. Structure of human plasminogen. Arrows on the left show cleavage sites of plasmin producing Lys-plg. The positions for the serine proteinase catalytic triad of His, Asp, Ser are shown as H,D,S. For other explanatory notes see text.

PURIFICATION

Plasminogen is readily purified from fresh human plasma by lysine-Sepharose affinity chromatography[19] in the presence of proteinase inhibitors (see modifications in [6]).

ACTIVITIES

Many assays exist for plasminogen, plasminogen activators and plasmin, which depend on immunoassay and/or proteolytic, amidolytic or esterolytic activity (e.g. see [6,17]).

ANTIBODIES

Rabbit and goat polyclonal antibodies, as well as several monoclonal antibodies to human plasminogen and its kringle domains are readily available commercially.

GENES

Human plasminogen has been cloned and sequenced[20]. GenBank numbers X05199 (2732 bp mRNA) and J05286.

REFERENCES

1. Sottrup-Jensen, L., Claeys, H., Zajdel, M., Petersen, T.E. and Magnusson, S. (1978) Progress in Chemical Fibrinolysis and Thrombolysis (Davidson et al. eds.) 3, 191-209, Raven Press, New York.
2. Lijnen, H.R. and Collen, D. (1982) Sem. Thromb. Hemostas. 8, 2-10.
3. Claeys, H. and Vermylen, J. (1974) Biochem. Biophys. Acta 342, 351-359.
4. McLean, J.W., Tomlinson, J.E., Kuang, W.-J., Eaton, D.L., Chen, E.Y., Fless, G.M., Scanu, A.M. and Lawn, R.M. (1987) Nature 300, 132-137.
5. Collen, D. (1980) Thromb. Haemostas. 43, 77-89.
6. Stephens, R.W., Pöllänen, J., Tapiovaara, H., Leung, K.-C., Sim, P.-S., Salonen, E.-M., Rønne, E., Behrendt, N., Danø, K. and Vaheri, A. (1989) J. Cell Biol. 108, 1987-1995.
7. Clemmensen, I., Petersen, L.C. and Kluft, C. (1986) Eur. J. Biochem. 156, 327-333.
8. Salonen, E.-M., Saksela, O., Vartio, T., Vaheri, A., Nielsen, L.S. and Zeuthen, J. (1985) J. Biol. Chem. 260, 12302-12307.
9. Preissner, K. (1990) Biochem. Biophys. Res. Communs. 168, 966-971.
10. Silverstein, R.L., Leung, L.L.K., Harpel, P.C. and Nachman, R.L. (1984) J. Clin. Invest. 74, 1625-1633.
11. Salonen, E.-M., Zitting, A. and Vaheri, A. (1984) FEBS Lett. 172, 29-32.
12. Lijnen, H.R., Hoylaerts, M. and Collen, D. (1980) J. Biol. Chem. 255, 10214-10222.
13. Miles, L. and Plow, E.F. (1988) Fibrinolysis 2, 61-71.
14. Summaria, L. and Robbins, K.C. (1976) J. Biol. Chem. 251, 5810-5813.
15. Ullberg, M., Kronvall, G., Karlsson, I. and Wiman, B. (1990) Infect. Immun. 58, 21-25.
16. Ellis, V., Scully, M.F. and Kakkar, V.V. (1989) J. Biol. Chem. 264, 2185-2188.
17. Danø, K., Andreasen, P.A., Grøndahl-Hansen, J., Kristensen, P., Nielsen, L.S. and Skriver, L. (1985) Adv. Cancer Res. 44, 139-266.
18. Pöllänen, J., Stephens, R.W. and Vaheri, A. (1991) Adv. Cancer Res. 57, 273-328.
19. Deutsch, D.G. and Mertz, E.T. (1970) Science 170, 1095-1097.
20. Petersen, T.E., Martzen, M.R., Ichinose, A. and Davie, E.W. (1990) J. Biol. Chem. 265, 6104-6111.

Ross W. Stephens and Antti Vaheri:
Department of Virology,
University of Helsinki,
SF-00290 Helsinki, Finland

Plasminogen Activator Inhibitor 1 (PAI-1)

PAI-1 is a specific, fast acting glycoprotein Arg-serpin inhibitor of both classes of plasminogen activating enzymes i.e. urokinase (u-PA) and tissue activator (t-PA). It is produced by vascular endothelium and platelets, as well as hepatocytes, fibroblasts, placental trophoblasts and cultured tumour cell lines of diverse origin. PAI-1 is considered to be the principal regulator of vascular fibrinolysis and may also contribute to regulation of pericellular proteolysis mediated by u-PA.

Human PAI-1 is a single-chain 52 kDa glycoprotein of 379 amino acids (43 kDa) after cleavage of a 23 amino acid signal peptide sequence and secretion[1]. PAI-1 belongs to the Arg-serpin class of proteinase inhibitors, with Arg-346 at the P_1 position of the reactive centre[2]. There are three potential asparagine-linked glycosylation sites, but no cysteine residues for disulphide linkages. The human gene for PAI-1 consists of 12.2 kB located on chromosome 7. PAI-1 is found in normal human plasma at 10-30 ng/ml, and at increased levels during sepsis and pregnancy and in patients subject to recurrent deep vein thrombosis and preeclampsia in pregnancy[3,4]. It is secreted by a diverse range of cultured tumour cell lines, often together with PAI-2, but under independent control[5]. Stimulators include TNF-α, IL-1 TGF-β, EGF, glucocorticoids and thrombin[6]. Probably due to its lack of disulphide linkages and therefore unstable secondary structure, PAI-1 rapidly loses activity after secretion, so that it may even appear to be absent in conditioned medium from producer cells, although bound and active in the extracellular matrix[7,8]. Vitronectin is the principal binding protein both in plasma[9,10] and in cell cultures[11], and this interaction stabilizes PAI-1. PAI-1 is a fast-acting inactivator of t-PA, u-PA (but not pro-u-PA) and activated protein C[6,12], but after binding to vitronectin its inhibition of thrombin is enhanced 200 fold while leaving its inhibition of t-PA unchanged[13]. PAI-1 reacts with these enzymes to form an equimolar and inactive covalent complex, with subsequent cleavage of a 33 amino acid sequence from Arg-346[14]. This inactivation reaction is slower but still functional when t-PA is bound to fibrin[15], and also occurs with u-PA when it is bound to the cell-surface u-PA receptor[16,17]. Thus PAI-1 is able to inhibit u-PA mediated extracellular matrix degradation by tumour cells *in vitro*[18].

PURIFICATION

PAI-1 may be purified by immunoaffinity chromatography from the culture medium of HT-1080 fibrosarcoma cells grown with 1 μM dexamethasone.

ACTIVITIES

PAI-1 is a specific, fast-acting ($K_1 = 1 \times 10^7\ M^{-1}s^{-1}$ for both u-PA and t-PA[19]) with very high affinity for plasminogen activators ($K_d = 1.5\text{-}5 \times 10^{-12}$ M for t-PA[20]).

ANTIBODIES

Rabbit and goat polyclonal antibodies to human PAI-1 are available commercially, as well as several monoclonal antibodies, including one which neutralizes inhibitory activity[16].

GENES

The human gene for PAI-1 has been isolated and cloned from endothelial cells[1] and placenta[2]. GenBank number M16006 (2876 bp mRNA, complete cds).

REFERENCES

1. Pannekoek, H., Veerman, H., Lambers, H., Diergaarde, P., Verweij, C.L., Van Zonneveld, A.-J. and Van Mourik, J.A. (1986) EMBO J. 5, 2539-2544.
2. Ny, T., Sawdey, M., Lawrence, D., Millan, J.L. and Loskutoff, D.J. (1986) Proc. Natl. Acad. Sci. (USA) 83, 6776-6780.
3. Pralong, G., Calendra, T., Glauser, M.P., Schellekens, J., Verhoef, J., Bachmann, F. and Kruithof, E.K.O. (1989) Thromb. Haemostas. 61, 459-462.
4. Estellés, A., Gilabert, J., Aznar, J., Loskutoff, D.J. and Schleef, R.R. (1989) Blood 74, 1332-1338.
5. Schleef, R.R., Wagner, N.V. and Loskutoff, D.J. (1988) J. Cell Physiol. 134, 269-274.
6. Andreasen, P.A., Georg, B., Lund, L.R., Riccio, A. and Stacey, S.N. (1990) Mol. Cell. Endocrinol. 68, 1-19.
7. Levin, E.G. (1986) Blood 67, 1309-1313.
8. Pöllänen, J., Saksela, O., Salonen, E.-M., Andreasen, P.A., Nielsen, L.S., Danø, K. and Vaheri, A. (1987) J. Cell Biol. 104, 1085-1096.
9. Wiman, B., Almquist, Å., Sigurdardottir, O. and Lindahl, T. (1988) FEBS Lett. 242, 125-128.
10. Declerck, P.J., Alessi, M.-C., Verstreken, M., Kruithof, E.K.O., Juhan-Vague, I. and Collen, D. (1988) J. Biol. Chem. 263, 15454-15461.
11. Salonen, E.-M., Vaheri, A., Pöllänen, J., Stephens, R.W., Andreasen, P.A., Mayer, M., Danø, K., Gailit, J. and Ruoslahti, E. (1989) J. Biol. Chem. 264, 6339-6343.
12. Fay, W.P. and Owen, W.G. (1989) Biochem. 28, 5773-5778.
13. Ehrlich, H.J., Gebbink, R.K., Keijer, J., Linders, M., Preissner, K. and Pannekoek, H. (1990) J. Biol. Chem. 265, 13029-13035.
14. Andreasen, P.A., Riccio, A., Welinder, K.G., Sartorio, R., Nielsen, L.S., Oppenheimer, C., Blasi, F. and Danø, K. (1986) FEBS Lett. 209, 213-218.
15. Røder, M., Philips, M., Suenson, E. and Thorsen, S. (1988) Fibrinolysis 2 (Suppl. 2) Abs. 225.
16. Stephens, R.W., Pöllänen, J., Tapiovaara, H., Leung, K.-C.,

Sim, P.-S., Salonen, E.-M., Rønne, E., Behrendt, N., Danø, K. and Vaheri, A. (1989) J. Cell Biol. 108, 1987-1995.
17.Cubellis, M.V., Wun, T.-C. and Blasi, F. (1990) EMBO J. 9, 1079-1085.
18.Cajot, J.F., Bamat, J., Bergonzelli, G.E., Kruithof, E.K.O., Medcalf, R.L., Testuz, J. and Sordat, B. (1990) Proc. Natl. Acad. Sci. (USA) 87, 6939-6943.
19.Thorsen, S., Philips, M., Selmer, J., Lecander, I. and Åstedt, B. (1988) Eur. J. Biochem. 175, 33-39.
20.Sprengers, E.D., Princen, H.M.G., Kooistra, T. and Van Hindsberg, V.W.M. (1985) J. Lab. Clin. Med. 105, 751-758.

Ross W. Stephens and Antti Vaheri:
Department of Virology,
University of Helsinki,
SF-00290 Helsinki,
Finland

Plasminogen Activator Inhibitor 2 (PAI-2)

PAI-2 is a specific, fast acting glycoprotein Arg-serpin inhibitor of plasminogen activating enzymes, primarily urokinase (u-PA). Normally undetectable in human plasma, it is produced by trophoblasts during pregnancy, by blood monocytes stimulated in vitro, and by leukemia cells of monocytic origin. It functions as an extracellular regulator of cell surface proteolytic activity.

PAI-2 exists in a 47 kDa intracellular and a 60 kDa (glycosylated) extracellular form[1], both consisting of 415 amino acids[2] and devoid of an N-terminal signal peptide sequence[3]. PAI-2 belongs to the Arg-serpin group of proteinase inhibitors, with Arg-380 at the P1 position of the reactive centre. The PAI-2 gene is located on chromosome 18[4] and is transcribed to a 1.8 kB mRNA. An untranslated 3' region of the cDNA contains an element responsive to several inflammatory mediators[2]. PAI-2 is not detectable in normal plasma or during sepsis, but appears in increasing concentration (up to 260 ng/ml) during pregnancy[5], due to secretion by placental trophoblastic epithelium[6]. It is also secreted by blood monocytes stimulated *in vitro* with e.g. muramyl dipeptide[7], and by leukemia cell lines of monocytic origin e.g. U937[1], but induction in cultured fibroblasts and melanoma cells by TNF-α leads to accumulation of the intracellular form without secretion[8]. Unlike PAI-1, secreted PAI-2 does not spontaneously lose activity, and remains active in solution; no binding proteins have been detected in the pericellular matrix. PAI-2 is primarily an inhibitor of u-PA[9]; its reaction with **tissue plasminogen activator (t-PA)** is slower, particularly in the physiological context of fibrinolysis, when the t-PA is bound to fibrin[10]. PAI-2 reacts with u-PA (but not prourokinase) to produce an equimolar and covalent inactive complex[9], with subsequent cleavage of a 35 residue C-terminal fragment from the PAI-2[11]. This inactivation reaction also occurs when the u-PA is bound to its specific cellular receptor[12,13], thus leading to abolition of urokinase-mediated monocyte invasion[14] and inhibition of matrix degradation by tumour cells[15]. Recent evidence suggests that PAI-2 inactivation of cell-surface u-PA is followed by internalization of the complex[16].

PURIFICATION

PAI-2 has been purified to homogeneity from two main sources, namely human placenta[9] and U937 monocytic leukemia cells[2,17], using sequential application of several conventional methods. Fully active recombinant PAI-2[13] can be produced in *E. coli*[2]. Antibodies suitable for purification by immunoaffinity chromatography[18] are also available.

ACTIVITIES

PAI-2 is a specific, fast acting ($K_1=3 \times 10^6$ $M^{-1}s^{-1}$)[19], high affinity ($K_i=10^{-11}$ M)[19] inactivator of u-PA.

ANTIBODIES

Rabbit and goat polyclonals to human PAI-2 are generally available, as well as monoclonal antibodies suitable for immunoaffinity purification[18].

GENES

The human gene for PAI-2 has been isolated and cloned from both placenta[20] and U937 leukemia[2,4]. GenBank numbers J03603 (1854 bp mRNA, complete cds) and J02685 (1884 bp mRNA, complete cds).

REFERENCES

1.Wohlwend, A., Belin, D. and Vassalli, J.-D. (1987) J. Exp. Med. 165, 320-339.
2.Antalis, T.M., Clark, M.A., Barnes, T., Lehrbach, P.R., Devine, P.L., Schevzov, G., Goss, N.H., Stephens, R.W. and Tolstoshev, P. (1988) Proc. Natl. Acad. Sci. 85, 985-989.
3.Ye, R.D., Wun, T.-C and Sadler, J.E. (1988) J. Biol. Chem. 263, 4869-4875.
4.Webb, A.C., Collins, K.L., Snyder, S.E., Alexander, S.J., Rosenwasser, L.J., Eddy, R.L., Shows, T.B. and Auron, P.E. (1987) J. Exp. Med. 166, 74-94.
5.Kruithof, E.K.O., Tran-Thang, C., Gudinchet, A., Hauert, J., Nicoloso, G., Genton, C., Welti, H. and Bachmann, F. (1987) Blood 69, 460-466.
6.Åstedt, B., Hägerstrand, I. and Lecander, I. (1986) Thromb. Haemostas. 56, 63-65.
7.Golder, J.P. and Stephens, R.W. (1983) Eur. J. Biochem. 136, 517-522.
8.Pytel, B.A., Peppel, K. and Baglioni, C. (1990) J. Cell Physiol. 144, 416-422.

9.Wun, T.-C. and Reich, E. (1987) J. Biol. Chem. 262, 3646-3653.
10.Leung, K.-C., Byatt, J.A. and Stephens, R.W. (1987) Thromb. Res. 46, 755-766.
11.Kiso, U., Kaudewitz, H., Henschen, A., Åstedt, B., Kruithof, E.K.O. and Bachmann, F. (1988) FEBS Lett. 230, 51-56.
12.Stephens, R.W., Pöllänen, J., Tapiovaara, H., Leung, K.-C., Sim, P.-S., Salonen, E.-M., Rønne, E., Behrendt, N., Danø, K. and Vaheri, A. (1989) J. Cell Biol. 108, 1987-1995.
13.Pöllänen, J., Vaheri, A., Tapiovaara, H., Riley, E., Bertram, K., Woodrow, G. and Stephens, R.W. (1990) Proc. Natl. Acad. Sci. 87, 2230-2234.
14.Kirchheimer, J.C. and Remold, H.G. (1989) J. Immunol. 143, 2634-2639.
15.Baker, M.S., Bleakley, P., Woodrow, G. and Doe, W. (1990) Cancer Res. 50, 4676-4684.
16.Estreicher, A., Mühlhauser, J., Carpentier, J.-L., Orci, L. and Vassalli, J.-D. (1990) J. Cell Biol. 111, 783-792.
17.Kruithof, E.K.O., Vassalli, J.-D., Schleuning, W.-D., Mattaliano, R.J. and Bachmann, F. (1986) J. Biol. Chem. 261, 11207-11213.
18.Åstedt, B., Lecander, I., Brodin, T., Lundblad, A. and Löw, K. (1982) Thromb. and Haemostas. 53, 122-125.
19.Christensen, U., Holmberg, L., Bladh, B. and Åstedt, B. (1982) Thromb. Haemostas. 48, 24-26.
20.Ye, R.D., Wun, T.-C. and Sadler, J.E. (1987) J. Biol. Chem. 262, 3718-3725.

Ross W. Stephens and Antii Vaheri:
Department of Virology,
University of Helsinki,
SF-00290 Helsinki,
Finland

Proteins Containing Ca^{2+}-Dependent Carbohydrate Recognition Domains (CRDs)

A common structural motif is found in proteins which share the ability to interact with carbohydrate ligands in a Ca^{2+}-dependent manner. These C-type carbohydrate recognition domains (CRDs) are found in association with a variety of other structural elements, including internal signal-anchor sequences, collagenous segments, complement homology sequences, EGF-like domains, and fibronectin type II repeats. Amongst these proteins, several proteoglycans found at the cell surface and in the extracellular matrix are those most likely to serve a structural role.

Proteins which contain C-type CRDs are summarized in the Figure. It is believed that the carbohydrate recognition function of these proteins represents a means of detecting the presence of specific complex sugar structures on glycoproteins, glycolipids, and larger particles. The effect of this sugar binding is then determined by the other "effector" domains associated with the CRD[1]. Two classes of membrane receptor have been identified. The first are type II transmembrane proteins (N-terminal cytoplasmic domains followed by internal signal anchors), which appear to function in receptor mediated endocytosis of carbohydrate containing ligands[2]. The low affinity IgE F_C receptor from lymphocytes is similar in overall structure[3], although it has not been shown to recognize a carbohydrate determinant. In contrast, the mannose receptor mediates phagocytosis of particulate ligands as well as endocytosis of soluble ligands. It is a transmembrane protein with the opposite orientation, and contains eight potential CRDs, as well as a fibronectin type II repeat[4]. The remaining group of receptors all contain N-terminal CRDs followed by EGF-like repeats and complement homology segments[5]. These proteins are collectively referred to as LEC-CAMs or **selectins**. One of the proteins (the MEL-14 antigen in mouse, corresponding to LAM-1 in humans) directs T-cell homing to peripheral lymph nodes.

A second class of proteins containing C-type CRDs are soluble, extracellular proteins. These include several proteins which are simply CRDs in isolation, found in vertebrate circulation (tetranectin[6]), various secretory ducts (pancreatic stone protein[7]), and in invertebrate secretions[8]. An additional class of proteins contains CRDs linked to collagenous sequences[9]. In serum, the mannose binding proteins mediate preimmune defense against pathogens, by functioning as opsonins and fixing complement[10,11]. The role of the similarly constructed pulmonary surfactant protein SP-A is less clear[12].

Probably of most interest in the context of the cytoskeleton and cell structure are the matrix proteoglycans which contain C-type CRDs. The most extensively studied of these proteins is the core protein of cartillage proteoglycan, which contains an N-terminal hyaluronic acid binding region, a central segment to which are attached keratan sulphate and chondroitin sulphate chains, followed by the C-type CRD and finally a complement homology region[13]. Although initially the CRD was recognized based on its sequence, Ca^{2+}-dependent sugar binding by this region of the molecular has been demonstrated experimentally[14]. It is possible that this activity is related to organization of the proteoglycan in the extracellular matrix. Finally, an additional proteoglycan from human fibroblasts, designated versican, resembles the cartillage protein in overall structure, but differs in the carbohydrate attachment region and includes an EGF-like repeat[15].

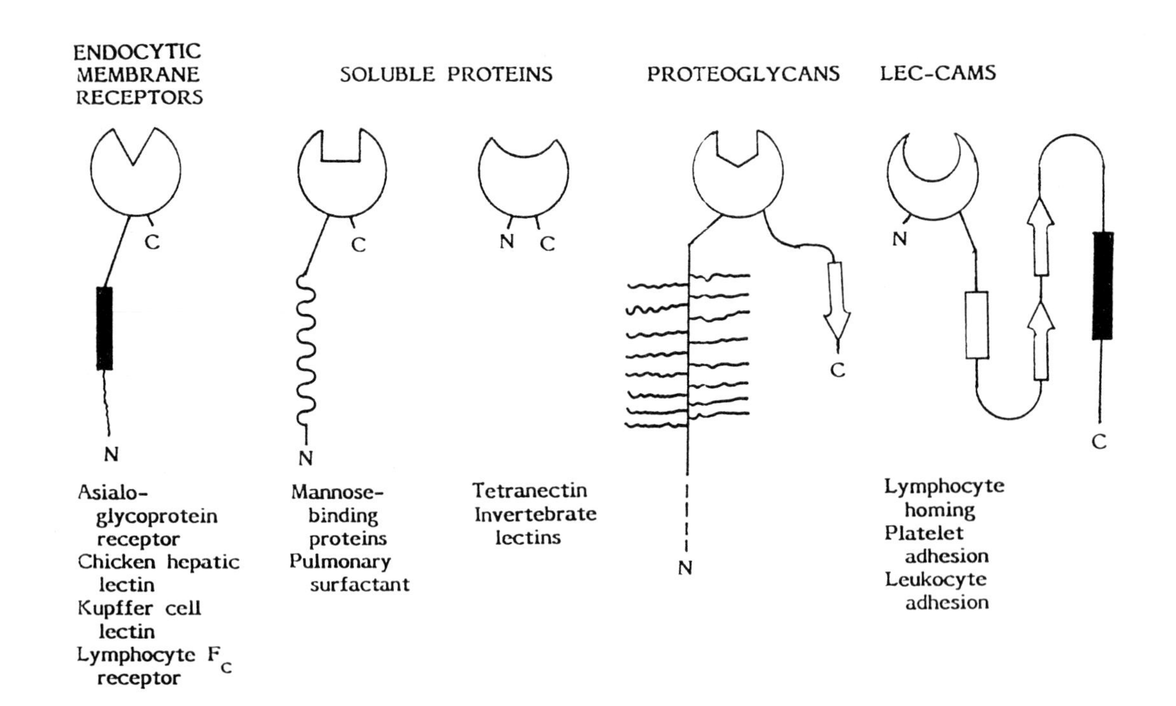

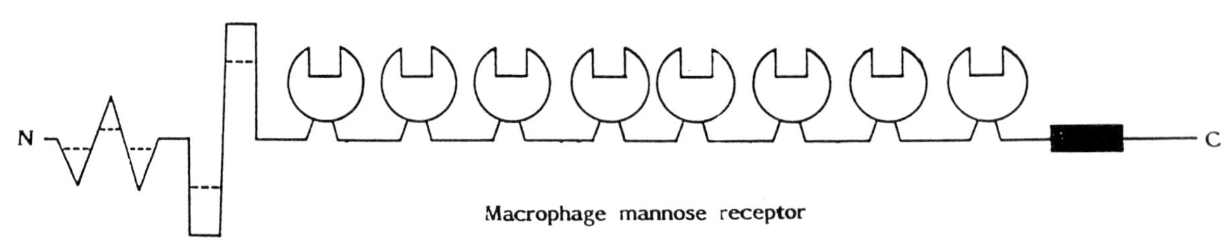

Figure. The structures of five classes of proteins which contain C-type carbohydrate-recognition domains are summarized. Solid rectangles represent membrane-spanning segments, the wavey line represents collagenous sequences, brush-like region denotes glycosamino glycan attachment, open box represents an EGF-like repeat, open arrows represent complement homology domains, and the folded portion of the mannose receptor represents a fibronectin type II repeat.

■ PURIFICATION

The purification schemes for each of the various proteins containing C-type CRDs are generally specific to the individual proteins. The endocytic receptors (extracted into Triton X-100) and the water soluble binding proteins can be effectively purified by affinity chromatography on immobilized sugars, usually in a single step[16,17]. Proteoglycans are generally purified by extraction in the presence of strong protein denaturant (such as 4 M guanidinium hydrochloride) followed by sequential CsCl density gradient centrifugation under associative and dissociative conditions[18].

■ ACTIVITIES

The carbohydrate binding activities of the CRDs have been measured using several precipitation/filtration assays[19], by affinity chromatography on immobilized sugars[14], and by binding of iodinated CRD to glycoproteins fixed onto nitrocellulose by electrotransfer from gels[20] or to neoglycolipids created from oligosaccharides and resolved by thin layer chromatography[21].

■ ANTIBODIES

Antibodies have been raised against many of the individual proteins containing C-type CRDs, but they have not generally been found to react with other proteins in the family. One serum, raised against a common peptide determinant, does appear to react with several different CRDs[22].

GENES

Full length cDNAs have been described for virtually all of the proteins described in the Figure. Specific references are generally those provided in the discussion above, or are immediately available therein.

REFERENCES

1. Drickamer, K. (1988) J. Biol. Chem. 263, 9557-9560.
2. Spiess, M. (1990) Biochemistry 29, 10009-10018.
3. Luden, C., Hofstetter, H., Sarfati, M., Levy, C.A., Suter, U., Alaimo, D., Kilchherr, E., Frost, H. and Delespesse, G. (1987) EMBO J. 6, 109-114.
4. Taylor, M.E., Conary, J.T., Lennartz, M.R., Stahl, P.D. and Drickamer, K. (1990) J. Biol. Chem. 265, 12156-12162.
5. Stoolman, L.M. (1989) Cell 56, 907-910.
6. Fuhlendorff, J., Clemmensen, I. and Magnusson, S. (1987) Biochemistry 26, 6757-6764.
7. DeCaro, A.M., Bonicel, J.J., Rouimi, P., DeCaro, J.D., Sarles, H. and Rovery, M. (1987) Eur. J. Biochem. 168, 201-207.
8. Takahashi, H., Komano, H., Kawaguchi, N., Kitamura, N., Nakanishi, S. and Natori, S. (1985) J. Biol. Chem. 260, 12228-12233.
9. Drickamer, K., Dordal, M.S. and Reynolds, L. (1986) J. Biol. Chem. 261, 6878-6886.
10. Ikeda, K., Sannoh, T., Kawasaki, N., Kawasaki, T. and Yamashina, I. (1987) J. Biol. Chem. 262, 7451-7454.
11. Kuhlman, M., Joiner, K. and Ezekowitz, R.A.B. (1989) J. Expt. Med. 169, 1733-1745.
12. Benson, B., Hawgood, S., Schilling, J., Clements, J., Damm, D., Cordell, B. and White, R.T. (1985) Proc. Natl. Acad. Sci. (USA) 82, 6379-6383.
13. Doege, K., Sasaki, M., Horigan, E., Hassell, J.R. and Yamada, Y. (1987) J. Biol. Chem. 262, 17757-17767.
14. Halberg, D.F., Proulx, G., Doege, K., Yamada, Y. and Drickamer, K. (1988) J. Biol. Chem. 263, 9486-9490.
15. Zimmermann, D.R. and Ruoslahti, E. (1989) EMBO J. 8, 2975-2981.
16. Ashwell, G. and Kawasaki, T. (1978) Methods Enzymol. 50, 287-288.
17. Mori, K., Kawasaki, T. and Yamashina, I. (1983) Arch. Biochem. Biophys. 222, 542-552.
18. Piez, K. and Reddi, A. (eds.) (1984) Extracellular Matrix Biochemistry (New York: Elsevier).
19. Grant, D.A. and Kaderbhai, N. (1986) Biochem. J. 234, 131-137.
20. Drickamer, K. (1989) Biochem. Soc. Trans. 17, 13-15.
21. Childs, R.A., Drickamer, K., Kawasaki, T., Thiel, S., Mizuochi, T. and Feizi, T. (1989) Biochem. J. 262, 131-138.
22. Blanck, O., Thibault, V., Ganier, C., van Rietschoten, J., Courageot, J. and Miquelis, R. (1990) Biochem. Biophys. Res. Commun. 169, 880-887.

Kurt Drickamer:
Department of Biochemistry and Molecular Biophysics,
Columbia University,
New York, NY 10032, USA

Restrictin

Restrictin is a neural extracellular matrix glycoprotein composed of four structural motifs. A cysteine-rich segment at the N-terminus is followed by four and a half EGF-like repeats and then by nine consecutive motifs that are similar to fibronectin type III repeats. At the C-terminus restrictin is related to the β- and γ-chains of fibrinogen including similarity to a Ca^{2+}-binding segment[1]. Restrictin is related to tenascin[2,3] including a substantial sequence similarity and an overall domain arrangement. Like tenascin it forms disulphide-linked oligomeric structures. It is expressed in the nervous system and implicated in neural cell attachment[4].

Restrictin is an extracellular matrix glycoprotein that can be isolated in different molecular mass forms. The native form of the major 180 kDa component appears to be disulphide-crosslinked and contains N-linked oligosaccharides[4]. The pI of restrictin ranges from 5.3 to 4.6. Restrictin copurifies with the axonal surface recognition molecule F11, a GPI-pI anchored member of the immunoglobulin superfamily[4-7]. Electron microscopic micrographs of purified restrictin show a mixture of three-armed oligomers and linear rodlike structures which are dimers and monomers of a restrictin subunit (Figure). The arms of the trimers are 50 nm long and a central nodule attached above the centre of each three-armed particle can be visualized[1]. This nodule is probably formed by the N-terminal cysteine-rich segment and links the three arms together. Each arm is of equal thickness and has like tenascin a small knob at the distal tip of each arm which corresponds to the C-terminal fibrinogen-related segment.

In contrast to many other adhesive proteins restrictin shows a very restricted expression pattern during development of the nervous system. For example, in the spinal cord restrictin is only found around motor neurons and in lower concentration on motor axons. In the embryonic *cerebellum*, restrictin appears to be primarily expressed in the prospective white matter and in the developing retina, restrictin is localized in the outer and inner plexiform layers and, at later stages also in the optic fibre layer[4]. Restrictin might be related to the mouse J1 glyco-

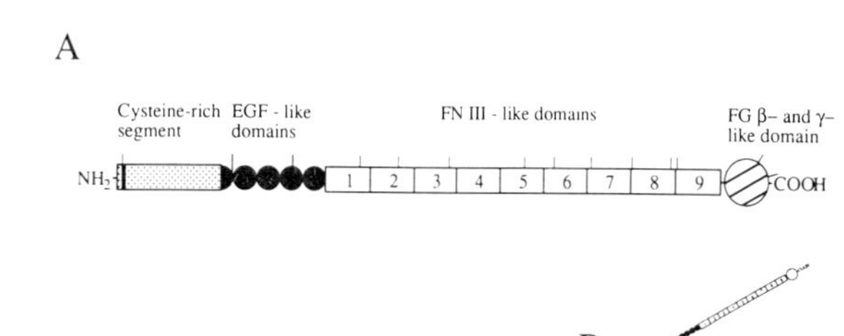

Figure. Domain organization of the restrictin polypeptide. (A) A long rectangle represents the N-terminal cysteine-rich segment and EGF-like repeats are indicated as filled circles. Boxes numbered 1 to 9 correspond to fibronectin type III-like domains and the C-terminal fibrinogen-like segment are shown by a hatched circle at the right. Potential N-glycosylation sites are indicated by vertical thin dashes. (B) Proposed model of a restrictin trimer. The N-terminal segments of each polypeptide are disulphide linked (taken from Reference 1).

proteins (J1-160/J1-180) which are expressed by oligodendrocytes[8,9].

■ PURIFICATION

Restrictin can be purified by immunoaffinity chromatography using anti-restrictin monoclonal antibodies from urea extracts of brain tissue[4].

■ ACTIVITIES

Restrictin has been shown to be implicated in neural cell attachment[4] which is not effected by the peptide GRGDSP, a sequence identified in the cell attachment site of other extracellular matrix glycoproteins and which is recognized by integrin heterodimers. The cell attachment site of restrictin is mapped by antibody perturbation experiments to the eighth or ninth fibronectin type III-like domain at the C-terminal region[1]. Restrictin does not stimulate the outgrowth of axons[4].

■ ANTIBODIES

Polyclonal and monoclonal antibodies to chicken restrictin have been described. They crossreact with other species including amphibians[4].

■ GENES

Complete cDNA sequences are available for chicken restrictin[1] (EMBL data library accession X64649).

■ REFERENCES

1. Nörenberg, U., Wille, H., Wolff, J.M., Frank, R. and Rathjen, F.G. (1992) Neuron, 8, 849-863.
2. Spring, J., Beck, K. and Chiquet-Ehrismann, R. (1989) Cell 59, 325-334.
3. Jones, F.S., Hoffman, S., Cunningham, B.A. and Edelman, G.M. (1989) Proc. Natl. Acad. Sci. (USA) 86, 1905-1909.
4. Rathjen, F.G., Wolff, J.M. and Chiquet-Ehrismann, R. (1991) Development 113, 151-164.
5. Rathjen, F.G., Wolff, J.M., Frank, R., Bonhoeffer, F. and Rutishauser, U. (1987) J. Cell Biol. 104, 343-353.
6. Wolff, J.M., Rathjen, F.G., Frank, R. and Roth, S. (1987) Eur. J. Biochem. 168, 551-561.
7. Brümmendorf, T., Wolff, J.M., Frank, R. and Rathjen, F.G. (1989) Neuron 2, 1351-1361.
8. Pesheva, P., Spiess, E. and Schachner, M. (1989) J. Cell Biol. 109, 1765-1778.
9. Morganti, M.C., Taylor, J., Pesheva, P. and Schachner, M. (1990) Exp. Neurology 109, 98-110.

■ *Fritz G. Rathjen:*
Zentrum für Molekulare Neurobiologie,
Hamburg, Germany

Serglycin

Serglycin (or PG 19) is a proteoglycan synthesized by cells of the myeloid lineage and by the yolk sac. In mast cells and other myeloid cells serglycin is present in secretory granules where it is believed to function as a matrix to which histamin and proteases are bound via ionic linkages.

Serglycin can be substituted with chondroitinsulphate or heparin glycosaminoglycan chains. Hybrid proteoglycans with both chondroitinsulphate and heparin have also been identified[1]. The rat serglycin core protein is synthesized as a precursor of 179 amino acids[2]. The precursor contains a 26 residue signal peptide as well as a 49 residue propeptide. The mature protein contains 104 amino acids which include a domain consisting exclusively of alternating serine and glycine residues repeated 24 times. The serine residues in this domain function as attachment sites for glycosaminoglycan chains. In human and mouse serglycin the repeat domain is shorter and contains eight and ten dipeptide repeats, respectively[3,4].

The mouse serglycin gene is divided into three exons[5]. Interestingly yolk sac carcinoma cells and cells of myeloid origin utilize different promoters in the gene. Serglycin mRNA produced by yolk sac carcinoma cells has an approximately 200 Bp longer 5'-nontranslated region than the message synthesized by myeloid cells[6]. Serglycin produced by yolk sac cells also has a different cellular distribution[7]. In these cells the proteoglycan is exported to the extracellular space whereas in myeloid cells the serglycin is exclusively found in intracellular granules.

PURIFICATION

The purification of serglycin is accomplished by using CsCl gradient centrifugation, ion exchange and molecular sieve chromatography[7].

ACTIVITIES

The serglycin is believed to function as a polyanionic matrix to which histamine and other secretory granule components are bound via ionic linkages. The serglycin core protein is essential for the biosynthesis of heparin, a well known anticoagulant[1].

ANTIBODIES

The proteoglycan is a poor antigen but antibodies have been raised against a mouse serglycin λ gt11 fusion protein[4].

GENES

The complete sequence of serglycin has been obtained and has the EMBL Data Bank number M14282.

REFERENCES

1. Kjellen, L. and Lindahl, U. (1991) Annu. Rev. Biochem. 60, 443-475.
2. Bourdon, M.A., Shiga, M. and Ruoslahti, E. (1986) J. Biol. Chem. 261, 12534-12537.
3. Stevens, R.L., Avraham, S., Gartner, M.C., Bruns, G.A., Austen, K.F. and Weis, J.H. (1988) J. Biol. Chem. 263, 7287-7291.
4. Kjellen, L., Pettersson, I., Lillhager, P., Steen, M.-L., Pettersson, U., Lehtonen, P., Karlsson, T., Ruoslahti, E. and Hellman, L. (1989) Biochem. J. 263, 105-113.
5. Angert, T., Huang, R., Aveskog, M., Pettersson, I., Kjellen, L. and Hellman, L. (1990) Gene 93, 235-240.
6. Bourdon, M.A., Shiga, M. and Ruoslahti, E. (1987) Mol. Cell. Biol. 7, 33-40.
7. Oldberg, Å., Hayman, E.G. and Ruoslahti, E. (1981) J. Biol. Chem. 256, 10847-10852.

Åke Oldberg:
Department of Physiological Chemistry,
University of Lund, P.O.Box 94,
S-22100 Lund, Sweden

SPARC/Osteonectin

SPARC/osteonectin, a 32 kDa Ca^{2+}-binding glycoprotein secreted by a variety of cells, is a prominent constituent of bone. Although the precise function of SPARC/osteonectin is unknown its expression has been linked to bone formation and to tissue differentiation and remodelling.

SPARC/osteonectin was first isolated from subperiosteal bone[1] and named "osteonectin" based on its ability to complex with type I collagen and hydroxylapatite[2]. However, subsequent studies showing that the same protein is expressed in many nonosteogenic tissues[3-8] have not supported the originally proposed role of SPARC/ osteonectin in connective tissue mineralization[2]. Indeed, the variable amounts of SPARC/osteonectin in different mineralized tissues[4] and in the same mineralized tissues of different species[9] indicate that the protein associates with preformed mineral in amounts that reflect accessibility to the mineral and the affinity of the protein[10]. That SPARC/osteonectin has a much broader functional role, in developmental processes[11], hemostasis[7], wound healing and tissue remodelling[3], is indicated from its expression in a variety of tissues including endodermal[5,6], epidermal[12] and soft connective tissues[3,4]. Consequently, the descriptive term "SPARC" (secreted protein, acidic and rich in cysteine)[5] is perhaps a more appropriate name for this protein.

The protein sequence for SPARC/osteonectin determined from cDNA sequences from several species including human[12], bovine[13] and mouse[5] is highly conserved. The mouse protein is encoded by ten exons[14] and has four distinct domains (I-IV, Figure 1)[15]. Exons three and four code for an N-terminal domain (I) comprising two glutamate rich segments that can bind more than eight Ca^{2+} ions. Ca^{2+}-binding in mouse SPARC is thought to induce a coil to helix transition[15]. However, the protein segment encoded by exon three, which is thought to bind to hydroxylapatite, is the least well conserved[16]. Thus, differences in this region could explain the high α-helical content of the porcine protein[10] in the absence of Ca^{2+}, and the low amount of SPARC/osteonectin bound to hydroxylapatite in rodent bones[9]. The second domain (II), encoded by exons five and six, shows some homology to ovomucoid[13] and is characteristically rich in disulphide bridges that stabilize the protein structure. The third domain (III), encoded by exons seven and eight, is predicted to be in an α-helical conformation and to be susceptible to proteolysis[15]. Exon nine codes for the fourth domain (IV) which contains a single high affinity EF-hand Ca^{2+}-binding site with the characteristic helix-loop-helix structure, stabilized by a disulphide bond. Since the Ca^{2+}-binding site (K_d=3x10^{-7}M) is expected to be fully occupied at physiological concentrations of Ca^{2+}, it is unlikely to be involved in Ca^{2+} regulation[15]. Although SPARC/osteonectin is the product of a single gene, an extracellular matrix glycoprotein (SC1) of brain tissue contains a C-terminal segment that is homologous to that part of SPARC/osteonectin encoded by exons five to nine, including the EF-hand motif[17].

PURIFICATION

SPARC/osteonectin is readily purified from foetal bone using demineralizing extractants (0.5 M EDTA) containing proteinase inhibitors to prevent proteolysis[1]. Following gel filtration on Sepharose CL-6B in 4 M guanidine hydrochloride, successive fractionations on fast Q and mono Q resins in the presence of 7 M urea yield pure protein[10]. Based on secondary structure analysis the protein renatures upon removal of the urea[10]. The platelet protein, which differs in glycosylation, has also been purified in native form by immunoaffinity chromatography[18].

ACTIVITIES

SPARC/osteonectin binds strongly to hydroxylapatite even in the presence of 4 M guanidine hydrochloride and 7 M urea[19] and inhibits hydroxylapatite crystal growth[20]. It also binds to extracellular matrix proteins including collagen types III and V and thrombospondin[21,22], but binding to type I collagen[2] is controversial[10,18,21]. SPARC/osteonectin will also bind to and inhibit spreading of endothelial and smooth muscle cells and fibroblasts[21] (Figure 2).

ANTIBODIES

Polyclonal antibodies[4,10,21] and monoclonal antibodies[18,23,24] raised against SPARC/osteonectin from various species have been described that crossreact with SPARC/osteonectin from most vertebrate species tested.

GENES

Full length cDNA sequences for human[12], mouse[5,12] (EMBL GenBank Y00755)[12] and bovine (J03233)[13] SPARC/osteonectin are available. Genomic clones used to study the bovine[16] and mouse (J03913)[14] gene structure and the respective promoter regions [(J04424)[25], (J04951)[26]] have also been isolated.

REFERENCES

1. Termine, J.D., Belcourt, A.B., Conn, K.M. and Kleinman, H.K. (1981) J. Biol. Chem. 256, 10403-10408.
2. Termine, J.D., Kleinman, H.K, Whitson, W.S., Conn, K.M., McGarvey, M.L. and Martin, G.R. (1981) Cell 26, 99-105.
3. Wasi, S., Otsuka, K., Yao, K.-L., Tung, P.S., Aubin, J.E., Sodek, J. and Termine, J.D. (1984) Can. J. Biochem. Cell Biol. 62, 470-478.
4. Tung, P.S., Domenicucci, C., Wasi, S. and Sodek, J. (1985) J. Histochem. Cytochem. 33, 531-540.
5. Mason, I.J., Taylor, A., Williams, J.G., Sage, H. and Hogan, B.L.M. (1986) EMBO J. 5, 1465-1472.
6. Sage, H., Johnson, C. and Bornstein, P. (1984) J. Biol. Chem. 259, 3993-4007.
7. Romberg, R.W., Werness, P.G., Lollar, P., Riggs, B.L. and Mann, K.G. (1985) J. Biol. Chem. 260, 2728-2736.
8. Mann, K., Deutzmann, R., Paulsson, M. and Timpl, R. (1987) FEBS Lett. 218, 167-172.
9. Zung, P., Domenicucci, C., Wasi, S., Kuwata, F. and Sodek, J. (1986) Can. J. Biochem. Cell. Biol. 64, 356-362.
10. Domenicucci, C., Goldberg, H.A., Hofmann, T., Isenman, D., Wasi, S. and Sodek, J. (1988) Biochem. J. 253, 139-151.
11. Nomura, S., Wills, A.J., Edwards, D.R., Heath, J.K. and Hogan, B.L.M. (1988) J. Cell Biol. 106, 441-450.
12. Lankat-Buttgereit, B., Mann, K., Deutzmann, R., Timpl, R. and Kreig, T. (1988) FEBS Lett. 236, 352-356.
13. Bolander, M.E., Young, M.F., Fisher, L.W., Yamada, Y. and Termine, J.D. (1988) Proc. Natl. Acad. Sci. 85, 2919-2923.
14. McVey, J.H., Nomura, S., Kelly, P., Mason, I.J. and Hogan, B.L.M. (1988) J. Biol. Chem. 263, 11111-11116.
15. Engel, J., Taylor, W., Paulsson, M., Sage, H. and Hogan, B.L.M. (1987) Biochemistry 26, 6958-6965.
16. Findlay, D.M., Fisher, L.W., McQuillan, C.L., Termine, J.D. and Young, M.F. (1988) Biochemistry 27, 1483-1489.

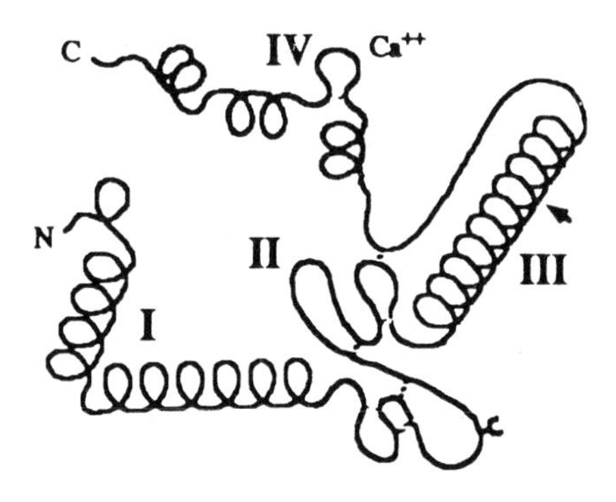

Figure 1. Representative structure for SPARC/osteonectin15.

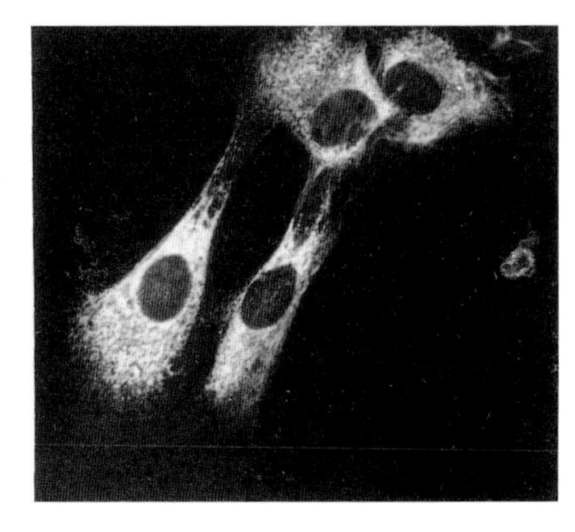

Figure 2. Immunofluorescence localization of SPARC/osteonectin in fibroblasts.

17.Johnston, I.G., Paladino, T., Gurd, J.W. and Brown, I.R. (1990) Neuron 2, 165-176.
18.Kelm Jr., R.J. and Mann, K.G. (1991) J. Biol. Chem. 266, 9632-9639.
19.Otsuka, K., Yao, K.-L., Wasi, S., Tung, P.S., Aubin, J.E., Sodek, J. and Termine, J.D. (1984) J. Biol. Chem. 259, 9805-9812.
20.Romberg, R.W., Werness, P.G., Riggs, B.L. and Mann, K.G. (1986) Biochemistry 25, 1176-1180.
21.Sage, H., Vernon, R.B., Funk, S.E., Everitt, E.A. and Angello, J. (1989) J. Cell Biol. 109, 341-356.
22.Clezardin, P., Malaval, L., Ehrensperger, A.-S., Delmas, P.D., Dechavanne, M. and McGregor, J.L. (1988) Eur. J. Biochem. 175, 275-284.
23.Bolander, M., Gehron-Robey, P., Fisher, L.W., Conn, K.M., Prabhakar, B.S. and Termine, J.D. (1989) Calc. Tiss. Int. 45, 74-80.
24.Malaval, L., Darbouret, B., Preaudat, C., Jolu, J.-P. and Delmas, P.D. (1991) J. Bone Min. Res. 6, 315-323.
25.Young, M.F., Findlay, D.M., Dominguez, P., Burbelo, P.D., McQuillan, C., Kopp, J.B., Gehron-Robey, P. and Termine, J.D. (1989) J. Biol. Chem. 264, 450-456.
26.Nomura, S., Hashmi, S., McVey, J.H., Ham, J., Parker, M. and Hogan, B.L.M. (1989) J. Biol. Chem. 264, 12201-12207.

Jaro Sodek:
MRC Group in Periodontal Physiology,
University of Toronto,
Toronto, Ontario,
Canada

Syndecan

Syndecan is an integral membrane proteoglycan, containing both heparan sulphate (HS) and chondroitin sulphate (CS) glycosaminoglycan (GAG) chains, that associates extracellularly with a variety of matrix molecules and to basic FGF, a heparin binding growth factor, and intracellularly with the actin cytoskeleton. Abundant on epithelia in mature tissues and on epithelial and induced mesenchymal cells during embryonic development, syndecan appears to be involved in regulating the shape and organization of epithelial cells.

Syndecan, originally isolated from cultured mouse mammary (NMuMG) epithelial cells[1], contains a 33 kDa core protein with three domains: an N-terminal 235 amino acid, cysteine-free, extracellular domain (ectodomain), containing five potential GAG attachment sites in two distinct regions and a dibasic protease susceptible site adjacent to the transmembrane domain, a hydrophobic transmembrane domain, and a 34 amino acid cytoplasmic domain[2] (Figure 1). Human mammary epithelial cell[3] and baby hamster kidney cell[4] syndecan are nearly identical (>90% sequence identity) in their cytoplasmic and transmembrane domains and in the site and sequence of their GAG-binding regions. Syndecan may represent a family of integral membrane proteoglycans: a human fibroblast HS proteoglycan core protein is highly similar to syndecan in its cytoplasmic (66%) and transmembrane (52%) domains but deviates substantially in its extracellular domain[5].

Via its HS chains, syndecan binds NMuMG cells to the interstitial matrix proteins fibronectin[6], types I, III and V collagen[7] and thrombospondin[8], as well as to the heparin binding growth factor, bFGF[9]. The significance of the binding to bFGF is unclear, but syndecan may alter the conformation, degradation or presentation of the peptide growth factor, modifying its interaction with the bFGF receptor[10].

Syndecan is involved in stabilizing epithelial shape (Figure 2). Binding of matrix proteins can crosslink syndecan in the plane of the membrane and lead to association of its cytoplasmic domain with the ***actin*** containing cytoskeleton[11]. Suspension of the cells leads to cleavage of the ectodomain at the protease susceptible site and rapid shedding ($t_{1/2}$ = 15 min); cell surface syndecan does not reappear while the cells are suspended[12]. Mammary epithelial cells made syndecan deficient with an antisense mRNA lose epithelial characteristics as demonstrated by

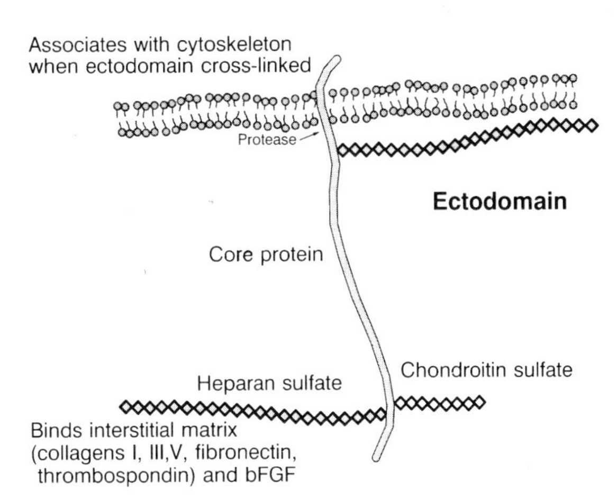

Figure 1. Model of syndecan at the cell surface. Note the intracellular, transmembrane and extracellular (ectodomain) domains of the core protein. The cytoplasmic domain can associate with the cytoskeleton. The ectodomain has a protease-susceptible site adjacent to the plasma membrane; cleavage at this site releases the ectodomain. The ectodomain bears both chondroitin and heparan sulphate chains, and the latter bind interstitial matrix components and bFGF. (Modified from Bernfield and Sanderson, 1990).

their fusiform shape, diminished cell and matrix adhesion, disorganized cortical actin filaments and reduced cell surface E-cadherin and β1 integrins[13].

Changes in syndecan localization and size correlate with tissue organization and cell type (Figure 2). On *stratified* epithelia, syndecan is over the entire cell surface and has a modal relative mass of ~92 kDa, on *simple* epithelia, syndecan localizes to basolateral surfaces and is ~160 kDa[14], whereas in mesenchymal cells, syndecan is predominantly intracellular and is ~300 kDa[15]. Differences in the number and size of GAG chains on a single core protein account for the polymorphism. In developing mouse embryos, syndecan is detected initially at the four cell stage and is expressed constitutively on epithelial sheets, transiently on epithelial induced mesenchymal condensations, and is lost when cells differentiate or undergo changes in shape[9]. Syndecan on precursor B lymphocytes in the bone marrow stromal matrix is lost when mature B cells are released into the circulation but is reexpressed on plasma cells residing in the matrices of lymphoid tissues[16].

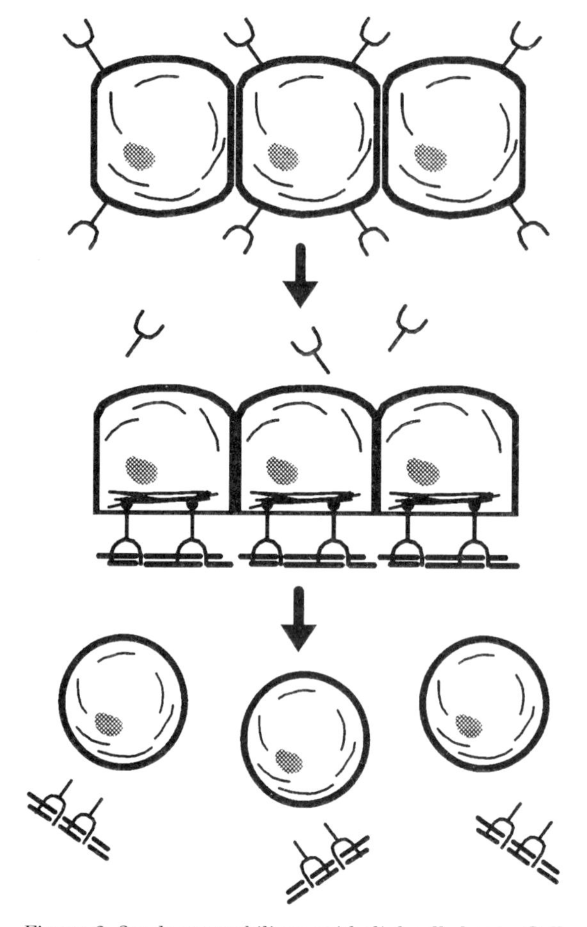

Figure 2. Syndecan stabilizes epithelial cell sheets. Cells are linked together by cell adhesion molecules. Syndecan is mobile in the plane of the membrane and surrounds subconfluent cultured cells (top). When cells become confluent, syndecan polarizes to the basolateral cell surface and binds to the underlying insoluble matrix. The matrix crosslinks syndecan in the plane of the membrane, causing it to associate intracellularly with the actin cytoskeleton which stabilizes the epithelial cell sheet (middle). Change in cell shape leads to cleavage of syndecan ectodomain, disrupting the linkage between the cells and the matrix (bottom) (After Bernfield and Sanderson, 1990).

PURIFICATION

The ectodomain is hydrophilic and is purified from the conditioned medium of cultured cells by anion exchange chromatography, isopycnic centrifugation in CsCl and immunoaffinity chromatography[12]. Medium from NMuMG cells yields ~10 μg per L of pure syndecan ectodomain. The intact, hydrophobic molecule is extracted and isolated from tissues and cells by anion exchange and immunoaffinity chromatography in the presence of detergent[14,16].

ACTIVITIES

Syndecan binds epithelial cells to a variety of immobilized matrix protein ligands[6-8] and this binding is duplicated by syndecan intercalated into liposomes[6]. Syndecan binds radiolabeled bFGF[9] and the cDNA for the hamster syndecan homologue was cloned by heparin inhibitable affinity to bFGF[4].

ANTIBODIES

A murine specific rat monoclonal antibody (281-2) raised against mouse syndecan recognizes the N-terminal region of the ectodomain[17]. Rabbit polyclonal antisera against the mouse ectodomain[18] and against the C-terminal peptide of the mouse cytoplasmic domain[2] crossreact with human, rat, bovine and canine syndecan.

GENES

Syndecan cDNAs prepared from mouse mammary epithelial cells (GenBank/EMBL X15487)[2], human mammary epithelial cells (GenBank/EMBL J05392)[3], and baby hamster kidney cells (GenBank/EMBL M29967)[4] have been cloned and sequenced. A single syndecan gene is on the proximal region of mouse chromosome 12 at a site syntenic with human chromosome 2p 24-25[19].

REFERENCES

1. Rapraeger, A. C. and Bernfield, M. (1983) J. Biol. Chem. 258, 3632-3636.
2. Saunders, S., Jalkanen, M., O'Farrell, S. and Bernfield, M. (1989) J. Cell Biol. 108, 1547-1556.
3. Mali, M., Jaakkola, P., Arvilommi, A.-M. and Jalkanen, M. (1990) J. Biol. Chem. 265, 6884-6889.
4. Kiefer, M. C., Stephans, J.C., Crawford, K., Okino, K. and Barr, P.J. (1990) Proc. Natl. Acad. Sci. 87, 6985-6989.
5. Marynen, P., Cassiman, J., Van den Berghe, H. and David, G. (1989) J. Biol. Chem. 264, 7017-7024.

6.Saunders, S. and Bernfield, M. (1988) J. Cell Biol. 106, 423-430.
7.Koda, J. E., Rapraeger, A. and Bernfield, M. (1985) J. Biol. Chem. 260, 8157-8162.
8.Sun, X., Mosher, D.F. and Rapraeger, A. (1989) J. Biol. Chem. 264, 2885-2889.
9.Bernfield, M. and Sanderson, R. D. (1990) Phil. Trans. R. Soc. (London) B 327, 171-186.
10.Klagsbrun, M. (1990) Curr. Opinion Cell Biol. 2, 857-863.
11.Rapraeger, A., Jalkanen, M., Bernfield, M. (1986) J. Cell Biol. 103, 2683-2696.
12.Jalkanen, M., Rapraeger, A., Saunders, S. and Bernfield, M. (1987) J. Cell Biol. 105, 3087-3096.
13.Kato, M. and Bernfield, M. (1990) J. Cell Biol. 111, 263a.
14.Sanderson, R. D. and Bernfield, M. (1988) Proc. Natl. Acad. Sci. 85, 9562-9566.
15.Kato, M. and Bernfield, M. (1989) J. Cell Biol. 109, 320a.
16.Sanderson, R. D., Lalor, P. and Bernfield, M. (1989) Cell Reg. 1, 27-35.
17.Jalkanen, M., Nguyen, H., Rapraeger, A., Kurn, N. and Bernfield, M. (1985) J. Cell Biol. 101, 976-984.
18.Jalkanen, M., Rapraeger, A.C., Bernfield, M. (1988) J. Cell Biol. 106, 953-962.
19.Oettinger, H. F., Streeter, H., Lose, E., Copeland, N.G., Gilbert, D.J., Justice, M.J., Jenkins, N.A., Mohandas, T. and Bernfield, M. (1991) Genomics 11, 334-338.

Michael T. Hinkes and Merton Bernfield:
Joint Program in Neonatology,
The Children's Hospital,
Harvard Medical School,
Boston, MA, USA

Tenascin

Tenascin[1,2] is a large disulphide linked hexameric extracellular matrix glycoprotein. It is a multidomain protein containing many repeated structural units such as heptad, EGF-like and fibronectin type III repeats, as well as a homology to the globular domain of β- and γ-fibrinogen. Tenascin is transiently expressed in many developing organs where it is often reexpressed in tumours. It is present in the central and peripheral nervous system as well as in smooth muscle and tendon. Proposed biological functions include effects on cell adhesion and cell morphology and consequently effects on cell differentiation.

Tenascin is most commonly isolated from conditioned media of either primary fibroblast cultures or from certain cell lines. The major form is a hexameric molecule consisting of disulphide linked subunits of 190, 200 and 230 kDa in the case of chicken tenascin. The high 230 kDa molecular weight splicing variants of tenascin from mammalian sources are larger. The isolated molecule consists of a central globule from which on opposite sides two short rods emanate. Each of these rods separates into three long arms. Each arm consists of a thinner inner part and a thicker outer region which is terminated by a globule. One tenascin molecule as it appears in the electron microscope after rotary shadowing is shown in Figure 1A. The model of one tenascin arm, as it was derived from the complete primary structure of the three splicing variants of chicken tenascin is shown in Figure 2.

The distinctive tissue distribution of tenascin has provoked much interest in this protein. During embryogenesis tenascin is transiently expressed in the dense mesenchyme surrounding many developing organs such as the mammary gland, tooth or kidney. It is present in embryonic cartilage and in the adult becomes confined to the perichondrium and periosteum, ligaments, tendons and myotendinous junctions. Also smooth muscle contains tenascin, whereas other organs or tissues such as the heart, skeletal muscle, or epithelial organs contain no or

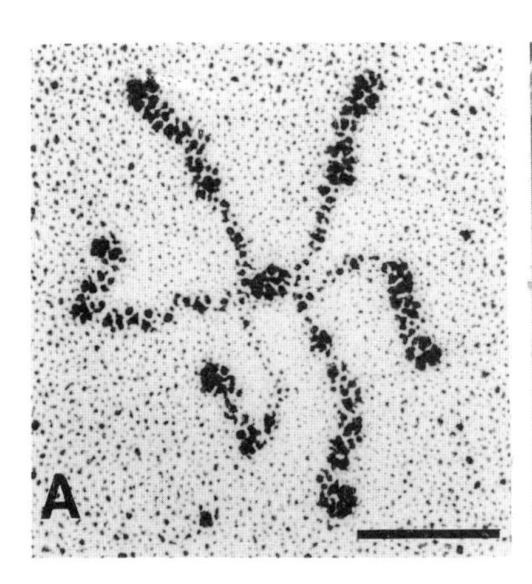

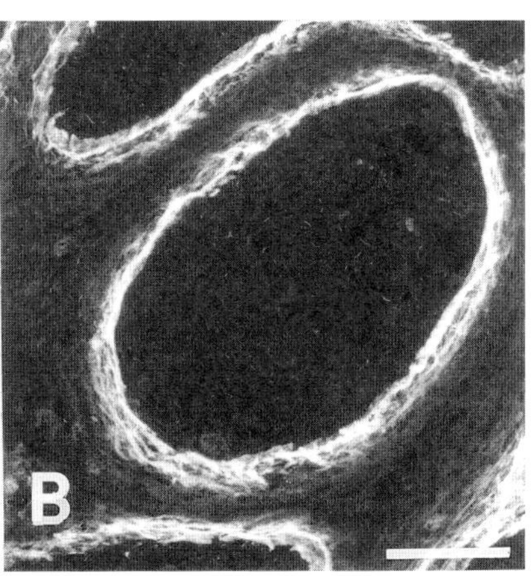

Figure 1. (A) Electron micrograph of a tenascin molecule after rotary shadowing. Bar 50 nm. (B) Immunofluorescence staining of tenascin in the stroma surrounding a rat mammary carcinoma. Bar 50 μm.

little tenascin. In contrast, epithelial tumours show very strong staining for tenascin in their surrounding tumour stroma, reminiscent of the embryonic situation (Figure 1B). In the central and peripheral nervous system tenascin is present during its development and also in regeneration.

Tenascin can be found in the literature under several names including myotendinous antigen, glioma mesenchymal extracellular matrix antigen, hexabrachion, J1 glycoproteins, cytotactin[1,2].

PURIFICATION

Tenascin can be purified from conditioned media of cell cultures by immunoaffinity chromatography using monoclonal anti-tenascin antibodies[3]. Alternatively conventional biochemical methods can be used, the most crucial step being the separation according to size in sucrose gradients, as tenascin is an unusually large protein of over 1000 kDa[1].

ACTIVITIES

The biological functions of tenascin are still partly controversial. Tenascin has been shown to be anti-adhesive for cells and to promote cell rounding[4,5]. On the other hand it has been proposed to mediate neuron glia adhesion and to promote neurite outgrowth[6]. Growth promotion, hemagglutination, immunosuppression of T-cells and the promotion of chondrogenesis are other reported effects mediated by tenascin[2].

ANTIBODIES

Polyclonal antisera against chick[7,8] and mouse[9] tenascin have been described. They generally crossreact with many species including amphibians[10]. Many monoclonal anti-chick tenascin antibodies have been described[3,4,11]. One of them (antibody M1) can be purchased from the "Developmental Hybridoma Bank" (The Johns Hopkins University, Baltimore MD). Other monoclonal anti-tenascin antibodies described are against human[12], mouse[13], or amphibian[10] tenascin respectively. A monoclonal antibody as well as a polyclonal antiserum against human tenascin is available from Telios Pharmaceuticals Incorporation (La Jolla, CA).

GENES

Complete cDNA sequences are available for chicken (GenBank M23121)[11], mouse (GenBank X56304)[13] and human tenascin (GenBank X56160)[12]. The human tenascin gene has been localized to chromosome 9q32-q34, the location of the hereditary disease *tuberous sclerosis*[14]. A tenascin-like transcript has been found to be encoded on the opposite strand of the human steroid 21-hydroxylase/complement component C4 locus on chromosome 6p21 (GenBank M25813 and X60189)[15,16].

REFERENCES

1. Erickson, H.P. and Bourdon, M.A. (1989) Ann. Rev. Cell Biol. 5, 71-92.
2. Chiquet-Ehrismann, R. (1990) FASEB J. 4, 2598-2604.
3. Chiquet, M. and Fambrough, D.M. (1984) J. Cell Biol. 98, 1937-1946.
4. Chiquet-Ehrismann, R., Kalla, P., Pearson, C.A., Beck, K. and Chiquet, M. (1988) Cell 53, 383-390.
5. Lotz, M.M., Burdsal, C.A., Erickson, H.P. and Clay, D.R. (1989) J. Cell Biol. 109,1795-1805.
6. Chiquet, M., Wehrle-Haller, B. and Koch, M. (1991) The Neurosciences 3, 341-350.
7. Chiquet, M. and Fambrough, D.M. (1984) J. Cell Biol. 98, 1926-1936.
8. Chiquet-Ehrismann, R., Mackie, E.J., Pearson, C.A. and Sakakura, T. (1986) Cell 47, 131-139.
9. Aufderheide, E. and Ekblom, P. (1988) J. Cell Biol. 107, 2341-2349.

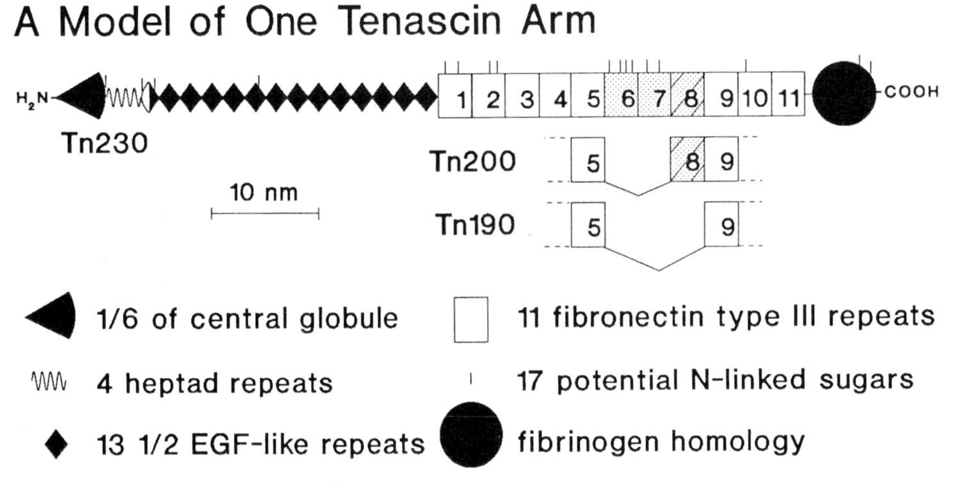

Figure 2. This model of chicken tenascin was derived from the primary sequence. It is drawn to scale and includes the three major splicing variants found in chick embryos.

10.Onda, H., Goldhamer, D.J. and Tassava, R.A. (1990) Development 108, 657-668.
11.Spring, J., Beck, K. and Chiquet-Ehrismann, R. (1989) Cell 59, 325-334.
12.Siri, A., Carnemolla, B., Saginati, M., Leprini, A., Casari, G., Baralle, F. and Zardi, L. (1991) Nucl. Acids Res. 19, 525-531.
13.Weller, A., Beck, S. and Ekblom, P. (1991) J. Cell Biol. 112, 355-362.
14.Gulcher, J.R., Alexakos, M.J., Le Beau, M.M., Lemons, R.S. and Stefansson, K. (1990) Genomics 6, 616-622.
15.Morel, Y., Bristow, J., Gitelman, S.E. and Miller, W.L. (1989) Proc. Natl. Acad. Sci. (USA) 86, 6582-6586.
16.Matsumoto, K.-I., Arai, M., Ishihara, N., Ando, A., Inoko, H. and Ikemura, T. (1992) Genomics 12, 485-491.

■ *Ruth Chiquet-Ehrismann:*
Friedrich Miescher Institute,
CH-4002 Basel, Switzerland

Thrombospondin

Thrombospondin is a 420 kDa adhesive glycoprotein which is composed of three equal molecular weight subunits[1,2]. Whereas, thrombospondin has a limited distribution in normal adult tissue, it is expressed at high levels in developing heart, muscle, bone and brain of the embryo and in response to injury and inflammation in adult tissue[3-5]. In cultured cells, thrombospondin has been shown to act synergistically with growth factors in promoting proliferation of smooth muscle cells[6]. These results have led to the hypothesis that thrombospondin is an extracellular matrix protein which is produced and utilized by the cell during proliferation.

Thrombospondin is organized into specific structural and functional domains[1,2,7]. A 25 kDa polypeptide can be proteolytically removed from the N-terminal domain which retains the high affinity heparin binding site. This region has a globular appearance in electron micrographs and is not homologous with any other proteins, which have been sequenced to date. The central portion of each chain contains a region which is homologous to procollagen, three type I repeats which are highly conserved and are homologous to complement factors and proteins produced by the malaria parasite and three type 2 or EGF-like repeats. The central portion of each subunit appears by electron microscopy to be thin and flexible. Each chain contains a globular domain at the C-terminal that contains seven type 3 repeats that form a tandem array of Ca^{2+}-binding sites. The last type 3 repeat contains the RGD sequence. The last 220 residues of each chain do not show significant homologies with other proteins, however, these residues and the type 3 repeats are highly conserved in human, mouse, chicken and frog.

Thrombospondin, like other adhesive proteins, has been shown to interact with multiple cell surface receptors via multiple sites in the protein. A cell surface heparin sulphate proteoglycan, designated syndecan, and sulphatides have been shown to bind the heparin-binding domain of the molecule[8,9]. The integrin $\alpha_v\beta_3$ has been reported to interact with the RGD sequence[10,11]. In addition, platelet glycoprotein IV (CD36) has been shown to function as a thrombospondin receptor on endothelial cells and monocytes[12]. Recent data suggests that this latter interaction may involve the VTCG sequence of the type 1 repeats[13]. In addition, antibodies which react with epitopes in the C-terminal domain inhibit platelet aggregation suggesting that this portion of the molecule is also capable of binding cell surfaces[14].

■ PURIFICATION

Thrombospondin can be purified from the supernatant of thrombin or ionophore A23187 treated platelets[7,9,10,14]. Most isolation procedures utilize heparin-Sepharose affinity chromatography and a step which separates on the basis of size (gel filtration or sucrose gradient centrifugation). MonoQ anion exchange chromatography is also very effective for purifying the released platelet protein and for purification from the media from cultured endothelial cells and fibroblasts[15]. Affinity chromatography on fibrinogen-Sepharose has also been used to purify thrombospondin[11].

■ ACTIVITIES

Thrombospondin, like other adhesive proteins, has been shown to modulate cell attachment, migration and proliferation. Thrombospondin is the endogenous platelet lectin which has been shown to agglutinate trypsinized, formalin fixed erythrocytes[16]. Platelet-monocyte interactions as measured by rosetting assays also involve thrombospondin[17]. Thrombospondin also supports attachment of normal and transformed cells, as well as malaria parasitized erythrocytes[18-20]. Whereas most cells do not spread on thrombospondin, some melanoma and carcinoma cell lines do spread on thrombospondin coated substrates[19,20]. Thrombospondin has been reported to support migration of human melanoma, carcinoma and granule cells[21,22]. It has also been reported to support smooth muscle cell proliferation[23]. In contrast, thrombospondin inhibits endothelial cell proliferation and angiogenesis[24-27].

■ ANTIBODIES

Many laboratories have produced polyclonal and monoclonal antibodies to human platelet thrombospondin and some of these reagents are commercially available (AMAC and Sigma). Very little information about species crossreactivity is available since most laboratories have focused on human tissue. One monoclonal fusion was performed with rat cells and at least one of the resulting antibodies, designated 5G11, reacts with mouse tissue[28].

GENES

Complete sequences for human endothelial cell[2] (GenBank X04665), human fibroblast[29] (GenBank X14787) and chick embryo[30] (GenBank M60858) thrombospondin are available. Two groups have reported sequences for the human promoter (GenBank J04447 and J04835)[31,32]. In addition, the structure and sequence of the mouse gene has been reported (GenBank J05605, J05606 and M62449 to M62470)[33,34]. The thrombospondin gene has been localized to human chromosome 15q15 and mouse chromosome 2 (region F)[35]. The positions of intron/exon boundaries for the human gene have been reported[36]. Recently, a cDNA clone that is apparently derived from a second mouse thrombospondin gene (Thbs2) has been described (GenBank M64860)[37].

REFERENCES

1. Lawler, J. (1986) Blood 67, 1197-1209.
2. Lawler, J. and Hynes R.O. (1986) J. Cell Biol. 103, 1635-1648.
3. O'Shea, K.S. and Dixit, J.M. (1988) J. Cell Biol. 107, 2737-2748.
4. Raugi, G.J., Olerud, J.E. and Gown, A.M. (1987) J. Invest. Dermatol. 89, 551-554.
5. Kreis, C., La Fleur, M., Menard, C., Paguin, R. and Beaulieu, A.D. (1989). J. Immunology 143, 1961-1968.
6. Majack, R.A. and Bornstein, P. (1987) Cell Membranes, Methods and Reviews 3, 55-77.
7. Galvin, N.J., Dixit, V.M., O'Rourke, K.M., Santoro, S.A., Grant, G.A. and Frazier, W.A. (1985) J. Cell Biol. 101, 1434-1441.
8. Sun, X., Mosher, D.F. and Rapraeger, A. (1989) J. Biol. Chem. 264, 2885-2889.
9. Roberts, D.D., Haverstick, D.M., Dixit, V.M., Frazier, W.A., Santoro, S.A. and Ginsburg, V. (1985) J. Biol. Chem. 260, 9405-9411.
10. Lawler, J., Weinstein, R. and Hynes, R.O. (1988) J. Cell Biol. 107, 2351-2361.
11. Tuszynski, G.P., Karczewski, J., Smith, L., Murphy, A., Rothman, V.L. and Knudsen, K.A. (1989) Exp. Cell Res. 182, 473-481.
12. Asch, A.S., Barnwell, J., Silverstein, R.L. and Nachman, R.L. (1987) J. Clin. Invest. 79, 1054-1061.
13. Rich, K.A., George, F.W. IV, Law, J.L. and Martin, W.J. (1990) Science 249, 1574-1577.
14. Dixit, V.M., Haverstick, D.M., O'Rourke, K.M., Hennessy, S.W., Grant, G.A., Santoro, S.A. and Frazier, W.A. (1985) Proc. Natl. Acad. Sci. (USA) 82, 3472-3476.
15. Clezardin, P., McGregor, J.L., Manach, M., Robert, F., Dechavanne, M. and Clemetson, K.J. (1984) J. Chromatogr. 296, 249-256.
16. Jaffe, E.A., Leung, L.L.K., Nachman, R.L., Levin, R.I. and Mosher, D.F. (1982) Nature 295, 246-248.
17. Silverstein, R.L. and Nachman, R.L. (1987) J. Clin. Invest. 79, 867-874.
18. Riser, B.L., Varani, J., O'Rourke, K. and Dixit, V.M. (1988) Exp. Cell Res. 174, 319-329.
19. Roberts, D.D., Sherwood, J.A. and Ginsburg, V. (1987) J. Cell Biol. 104, 131-139.
20. Howard, R.J. and Gilladoga, A.D. (1989) Blood 74, 2603-2618.
21. Taraboletti, G., Roberts, D.D. and Liotta, L.(1987) J. Cell Biol. 105, 2409-2415.
22. O'Shea, K.S., Rheinheimer, J.S.T. and Dixit, V.M. (1990) J. Cell Biol. 110, 1275-1283.
23. Majack, R.A., Goodman, L.V. and Dixit, V.M. (1988) J. Cell Biol. 106, 415-422.
24. Good, D.J., Polverini, P.J., Rastinejad, F., LeBeau, M.M., Lemons, R.S., Frazier, W.A. and Bouck, N.P. (1990) Proc. Natl. Acad. Sci. (USA) 87, 6624-6628.
25. Bagavandoss, P. and Wilks, J.W. (1990) Biochem. Biophys. Res. Comun. 170, 867-872.
26. Taraboletti, G., Roberts, D., Liotta, L. and Giavazzi, R. (1990) J. Cell Biol. 111, 765-772.
27. Iruela-Arispe, M.L., Bornstein, P. and Sage, H. (1991) Proc. Natl. Acad. Sci. (USA) 88, 5026-5030.
28. Legrand, C., Dubernard, V., Kieffer, N. and Narden, A.T. (1988) Eur. J. Biochem. 171, 393-399.
29. Hennessy, S.W., Frazier, B.A., Kim, D.D., Deckwerth, T.L., Baumgartel, D.M., Rotwein, P. and Frazier, W.A. (1989) J. Cell Biol. 108, 729-736.
30. Lawler, J., Duquette, M. and Ferro, P. (1991) J. Biol. Chem. 266, 8039-8043.
31. Donoviel, D.B., Framson, P., Edlridge, C.F., Cooke, M., Kobayashi, S. and Bornstein, P. (1988) J. Biol. Chem. 263, 18590-18593.
32. Laherty, C.D., Gierman, T.M. and Dixit, V.M. (1989) J. Biol. Chem. 264, 11222-11227.
33. Bornstein, P., Alfi, D., Davarayala, S., Framson, P. and Li, P. (1990) J. Biol. Chem. 265, 16691-16698.
34. Lawler, J., Duquette, M., Ferro, P., Copeland, N.G., Gilbert, D.J. and Jenkins, N.J. (1991) Genomics 11, 587-600.
35. Jaffe, E., Bornstein, P. and Disteche, C.M. (1990) Genomics 7, 123-126.
36. Wolf, F.W., Eddy, R.L., Shows, T.B. and Dixit, V.M. (1990) Genomics 6, 685-691.
37. Bornstein, P., O'Rourke, K., Wikstrom, K., Wolf, F.W., Katz, R., Li, P. and Dixit, V.M. (1991) J. Biol. Chem. 266, 12821-12824.

Jack Lawler, Ph.D.:
Vascular Research Division,
Department of Pathology,
Brigham and Women's Hospital
and Harvard Medical School, Boston, MA, USA

Tissue-Type Plasminogen Activator (t-PA)

t-PA is a serine proteinase of high specificity for the plasma zymogen plasminogen. It activates plasminogen to form the active serine proteinase plasmin. It has also binding sites for fibrin which enable localization of plasminogen activation on this plasmin substrate in thrombi. A product of endothelial cells, t-PA thus functions as the principal activator in vascular fibrinolysis. In vitro, t-PA is produced by endothelial cells, but also by melanoma and neuroblastoma cells.

t-PA is a highly specific serine proteinase, acting on a single peptide bond in plasminogen to form plasmin. It is released into the circulation from the vascular endothelium, and rapidly removed from the circulation by receptors on hepatocytes, so that the normal concentration in plasma is 3.4 ng/ml[1] with a half-life of 5-10 min. t-PA is also produced by several tumour cell lines *in vitro*, notably melanoma and neuroblastoma. The gene for human t-PA is located on chromosome 8 and is transcribed to a 2.7 kB mRNA. t-PA is secreted as a single chain 70 kDa glycoprotein which binds to fibrin, heparin[2], fibronectin, laminin[3] and Thy-1[4]. This single chain enzyme (sct-PA) has functional catalytic activity[5] which is increased by plasmin cleavage at Arg-275 and formation of a two-chain enzyme (tct-PA) held together by a single disulphide bond. The activities of both forms of the enzyme are markedly stimulated after binding to fibrin[6] and furthermore, the activity of fibrin bound t-PA is protected against rapid inactivation by **PAI-1** and **PAI-2**[7,8]. t-PA consists of 527 amino acids arranged in five main domains (see Figure); finger (homologous to the type 1 repeats of fibronectin[9]), growth factor, kringle 1, kringle 2, and catalytic domain containing the serine proteinase triad His-322, Asp-371 and Ser-478[10]. The kringles each consist of triple loops held together by three disulphide bridges (see Figure). The fibrin affinity is endowed by the finger and kringle 2[11], and kringle 2 may also be involved in the initial binding of t-PA to PAI-1[12]. t-PA inactivation by PAI-1 further requires Lys-296-Arg-304 of the catalytic domain[13]. Secretion and subsequent cell binding of t-PA has been reported for endothelial[14], neural[15] and melanoma[16] cells, but a cell receptor for t-PA has not been defined, nor is there evidence for involvement of the growth factor domain, the fibrin binding sites, the carbohydrate moieties or the catalytic domain[17] (see **u-PA**). Cell-bound t-PA retains activity[14-16] and may be protected from inactivation by PAI-1[14]. Hepatocytes bind and internalize t-PA by a PAI-1 dependent mechanism[18].

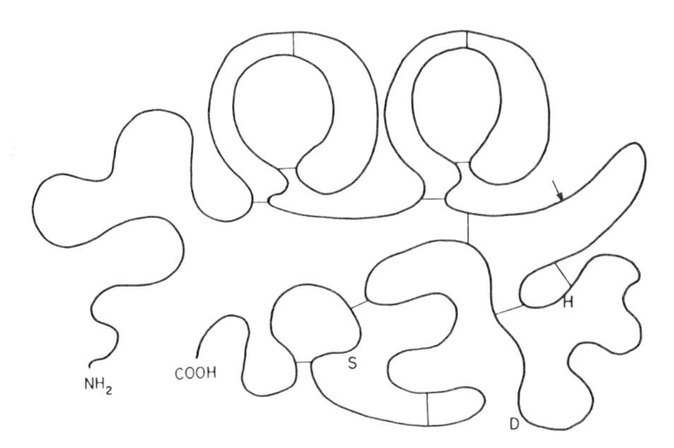

Figure. Structure of human t-PA. The domains consist of finger (N-terminal region), growth factor (Cys-56-Cys-75), kringle 1 (Cys-92-Cys-173), kringle 2 (Cys-180-Cys-261) and catalytic (Cys-264-Cys-502). The arrow indicates the plasmin cleavage site for formation of tct-PA from sct-PA. The serine proteinase catalytic triad is also indicated (H,D,S).

PURIFICATION

t-PA is conveniently purified from culture medium of the Bowes human melanoma cell line by immunoaffinity chromatography[19]. Recombinant t-PA, intended for thrombolytic therapy, is widely available.

ACTIVITIES

t-PA may be assayed directly as amidolytic activity or indirectly by plasmin generation, or as t-PA antigen by ELISA methods[1,16,20]. The standard preparations of t-PA generally available have a specific activity of approximately 800 IU/μg protein.

ANTIBODIES

Several preparations of rabbit and goat polyclonal antibodies to t-PA are commercially available[3,16] and murine monoclonal antibodies to several different epitopes of the A and B chains. Some of these antibodies are anti-catalytic[16], to the kringles and others distinguish between single and two-chain t-PA.

GENES

Since the first cloning of t-PA from human melanoma cells[10], the t-PA gene has been extensively studied and engineered with the aim of producing therapeutic t-PA variants with longer half-life *in vivo*, higher fibrin affinity and specificity, and resistance to inactivation by PAI-1[13]. GenBank numbers K03021 (human gene 36,594 bp complete cds) and M15518 (human mRNA 2,519 bp complete cds).

■ REFERENCES

1.Holvoet, P., Cleemput, H. and Collen, D. (1985) Thromb. Haemostas. 54, 684-687.
2.Andrade-Gordon, P. and Strickland, S. (1986) Biochem. 25, 4033-4040.
3.Salonen, E.-M., Saksela, O., Vartio, T., Vaheri, A., Nielsen, L.S. and Zeuthen, J. (1985) J. Biol. Chem. 260, 12302-12307.
4.Liësi, P., Salonen, E.-M., Dahl, D., Vaheri, A. and Richards, S.-J. (1990) Exp. Brain Res. 79, 642-650.
5.Boose, J.A., Kuismanen, E., Gerard, R., Sambrook, J. and Gething, M.-J. (1989) Biochem. 28, 635-643.
6.Hoylaerts, M., Rijken, D.C., Lijnen, H.R. and Collen, D. (1982) J. Biol. Chem. 257, 2912-2919.
7.Røder, M., Philips, M., Suenson, E. and Thorsen, S. (1988) Fibrinolysis 2 (Suppl. 2) Abs. 225.
8.Leung, K., Byatt, J.A. and Stephens, R.W. (1987) Thromb. Res. 46, 755-766.
9.Bányai, L., Váradi, A. and Patthy, L. (1983) FEBS Lett. 163, 37-41.
10.Pennica, D., Holmes, W.E., Kohr, W.J., Harkins, R.N., Vehar, G.A., Ward, A.A., Bennett, W.F., Yelverton, E., Seeburg, P.H., Heyneker, H.L. and Goeddel, D.V. (1983) Nature 301, 214-221.
11.Van Zonneveld, A.-J., Veerman, H., MacDonald, M.E., Van Mourik, J.A. and Pannekoek, H. (1986) J. Cell Biochem. 32, 169-178.
12.Wilhelm, O.G., Jaskunas, S.R., Vlahos, C.J. and Bang, N. (1990) J. Biol. Chem. 265, 14606-14611.
13.Madison, E.L., Goldsmith, E.J., Gerard, R.D., Gething, M.-J. and Sambrook, J.F. (1989) Nature 339, 721-724.
14.Hajjar, K.A., Hamel, N.M., Harpel, P.C. and Nachman, R.L. (1987) J. Clin. Invest. 80, 1712-1719.
15.Verrall, S. and Seeds, N.W. (1988) J. Neurosci. Res. 21, 420-425.
16.Bizik, J., Lizonová, A., Stephens, R.W., Grófová, M. and Vaheri, A. (1990) Cell Regulation 1, 895-905.
17.Reilly, T.M., Whitfield, M.D., Taylor, D.S. and Timmermans, P.B.M.W.M. (1989) Thromb. Haemostas. 61, 454-458.
18.Morton, P.A., Owensby, A.A., Sobel, B.E. and Schwartz, A.L. (1989) J. Biol. Chem. 264, 7228-7235.
19.Andreasen, P.A., Nielsen, L.S., Grøndahl-Hansen, J., Skriver, L., Zeuthen, J., Stephens, R.W. and Danø, K. (1984) EMBO J. 3, 51-56.
20.Rånby, M. (1990) Thromb. Haemostas. 63,139.

■ *Ross W. Stephens and Antti Vaheri:*
Department of Virology,
University of Helsinki,
SF-00290 Helsinki,
Finland

Urokinase Type Plasminogen Activator (u-PA)

Urokinase is a serine proteinase of high specificity for the plasma zymogen plasminogen, which it activates to form the active serine proteinase plasmin. It also has a receptor binding domain which enables localization of the plasminogen activating system on the surface of cells expressing the specific high affinity u-PA receptor. u-PA functions in the destruction of extracellular matrix which accompanies cell migration and invasion (especially tumour invasion), tissue remodelling and inflammation.

u-PA is a highly specific serine proteinase, acting specifically on a single peptide bond in plasminogen to form plasmin. It occurs at very low levels in normal human plasma (0.05-0.10 IU/ml), but at relatively high concentrations in urine, which has been used extensively as the source for purification. The gene for human u-PA is located on chromosome 10 and is transcribed to a 2.5 kB mRNA. Increased expression and secretion of u-PA is a strong correlate of the malignant phenotype of many types of cells[1,2]. An immature form of u-PA is secreted as a 52 kDa single chain glycoprotein (prourokinase, pro-u-PA) which binds to thrombospondin[3], heparin[4] and fibrin[5], but has less than 1/250 the activity of u-PA enzyme[6]. Pro-u-PA contains 411 amino acids, with four distinct domains (Figure 1): growth factor (receptor binding), kringle, connecting peptide, and catalytic region containing the serine proteinase catalytic triad (His-204, Asp-255 and Ser-356[7]). The growth factor domain has considerable homology with EGF and several components of the coagulation cascade, while the kringle is homologous to those in plasminogen and t-PA. Following secretion, pro-u-PA is avidly (K_d=10^{-10} M) bound by the specific cell surface

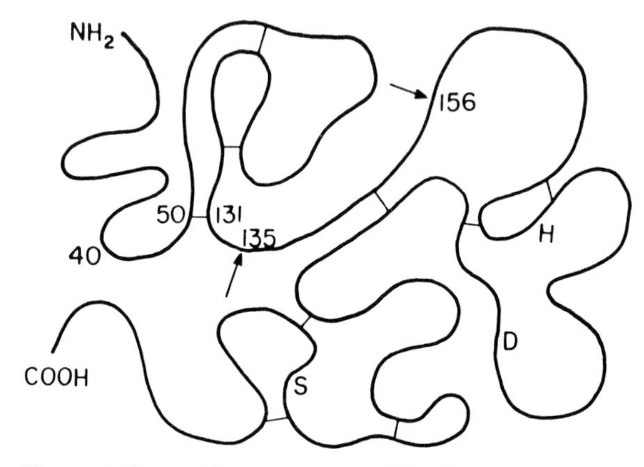

Figure 1. Prourokinase structure. The domains consist of growth factor (Ser_1-Gln_{40}), kringle (Cys-50 - Cys-131), connecting peptide (Lys-135 - Arg-156), catalytic (Cys-189 - Cys-380). Arrows indicate plasmin cleavage sites for proenzyme activation (Lys-158) and separation of catalytic domain from the cell binding domain (Lys-136).

receptor (u-PAR)[8,9], a 55 kDa glycoprotein which has a laterally mobile lipid anchor instead of a transmembrane domain[10]. In adherent cells, u-PA antigen is largely confined to focal adhesions and cell-cell contact sites[11] (Figure 2). Bound pro-u-PA is not internalized, but favourably acted upon by cell surface plasmin to form active u-PA[8,12]. The plasmin cleavage is at Lys-158 - Ile-159, thus forming a two chain structure held together by one disulphide bridge (Cys-148 - Cys-279). There is considerable evidence that cell surface u-PA catalyzed plasminogen activation leads to degradation of pericellular matrix, loss of adhesion of cells to matrix, promotion of cellular migration and invasion of tissue structures[13-15]. Proliferation may also be stimulated[16]. Cell surface u-PA is subject to regulation by the specific plasminogen activator inhibitors (PAI's)[11,12,17-20], which inactivate the receptor bound u-PA activity and promote its internalization[19,20]. u-PA activity may also be lost from the cell surface by plasmin cleavage at Lys-135 or Lys-136 within the connecting peptide region, so that the catalytic region is cleaved off and only the so-called N-terminal fragment (ATF) remains bound via its growth factor domain to the u-PAR.

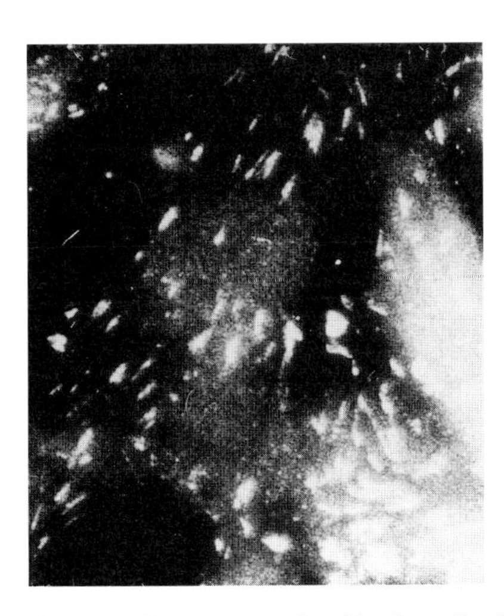

Figure 2. Immunofluorescence localization of cell surface bound u-PA at the focal adhesions of human RD (rhabdomyosarcoma) cells (from reference[11]).

■ PURIFICATION

u-PA is readily purified from either urine or tumour cell culture supernatants by immunoaffinity chromatography[1]. Recombinant u-PA can also be produced in *E. coli* .

■ ACTIVITIES

Both direct assays, based on amidolytic activity, and indirect assays, based on generation of plasmin activity, are used for assay of u-PA[1,12]. The activity standards available for u-PA have a specific activity ≈80,000 IU/mg protein. Sensitive ELISA assays of u-PA protein are also used[2].

■ ANTIBODIES

Several preparations of rabbit and goat polyclonal antibodies to u-PA are available commercially, as well as murine monoclonal antibodies to several different epitopes[2,12], e.g. some monoclonal antibodies are anti-catalytic, some prevent binding to the cell receptor, and others distinguish between the single-chain and two chain forms.

■ GENES

The u-PA gene has been cloned from human, pig, and mouse sources[7]. GenBank number M18182 contains complete human coding sequence.

■ REFERENCES

1. Danø, K., Andreasen, P.A., Grøndahl-Hansen, J., Kristensen, P., Nielsen, L.S. and Skriver, L. (1985) Adv. Cancer Res. 44, 139-266.
2. Jänicke, F., Schmitt, M., Hafter, R., Hollrieder, A., Babic, R., Ulm, K., Gössner, W. and Graeff, H. (1990) Fibrinolysis 4, 69-78.
3. Harpel, P.C., Silverstein, R.L., Pannell, R., Gurewich, V. and Nachman, R.L. (1990) J. Biol. Chem. 265, 11289-11294.
4. Andrade-Gordon, P. and Strickland, S. (1986) Biochem. 25, 4033-4040.
5. Pannell, R. and Gurevich, V. (1986) Blood 67, 1215-1223.
6. Petersen, L.-C., Lund, L.R., Nielsen, L.S. and Danø, K. (1988) J. Biol. Chem. 263, 11189-11195.
7. Blasi, F. (1988) Fibrinolysis 2, 73-84.
8. Stoppelli, M.P., Tacchetti, C., Cubellis, M.V., Corti, A., Hearing, V.J., Cassani, G., Appella, E. and Blasi, F. (1986) Cell 45, 675-684.
9. Behrendt, N., Rønne, E., Ploug, M., Petri, T., Løber, D., Nielsen, L.S., Schleuning, W.-D., Blasi, F., Appella, E. and Danø, K. (1990) J. Biol. Chem. 265, 6453-6460.
10. Ploug, M., Rønne, E., Behrendt, N., Jensen, A.L., Blasi, F. and Danø, K. (1990) J. Biol. Chem. 266, 1926-1933.
11. Pöllänen, J., Vaheri, A., Tapiovaara, H., Riley, E., Bertram, K., Woodrow, G. and Stephens, R.W. (1990) Proc. Natl. Acad. Sci. (USA) 87, 2230-2234.
12. Stephens, R.W., Pöllänen, J., Tapiovaara, H., Leung, K.-C., Sim, P.-S., Salonen, E.-M., Rønne, E., Behrendt, N., Danø, K. and Vaheri, A. (1989) J. Cell Biol. 108, 1987-1995.
13. Ossowski, L. (1988) J. Cell Biol. 107, 2437-2445.
14. Hearing, V.J., Law, L.W., Corti, A., Appella, E. and Blasi, F. (1988) Cancer Res. 48, 1270-1278.
15. Kirchheimer, J.C. and Remold, H.G. (1989) J. Immunology 143, 2634-2639.
16. Kirchheimer, J.C., Wojta, J., Christ, G. and Binder, B.R. (1989) Proc. Natl. Acad. Sci. (USA) 86, 5424-5428.
17. Cubellis, M.V., Andreasen, P.A., Ragno, P., Mayer, M., Danø, K. and Blasi, F. (1989) Proc. Natl. Acad. Sci. (USA) 86, 4828-4832.
18. Ellis, V., Wun, T.-C., Behrendt, N., Rønne, E. and Danø, K. (1990) J. Biol. Chem. 265, 9904-9908.
19. Cubellis, M.V., Wun, T.-C. and Blasi, F. (1990) EMBO J. 9, 1079-1085.
20. Estreicher, A., Mühlhauser, J., Carpentier, J.-L., Orci, L. and Vassalli, J.-D. (1990) J. Cell Biol. 111, 783-792.

Ross W. Stephens and Antti Vaheri:
Department of Virology,
University of Helsinki,
SF-00290 Helsinki, Finland

Versican

Versican[1,2] is a large chondroitin sulphate proteoglycan expressed by fibroblast cells. The multi-domain structure of its core protein includes a putative hyaluronic acid binding domain, a glycosaminoglycan attachment region, two EGF-like repeats, a lectin-like domain and a complement regulatory protein-like element. The N-terminal domain of versican is most likely identical to the glial hyaluronate binding protein (GHAP)[3]. The structural similarities of versican domains with hyaluronic acid binding proteins on one end of the core protein and with LEC-CAMs (or selectins)[4] on the other end suggest an involvement of versican in the interaction between extracellular matrix components and cell surface glycoproteins.

Versican, originally identified in the culture medium of human IMR-90 lung fibroblasts, has an apparent molecular mass of ≥ 1000 kDa. It consists of a large core protein (~400 kDa) and covalently attached chondroitin sulphate side chains[2]. The entire primary structure of the core protein has been determined based on overlapping cDNA clones isolated from IMR-90 fibroblast and human placental libraries[1,2]. The cDNAs code for a 2409 amino acids long polypeptide chain including a 20 residues long secretory signal sequence (Figure). The sequence contains putative attachment sites for 12-15 chondroitin sulphate side chains as well as for a number of N- and O-linked oligosaccharides. Versican includes at the N-terminus a domain with high similarities to the three-loop structure of cartilage **link protein**[5], to the G1 and G2 domains of the cartilage proteoglycan[6] (now named **aggrecan**), and to the cell surface hyaluronate receptor **CD44**[7]. In addition, partial amino acid sequences of the glial hyaluronate binding protein (GHAP)[3] are virtually identical to these versican sequences suggesting that GHAP is a proteolytical fragment of versican. The over 1700 amino acids long middle portion of the versican core protein contains a polyglutamic acid stretch (EENEEEEE) and the putative glycosaminoglycan attachment domain. In the C-terminal part, versican includes two EGF-repeats, a lectin-like sequence as well as a complement regulatory protein-like domain. A similar set of structural elements, although in different order, has been identified in a family of cell surface proteins termed LEC-CAMs (or **selectins**)[4]. The closest similarities in this region, however, have been observed with the cartilage derived aggrecan. Although versican and aggrecan have a high resemblance in both N- and C-terminal portions, the two proteoglycans are completely different in the large glycosaminoglycan carrying middle part of the core proteins and only an alternatively spliced isoform of aggrecan[8] contains an EGF-like repeat.

To date, little is known about the spatial and temporal distribution of versican. Nevertheless, it seems likely that versican is similar, if not identical, to a number of large chondroitin sulphate proteoglycans isolated from various cell cultures and tissue sources.

PURIFICATION

Versican has been partially purified from spent culture medium of human IMR-90 lung fibroblast cells through

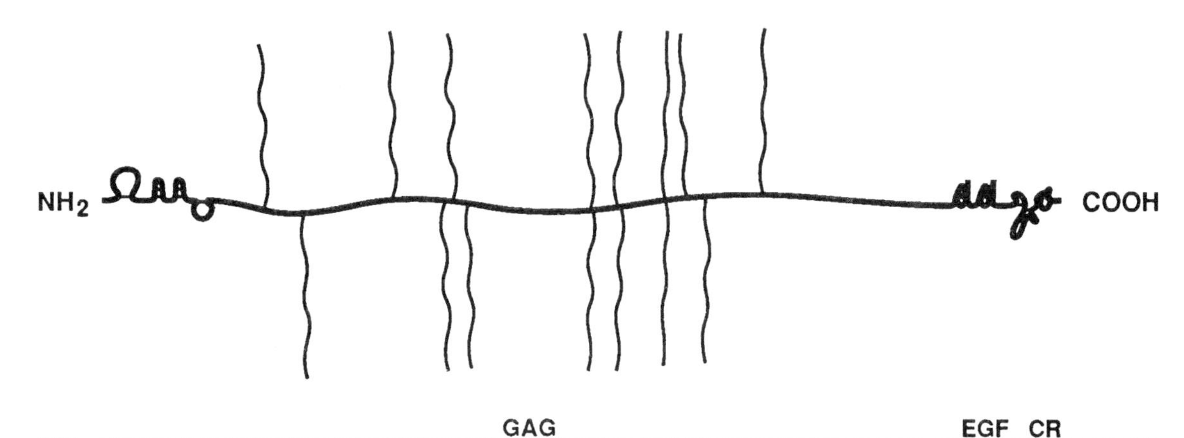

Figure. Structural model of versican based on the cDNA sequence. The core protein is depicted by the thick line and the wavy lines illustrate the chondroitin sulphate side chains. The schematic representation below shows the organization of the various structural elements in versican. HABR: hyaluronic acid binding region; GAG: glycosaminoglycan attachment domain; EGF: epidermal growth factor-like repeat; Lectin: lectin-like domain; CR: complement regulatory protein-like domain.

ammonium sulphate precipitations and DEAE-cellulose chromatography[2]. Similar proteoglycans have been extracted with guanidine-hydrochloride and purified by a combination of CsCl gradient centrifugation, anion exchange and/or gel filtration chromatography[9].

ACTIVITIES

Based on the identity of the N-terminal domain of versican with GHAP it seems likely that also the intact versican molecule binds to hyaluronic acid. The function of the other structural domains are unknown. PG-M, a similar proteoglycan isolated from chicken fibroblast cultures, has a negative effect on cell adhesion[10].

ANTIBODIES

Peptide antibodies against versican were prepared as described[2].

GENES

The complete cDNA sequence of human versican (EMBL/GenBank/DDBJ X15998) has been published[1].

REFERENCES

1. Zimmermann, D.R. and Ruoslahti, E. (1989) EMBO J. 8, 2975-2981.
2. Krusius, T., Gehlsen, K.R. and Ruoslahti, E. (1987) J. Biol. Chem. 262, 13120-13125.
3. Perides, G., Lane, W.S., Andrews, D., Dahl, D. and Bignami, A. (1989) J. Biol. Chem. 264, 5981-5987.
4. Springer, T.A. (1990) Nature 346, 425-434.
5. Deák, F., Kiss, I., Sparks, K.J., Argraves, W.S., Hampikian, G. and Goetinck, P.F. (1986) Proc. Natl. Acad. Sci. 83, 3766-3770.
6. Doege, K., Sasaki, M., Horigan, E., Hassell, J.R. and Yamada, Y. (1987) J. Biol. Chem. 262, 17757-17767.
7. Stamenkowic, I., Amiot, M., Pesando, J.M. and Seed, B. (1989) Cell 56, 1057-1062.
8. Baldwin, C.T., Reginato, A.M. and Prockop, J.P. (1989) J. Biol. Chem. 264, 15747-15750.
9. Heinegård, D. and Sommarin, Y. (1987) Methods Enzymol. 144, 319-372.
10. Yamagata, M., Suzuki, S., Akiyama, S.K., Yamada, K.M. and Kimata, K. (1989) J. Biol. Chem. 264, 8012-8018.

Dieter R. Zimmermann:
Institute of Pathology,
University of Zürich,
Schmelzbergstrasse 12,
8091 Zürich, Switzerland

Vitronectin

Vitronectin is an abundant blood plasma protein that has been studied because of potent cell adhesive activity, inhibition of complement mediated cell lysis, anti-heparin activity, interactions with plasminogen activator inhibitor and presence in atherosclerotic plaques. It probably functions as a modulator of the products of proteolytic cascades.

The concentration of vitronectin in plasma is approximately 300 μg/ml (4 μM)[1,2]. Most, if not all, circulating vitronectin is produced by the liver. Vitronectin is synthesized as a single chain and cleaved by a signal peptidase, N-glycosylated, phosphorylated, and sulphated prior to secretion. Human vitronectin circulates as one-chain and two-chain forms, the latter due to proteolytic cleavage, probably by an intracellular processing protease. The one-chain form migrates as a 75 kDa band on SDS-PAGE unreduced or reduced whereas the two-chain form dissociates into 65 and 10 kDa bands upon reduction. The relative amount of one-chain and two-chain forms is determined by a polymorphism near the cleavage site[3,4]. Vitronectin migrates as an α_2-globulin on agarose gel electrophoresis.

Vitronectin is composed of a disulphide rich domain homologous to modules of the PC-1 glycoprotein of lymphocytes, an acidic stretch that includes the RGD cell adhesion sequence and sites for sulphation and transglut-

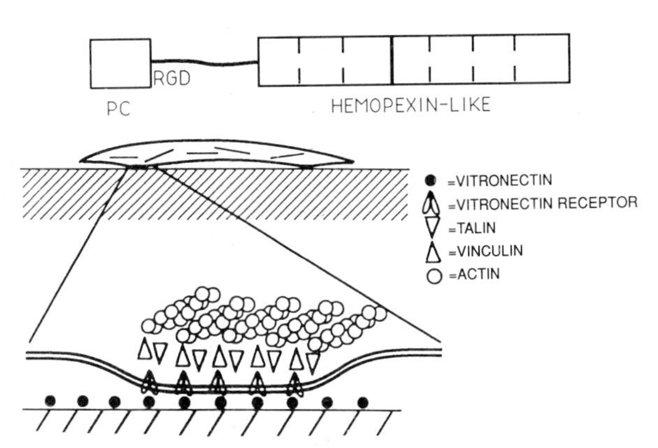

Figure. *Top* - Schematic of domains of vitronectin: PC, PC-1 domain; RGD, ArgGlyAsp cell attachment sequence; dotted lines, tandem repeats in hemopexin domains. *Bottom* - Schematic of how vitronectin may mediate cell adhesion.

amination and two copies of a homology unit also found in hemopexin, interstitial collagenase, and stromelysin[5-8] (Figure). It undergoes a conformational change upon binding of thrombin-antithrombin complexes, treatment with denaturants, or binding to surfaces[9]. Consequences of the change include increased affinity for heparin and increased lability of free sulfydryl groups. The change is probably similar to the conformational change induced in hemopexin by heme.

Vitronectin has been identified by immunofluorescence in supporting stroma of a variety of normal tissues but not in basement membranes *per se*[10]. There is an age dependent increase in deposition of vitronectin about elastic fibres in skin[11]. Atherosclerotic lesions are rich in vitronectin[12]. Pathological lesions containing deposits of terminal complement components also contain vitronectin. Detailed ontogenetic and phylogenetic analyses of vitronectin have not been reported.

PURIFICATION

Native vitronectin can be purified from plasma by differential precipitation with polyethylene glycol[13] or ammonium sulphate[14] followed by sequential chromatography on DEAE-Sephacel, Blue-Sepharose, and Sephacryl S-200. The final product is contaminated with albumin and a number of other proteins. A further step in which vitronectin is bound to heparin-Sepharose in 50 mM sodium chloride, pH 7.4, and eluted with gradient of 50-400 mM sodium chloride eliminates these contaminants[15]. Reduced glutathione must be included in buffers to prevent formation of disulphide bonded multimers. Denatured vitronectin can be purified from plasma or serum by a two step procedure based on the fact that, uniquely among plasma proteins, vitronectin binds to heparin at physiological salt concentration only after treatment with urea[16]. Untreated plasma or serum is applied to a heparin affinity column to remove proteins with intrinsic affinity for heparin, the pass-through is treated with urea and applied to a second column, and vitronectin is eluted from the second column with 500 mM NaCl.

ACTIVITIES

Vitronectin has diverse activities[1,2]. It adsorbs strongly to glass, tissue culture plastic, and a variety materials, even in the presence of high concentrations of other proteins. Adsorbed vitronectin is recognized by α_V- and β_3-containing **integrin** cell adhesion receptors and thus mediates adhesion and spreading of cells (Figure). Vitronectin also prevents heparin from catalyzing the interaction between thrombin and antithrombin III (anti-heparin activity), inhibits cell lysis by the C5b-9 complement complex, and stabilizes the activity of **plasminogen activator inhibitor**.

ANTIBODIES

Human vitronectin is highly antigenic and high titered, broadly reactive antisera can be produced in rabbits. A number of monoclonal antibodies have been produced to human native vitronectin (sometimes called S-protein) and to denatured vitronectin. One of the first to be described, 8E6[10], is interesting because it recognizes denatured or complexed vitronectin in preference to native vitronectin[9].

GENES

cDNAs for human vitronectin were sequenced by two groups[5,6] and can be accessed in GenBank (XO3168). The cDNA sequence of a rabbit homolog is also available[12]. The gene for human vitronectin is small, 5.3 kB (GenBank XO5006)[7]. Its chromosomal location has not been reported.

REFERENCES

1. Tomasini, B.R. and Mosher, D.F. (1991) Prog. Hemost. Thromb. 10, 269-305.
2. Preissner, K.T. (1989) Blut, 59, 419-431.
3. Tollefsen, D.M., Weigel, C.J. and Kabeer, M.H. (1990) J. Biol. Chem. 265, 9778-9781.
4. Kubota, K., Hayashi, M., Oishi, N. and Sakaki, Y. (1990) Biochem. Biophys. Res. Commun., 167, 1355-1360.
5. Jenne, D. and Stanley, K.K. (1985) EMBO J. 4, 3153-3157.
6. Suzuki, S., Oldberg, Å., Hayman, E.G., Pierschbacher, M.D. and Ruoslahti, E. (1985) EMBO J. 4, 2519-2524.
7. Jenne, D. and Stanley, K.K. (1987) Biochemistry 26, 6735-6742.
8. Hunt, L.T., Barker, W.C. and Chen, H.R. (1987) Prot. Seq. Data Anal. 1, 21-26.
9. Tomasini, B.R. and Mosher, D.F. (1988) Blood 72, 903-912.
10. Hayman, E.G., Pierschbacher, M.D., Ohgren, Y. and Ruoslahti, E. (1983) Proc. Natl. Acad. Sci. (USA), 80, 4003-4007.
11. Dahlbäck, K., Löfberg, H. Alumets, J. and Dahlbäck, B. (1989) J. Invest. Dermatol. 92, 727-733.
12. Sato, R., Komine, Y., Imanaka, T. and Takano, T. (1990) J. Biol. Chem. 265, 21232-21236.
13. Dahlbäck, B. and Podack, E.R. (1985) Biochemistry 24, 2368-2374.
14. Preissner, K.T., Wassmuth, R. and Müller-Berghaus, G. (1985) Biochem. J. 231, 349-355.
15. Mosher, D.F. (1990) Unpublished observation.
16. Yatohgo, T., Izumi, M., Kashiwagi, H. and Hayashi, M. (1988) Cell Struct. Function 13, 281-292.

Deane F. Mosher:
Departments of Medicine and Physiological Chemistry,
University of Wisconsin,
1300 University Avenue,
Madison, WI 53706, USA

von Willebrand Factor

von Willebrand factor (vWF) is a large multimeric glycoprotein that mediates platelet adhesion and thrombus formation at sites of vascular injury, and also serves as carrier for procoagulant factor VIII. Both functions are essential for the normal arrest of bleeding (hemostasis). In pathological conditions, vWF is involved in the processes leading to thrombotic arterial occlusion. Congenital abnormalities of vWF result in the most common inherited human disorder of hemostasis, von Willebrand disease.

vWF is synthesized in endothelial cells[1] and megakaryocytes[2] as a series of disulphide-linked oligomers ranging in molecular mass from ~400 kDa to in excess of (estimated) 10,000 kDa[3]. Large, fully extended multimers can reach lengths of >1,300 nm[4], making vWF the largest known soluble protein. Secretion by the vascular endothelium can be either into the circulation or abluminal, in which case vWF becomes part of the subendothelial matrix. Only circulating vWF forms a noncovalent complex with factor VIII. Endothelial secretion follows either a constitutive or a regulated pathway[5], the latter involving storage organelles known as Weibel-Palade bodies. vWF synthesized in megakaryocytes is not secreted into the circulation, but stored in the α-granules; these become part of platelets from which vWF is released when they are activated[6].

The vWF gene is ~178 kB and contains 52 exons. All exons and intron/exon boundaries have been sequenced[7]. The vWF mRNA is ~9 kB and codes for a 2813 amino acid residue protein. This protein represents prepro-vWF[8] (Figure). The vWF precursor sequence contains four distinct types of repeated domains (A-D) accounting for >90% of the sequence[9] (Figure). Several other proteins contain homologous A domains, including proteins of the

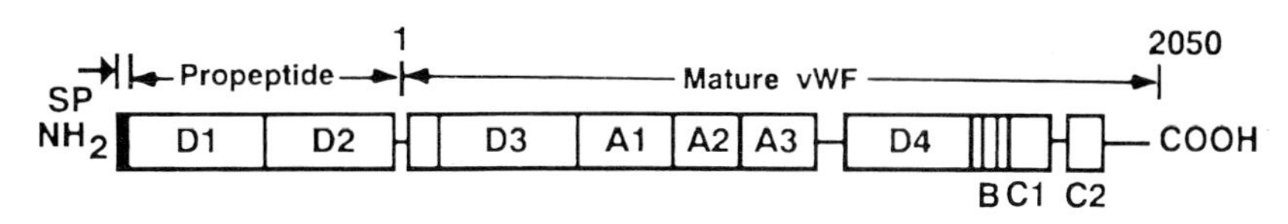

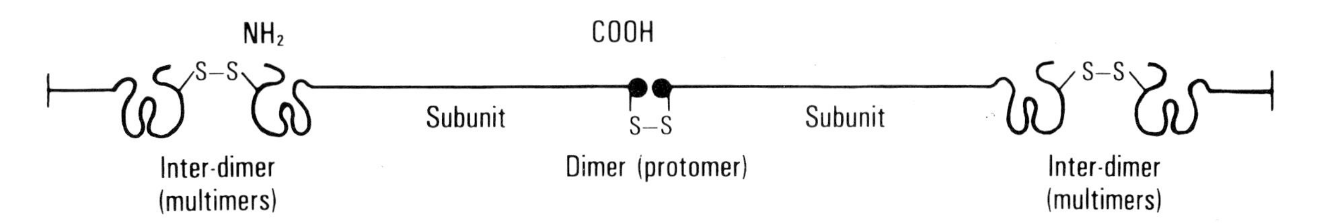

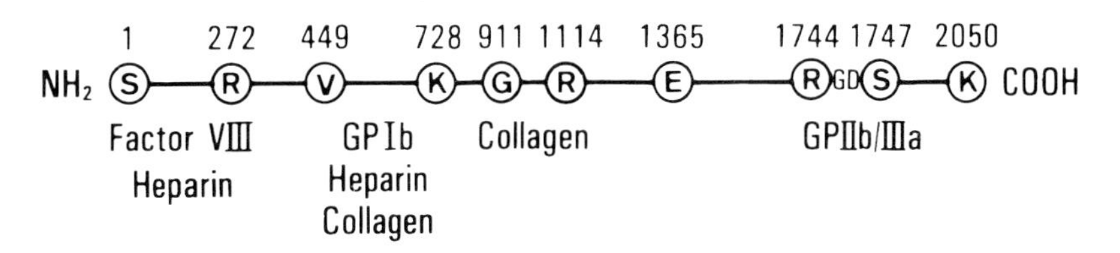

Figure. Schematic representation of the vWF molecule. *Top.* Distribution of homologous repeat domains (A-D) in the pre-pro-vWF cDNA. *Middle.* Assembly of the vWF subunit into dimers by S-S bonding at two subunit C-termini. Dimers are the protomeric unit of vWF oligomers, and assemble into higher order structures by S-S bonding at N-termini. *Bottom.* Location of functional domains within the 2050 residue mature subunit.

immune system (**LFA-1**, Mac-1, p150-95, factor B and component C2 of the complement activation pathway); cell adhesion molecules (VLA-1 and VLA-2); and extracellular matrix proteins (**cartilage matrix protein**, type VI **collagen**). Type A domains may represent a common structure for distinct ligand interactions. Type D domains are homologous to invertebrate vitellogenin.

The initial vWF translation product forms dimers in the endoplasmic reticulum by interchain disulphide bonding, presumably between C-termini of two subunits (Figure)[10], and undergoes posttranslational processing, including removal of a 22 residue signal peptide and glycosylation[11]. vWF dimers form multimers of increasing size by interchain disulphide bonds, presumably between N-termini (Figure)[10]. This process may require a cellular environment with acidic pH and is directed by the 741 residue propolypeptide sequence, which is normally cleaved after multimer assembly[12]. Mature vWF multimers contain a 2050 residue subunit[13] and no propeptide; this, also known as von Willebrand antigen II, is secreted as a distinct protein, or is found in the α-granules of platelets; it has no known function. The subunit of circulating vWF undergoes proteolytic cleavage under physiological conditions, while endothelial and platelet vWF are not proteolyzed. As a result, nonidentical subunits are present in plasma vWF multimers and cause their electrophoretic heterogeneity[14]. Approximately 18.7% (w/w) of the total mass of the vWF molecule consists of carbohydrate. There are 22 probable glycosylation sites within the mature subunit, of which 12 are N-linked and ten are O-linked[13].

Abnormalities of vWF characterize von Willebrand disease. In the most common cases (~70% of patients), decreased amounts of what appears to be a structurally and functionally normal molecule can be detected. Rare patients have no demonstrable vWF synthesis, and in some cases complete deletions of the vWF gene have been reported[15]. The remainder of the patients have various structural abnormalities resulting in a dysfunctional molecule. Several different variant forms have been recognized and classified[3].

PURIFICATION

vWF is purified from plasma by cryoprecipitation, differential polyethylene glycol precipitation, and size exclusion chromatography. This method allows isolation of the larger molecular forms present in plasma, with a relative depletion of the smaller ones[16].

ACTIVITIES

vWF links platelets to the subendothelium (adhesion) and to one another (aggregation) by means of distinct domains that mediate specific interactions with collagen, proteoglycans, and two platelet receptors, **GP Ib-IX** and **GP IIb-IIIa**; factor VIII binding also involves a distinct site of the molecule (Figure). Each function can be measured with specific binding assays. A commonly used test, measuring vWF mediated platelet agglutination in the presence of the antibiotic ristocetin, reflects vWF interaction with GP Ib-XI and is not, as usually stated, a method to measure globally vWF function.

ANTIBODIES

Polyclonal antibodies against human vWF (commercially available from Dako Corporation; Carpenteria, CA) show good crossreactivity with vWF from other mammals, including porcine, bovine, canine, and feline vWF; monoclonal antibodies usually exhibit more restricted species specificity. Numerous monoclonal antibodies have been prepared against distinct domains of the molecule and have provided relevant information on structure-function relationships, particularly with regard to the binding sites for GP Ib[17], collagen[18], and factor VIII[19]. Epitope mapping of vWF has been performed in significant detail[20]. Anti-vWF antibodies are commonly used to identify vascular structures by immunofluorescent staining of endothelial cells.

GENES

The vWF gene is located on the short arm of chromosome 12, region 12p12-12pter (GenBank EMBL M25716)[21]. A homologous sequence has been located on chromosome 22 and most likely represents a partial pseudogene[22]. The complete cDNA sequence of pre-pro vWF is also published[8].

REFERENCES

1. Jaffe, E.A., Hoyer, L.W. and Nachman, R.L. (1974) Proc. Natl. Acad. (USA), 71, 1906-1909.
2. Sporn, L.A., Chavin, S.I., Marder, V.J. and Wagner, D.D. (1985) J. Clin. Invest. 76, 1102-1106.
3. Ruggeri, Z.M. and Zimmerman, T.S. (1987) Blood 70, 895-904.
4. Fowler, W.E. and Fretto, L.J. (1989) Coagulation and Bleeding Disorders 181-193.
5. Lynch, D.C. (1989) Coagulation and Bleeding Disorders 137-159.
6. Wagner, D.D. (1989) Coagulation and Bleeding Disorders 161-180.
7. Mancuso, D.J., Tuley, E.A., Westfield, L.A., Worrall, N.K., Shelton-Inloes, B.B., Sorace, J.M., Alevy, Y.G. and Sadler, J.E. (1989) J. Biol. Chem. 264, 19514-19527.
8. Bonthron, D., Orr, E.C., Mitsock, L.M., Ginsburg, D., Handin, R.I. and Orkin, S.H. (1986) Nucleic Acids Res. 14, 7125-7127.
9. Shelton-Inloes, B.B., Titani, K. and Sadler, J.E. (1986) Biochemistry 25, 3164-3171.
10. Wagner, D.D., Lawrence, S.O., Ohlsson-Wilhelm, B.M., Fay, P.J. and Marder, V.J. (1987) Blood 69, 27-32.
11. Bonthron, D.T., Handin, R.I., Kaufman, R.J., Wasley, L.C., Orr, E.C., Mitsock, L.M., Ewenstein, B., Loscalzo, J., Ginsburg, D. and Orkin, S.H. (1986) Nature 324, 270-273.
12. Fay, P.J., Kawai, Y., Wagner, D.D., Ginsburg, D., Bonthron, D., Ohlsson-Wilhelm, B.M., Chavin, S.I., Abraham, G.N., Handin, R.I., Orkin, S.H., Montgomery, R.R. and Marder, V.J. (1986) Science 232, 995-998.
13. Titani, K., Kumar, S., Takio, K., Ericsson, L.H., Wade, R.D., Ashida, K., Walsh, K.A., Chopek, M.W., Sadler, J.E. and Fujikawa, K. (1986) Biochemistry 25, 3171-3184.
14. Ruggeri, Z.M. and Zimmerman, T.S. (1981) Blood 57, 1140-1143.

15.Ngo, K.Y., Glotz, V.T., Koziol, J.A., Lynch, D.C., Gitschier, J., Ranieri, P., Ciavarella, N., Ruggeri, Z.M. and Zimmerman, T.S. (1988) Proc. Natl. Acad. (USA), 85, 2753-2757.
16.De Marco, L. and Shapiro, S.S. (1981) J. Clin. Invest. 58, 321-328.
17.Mohri, H., Yoshioka, A., Zimmerman, T.S. and Ruggeri, Z.M. (1989) J. Biol. Chem. 264, 17361-17367.
18.Roth, G.J., Titani, K., Hoyer, L.W. and Hickey, M.J. (1986) Biochemistry 25, 8357-8361.
19.Foster, P.A., Fulcher, C.A., Marti, T., Titani, K. and Zimerman, T.S. (1987) J. Biol. Chem. 262, 8443-8446.
20.Berkowitz, S.D., Dent, J., Roberts, J., Fujimura, Y., Plow, E.F., Titani, K., Ruggeri, Z.M. and Zimmerman, T.S. (1987) J. Clin. Invest. 79, 524-531.
21.Ginsburg, D., Handin, R.I., Bonthron, D.T., Donlon, T.A., Bruns, G.A., Latt, S.A., Orkin, S.H. (1985) Science 228, 1401-1406.
22.Shelton-Inloes, B.B., Chehab, F.F., Mannucci, P.M., Federici, A.B. and Sadler, J.E. (1987) J. Clin. Invest. 79, 1459-1465.

*Zaverio M. Ruggeri and Jerry Ware:
Roon Research Center for Arteriosclerosis and Thrombosis, Department of Molecular and Experimental Medicine,
Committee on Vascular Biology,
Research Institute of Scripps Clinic,
La Jolla, California, USA*

2

Cell Adhesion and Cell-Cell Contact Proteins

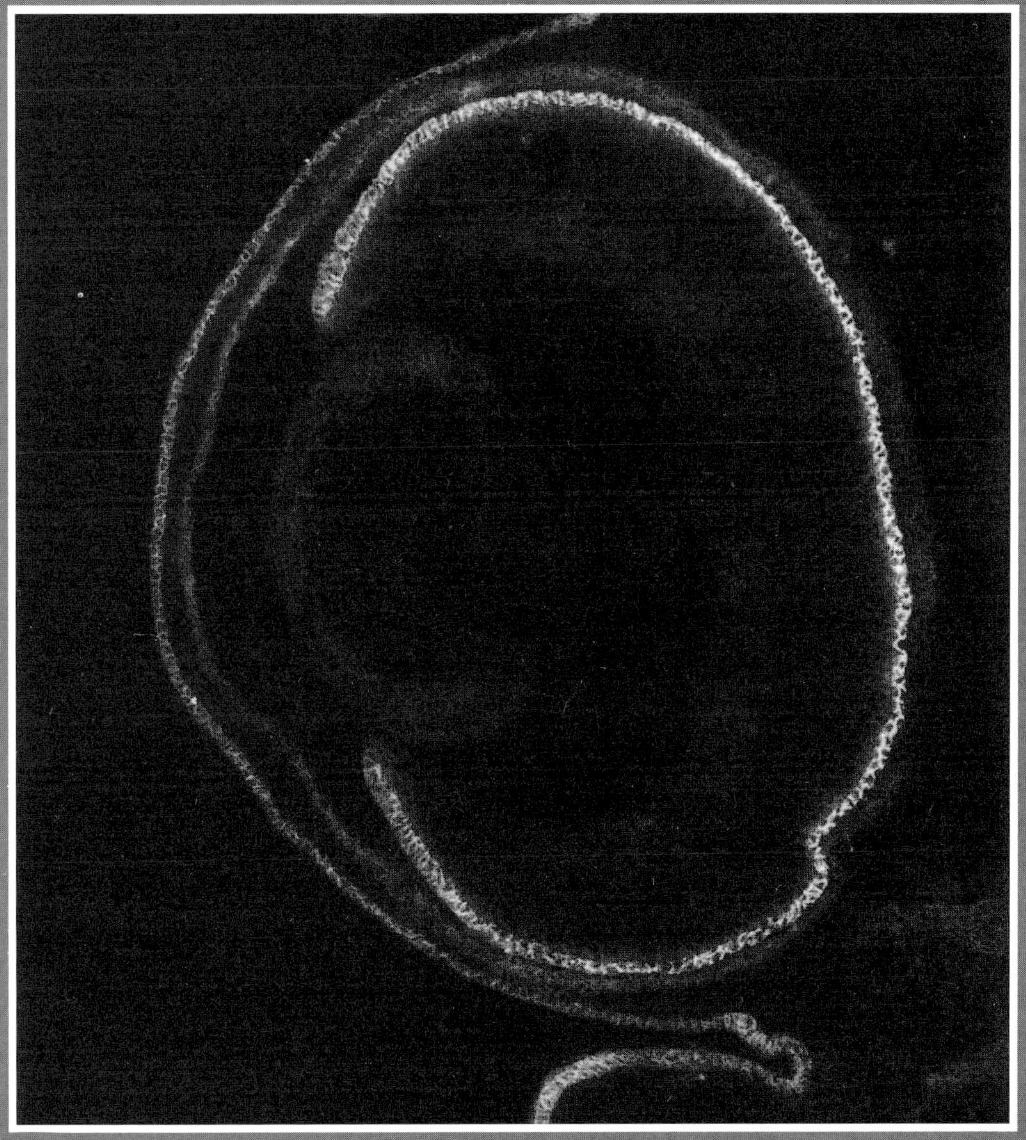

Localization of P-cadherin in the eye region of a foetal mouse (for more details see 'Cadherins' by M. Takeichi).

Cell Adhesion and Cell-Cell Contact Proteins

Cell adhesion is required at all stages of development and is of fundamental importance in the establishment and maintenance of the organized structure and function of multicellular organisms. In recent years a number of specific cell adhesion molecules (CAMs) have been identified which mediate this process. Based upon sequence analyses they can be organized into different families, the major of which are the immunoglobulin superfamily, the cadherin family, the integrin family, the selectins and the H-CAM family. In addition to mediating stable intercellular bonds CAMs participate in transmembrane signalling. Several induction molecules have been demonstrated to have cell adhesion activity. Both during embryogenesis and in mature organisms CAMs are expressed in characteristic spatiotemperal sequences and it is believed that cell adhesion selectivity is acquired by concerted actions of several CAMs and adhesive systems.

Cell adhesion is of prime importance for the formation and functional maintenance of multicellular organisms. The classical experiments on marine sponges by Wilson[1], and on amphibian embryos by Holtfreter[2], early this century demonstrated that cells possess intrinsic adhesiveness that guide their association into specific tissue architectures. However, for a long period it remained unclear what the molecular basis for these adhesion activities could be.

A major breakthrough in this field came when immunological approaches were combined with modern cell biological and molecular biological techniques. This was first introduced in the study of cellular slime molds by Gerisch and coworkers[3], and then in vertebrate systems by Edelman and coworkers[4]. The principle of this approach was to first inhibit cell adhesion with monovalent antibodies (Fab) against cell surface antigens and then to purify the antigens that were able to neutralize this inhibition. The first vertebrate molecule to be identified in this way was **NCAM**[5], soon followed by L-CAM[6], uvomorulin[7], cell-CAM 120/80[8] and **C-CAM** (cell-CAM 105)[9]. Somewhat later the ability of monoclonal antibodies to interfere with various adhesion reactions was utilized in the identification of several new adhesion molecules. Polyclonal and monoclonal antibodies have also been used to interfere with adhesion-related processes in more sophisticated *in vitro* systems, or even in *in vivo* situations[10-13].

Another approach to identify molecules that are involved in cell contact phenomena has been to isolate intercellular junctions, and to subject them to biochemical analyses. In this way proteins mediating cell-cell binding in gap junctions (**connexins**, for references, see also **14**), adherens junctions (A-CAM)[15] and desmosomes (**desmoglein** and **desmocollins**)[16,17] have been identified.

When some of the molecules identified by the procedures described above were cloned and sequenced it became apparent that many of the CAMs could be grouped into various protein superfamilies. Other molecules that have been cloned and sequenced have then been described as putative CAMs on the basis of their sequence similarity with one of these CAM-superfamilies[18-20]. Five major superfamilies can be recognized today, but several more will probably be revealed in the near future. The five CAM superfamilies are the immunoglobulins, **cadherins**, **integrins**, H-CAMs and the other cell adhesion molecules.

THE IMMUNOGLOBULIN SUPERFAMILY

This is a large family that, in addition to the immunoglobulins (Ig)[19], contains many other proteins, the majority of which are involved in cellular recognition phenomena[21]. The common building block is the Ig domain of about 100 amino acid residues that is arranged as a sandwich of two sheets of anti-parallel β-strands. Different members of the superfamily have varying numbers of Ig-like domains. The prototype for a CAM is N-CAM which has five Ig-like domains in the extracellular portion of the molecule. N-CAM and several other members of this superfamily appear in several isoforms as a result of alternative splicing. Some of these molecules are single-pass transmembrane proteins, whereas others are linked to the membrane via a glycan-phosphatidylinositol link[18]. CAMs belonging to the Ig-superfamily have important functions in the embryonic development and appear in a number of different cell types in adult animals, including cells of the nervous system, various epithelia and leukocytes.

THE CADHERIN SUPERFAMILY

This family comprises strictly Ca^{2+}-dependent, single-pass transmembrane proteins. The original four members of this family, L-CAM, E-cadherin (uvomorulin, cell-CAM 120/80), N-cadherin, P-cadherin, were demonstrated by direct functional assays to be involved in cell-cell adhesion[12,13]. Recently sequence-analysis of several newly cloned proteins has led to expansion of this superfamily[13,20], which today contains well over 20 different members. Interestingly, the adhesive molecules of two different types of intercellular junctions, adherens junctions (E-cadherin, N-cadherin = A-CAM)[12] and desmosomes (desmoglein and desmocollins)[22-25], belong to the cadherin superfamily. A characteristic feature of the cadherins is that they interact with two members of the cytoskeletal system, the microfilaments[12,13] or the intermediate filaments[22-25]. Cadherins are essential in tissue formation during embryonic development[12,13]. In adult organisms various cadherins are selectively distributed in neural, muscular and epithelial tissues[12].

THE INTEGRIN SUPERFAMILY

The integrin superfamily is a growing superfamily of proteins that are involved in various aspects of cell-cell adhesion and cell adhesion to extracellular matrix components (see also article by Reichardt on **Extracellular Matrix Proteins** in this volume)[26,27]. Integrins are heterodimers consisting of one α-chain and one β-chain, which both are single-pass transmembrane proteins. They are divided into subclasses according to the identity of the β-chain. Each β-chain can associate with several different α-chains, and some α-chains can interact with different β-chains. Thus, a huge repertoir of dimer combinations exists, which results in a broad spectrum of binding specificities. The binding activity of many integrins is dependent on divalent cations, and like the cadherins, integrins also can interact with the cytoskeletal system, e.g. β-1 chains associate with microfilaments in focal adhesion contacts[28] and β-4 chains associate with intermediate filaments in hemi-desmosomes[29]. Various types of integrins are found in fibroblasts, epithelial cells, muscular cells and leukocytes.

THE SELECTIN (LEC-CAM) SUPERFAMILY

Proteins in the **selectin** family are found in leukocytes, platelets and vessel endothelial cells[19,27,30]. The first member to be identified in this family was Mel-14, which was characterized as a lymphocyte homing receptor. Other members like ELAM-1 and GMP-140 mediate binding of polymorphonuclear leukocytes to vessel endothelia and participate in platelet interactions. These molecules are single-pass transmembrane proteins, which in their extracellular portions have a lectin-like domain (L), an EGF-like domain (E) and a varying number of complement binding protein-like domains (C). Hence, they are called LEC-CAM, but recently it has been agreed to use the name selectins for this superfamily[30]. It has been demonstrated that they indeed do bind specific carbohydrate structures via their lectin domains.

THE H-CAM SUPERFAMILY

Members of the H-CAM family were also originally identified as lymphocyte homing receptors[19]. Sequence analysis then demonstrated that related proteins are found in several cell types, including various epithelia and a subset of glia cells. These molecules are single-pass transmembrane proteins which have motifs homologous to the proteoglycan core protein (see also article by Lander on **Proteoglycans** in this volume) and the link protein of cartilage in their extracellular portion. Some of the members of the H-CAM family are proteoglycans themselves and carry chondroitin sulfate chains. One of the H-CAMs, **CD44**, is the major receptor for **hyaluronan**[31].

OTHER CELL ADHESION MOLECULES

Several other proteins that are involved in various cellular adhesive interactions have been identified. Some, like **fasciclin I**[32] occurring in the nervous system of insects do not show sequence homology with any other proteins, while others do. **AMOG** which is present on glia cells, for instance is highly homologous to the β-chain of the Na^+,K^+-ATPase[33] and **neurotactin**[34,35], which appears early in the embryogenesis and later is expressed in the nervous system is homologous to serine esterases. In the next few years we are likely to see the birth of new superfamilies containing proteins with cell adhesion activities.

BIOLOGICAL FUNCTIONS OF CELL ADHESION MOLECULES

General considerations

The functional roles of several CAMs and cell contact molecules seem obvious. Thus, connexins form gap junction channels and mediate intercellular communication. Desmoglein and desmocollins mediate cell-cell binding in desmosomes and contribute to the tensile strength of epithelial tissues. Cadherins that are components of adherens junctions might participate in regulation of cellular motility. Integrins contribute both to proper tissue organization and cellular migration. It has also been clearly demonstrated that cadherins have vital roles during embryogenesis, and are master molecules in various form-shaping events and formation of tissues. Variations in abundance and type of cadherin expression are driving forces in cell sorting events that are crucial in tissue segregation. In recent years it has also been extensively demonstrated how selectins, integrins and members of the Ig-superfamily interact and participate in adhesive, invasive and migrational events in leukocytes in inflammatory processes. Thus, it seems clear that CAMs have more sophisticated functions than just forming mechanical bonds between cells, and that different CAMs have different functions.

In trying to elucidate and understand the functional roles of various CAMs one important observation is that several distinct CAMs can be expressed simultaneously in one and the same cell type. This apparent redundancy might be part of a fail-safe system that guarantees proper cellular organization even if one of the components fails. A perhaps more likely situation is that it is the various combinations of different CAMs that confer tissue-specific adhesive properties. This has been discussed by Butcher[36] and by Hynes and Lander[37]. It is also possible that the major task of some of the CAMs is not to form physical bonds between cells, but rather involvement in other kinds of cellular interactions such as transmembrane signalling. Since signalling between cells must involve molecular contacts, it follows that if the binding energy in such contacts is high enough to withstand the disruptive forces in the adhesion assay used, the molecules involved in such interactions will be recognized in the screening for adhesion molecules. Thus, the definition of CAMs, as we know them today, is an operational one, based on the adhesion assay.

That signalling molecules can function as CAMs has recently been beautifully demonstrated for some cell surface molecules that are essential for developmental events in *Drosophila melanogaster*. In the development of the compound eye of *Drosophila* proper differentiation of cell number 7 in the ommatidium into a photoreceptor cell requires the gene *Sevenless* that codes for the protein sev[38]. Sev is a transmembrane cell surface protein with a cytoplasmic tyrosine kinase domain whose activity is essential for differentiation[39]. Sev binds to another transmembrane cell surface protein called boss (bride of sevenless) on the neighbouring cell number 8[40]. This interaction triggers or induces the differentiation of cell 7. There is no evidence to suggest that either sev or boss are involved in the topological organization of the cells within the ommatidium or in the maintenance of physical tensile strength of the tissue. However, it was recently demonstrated that sev and boss can mediate cell adhesion between nonadhesive cells that have been transfected with complementary DNA coding for these two proteins[40]. Thus, in this assay, sev and boss clearly are heterophilic cell adhesion molecules, although their prime function is inductive transmembrane signal transduction and not formation of cell-cell bonds important for pattern formation or physical integrity of the eye tissue. A similar type of interaction occurs between the gene products of *notch* and *delta*[41] These transmembrane proteins are involved in inductive interactions important for the development of the nervous system in *Drosophila*. Similar transfection experiments have demonstrated that the proteins coded for by *Notch* and *Delta* can bind to each other and mediate heterophilic cell adhesion. Notch and delta both belong to a protein superfamily whose members are characterized by having a large number of EGF like repeats in their extracellular domains[41].

It would not be surprising if several of the CAMs that we know of today are involved in signal transduction between cells and between the extracellular matrix and cells. In fact this has already been demonstrated for several integrins and for some of the members of the Ig-superfamily in lymphocytes. There is for instance some evidence that interactions between T-lymphocytes mediated by the heterophilic pair of adhesion molecules **CD2-LFA-3** (members of the Ig-superfamily) can affect and strengthen the adhesion mediated by **LFA-1-ICAM-1** (an integrin-IG-superfamily heterophilic pair)[42]. Furthermore, the accessory T-lymphocyte adhesion molecules **CD4** and **CD8** (Ig-superfamily members) interact specifically with the tyrosine kinase lck[43], and it has been demonstrated that activation of the T cell receptor activates the LFA-1 integrin which binds ICAM-1 on the target cell[27,44]. It has also been reported that antibody-mediated perturbation of **L1** or NCAM (members of the Ig-superfamily) elicit transient changes in intracellular levels of phosphoinositides and calcium concentration[45]. Furthermore both NCAM and N-cadherin can induce neurite extension via regulation of neuronal Ca^{2+} channels[46]. An appealing possibility is that a primary interaction between one pair of adhesion molecules would signal to the cells to express new combinations of adhesion molecules. In this way cell adhesion interactions would be involved in regulating the dynamic variations in expression of adhesion molecules seen in many processes both in embryonic development and in mature animals, e.g. in leukocyte interactions in inflammation.

Specific examples

During the last few years a vast body of information has accumulated concerning the involvement and contribution of various CAMs and ECM-receptors in several types of cellular organizations (for reviews see **11-13,18,27,36,37,47,48**), and a few of them are discussed below. Most of these examples concern events during embryonic development, but valuable information has also emerged from studies of tumour invasion and leukocyte-endothelial cell interactions. Especially studies of leukocyte adhesion have contributed significantly to our understanding of cooperation between various adhesion molecules.

Leukocyte-endothelial cell interactions

Leukocyte trafficking is fundamental for their participation in the defense against various infectious agents. Thus, leukocytes can leave the blood stream and migrate out into the tissues where they fulfil their functions. In this extravasation process members of the selectin family, the integrin family, the Ig-superfamily and the H-CAM family have been found to participate[19,27,37]. So far, there is no evidence that cadherins are involved in extravasation or other leukocyte cellular interactions. The same types of molecules are used both in extravasation of polymorphonuclear cells and homing of lymphocytes. None of the molecules seem to be highly selective in spite of the specific patterns of tissue targeting seen in e.g. lymphocyte homing. As a solution to this dilemma Butcher[36] has suggested that it is various combinations of these adhesion molecules that give rise to the selectivity observed in leukocyte-endothelial cell interactions.

This is best understood in extravasation of neutrophils[36,37]. Their initial targeting is mediated by selectins. Inflammatory agents initiate a rapid exposure of P-selectin on the endothelial cells. P-selectin binds to specific carbohydrates that are present on the neutrophil surfaces. This slows down the neutrophils which start to roll on the endothelial lining. Other cellular interactions then activate the integrin LFA-1 on the neutrophils which will bind to the Ig-superfamily member ICAM-2 present on the endothelial cells. This causes a strong binding of the neutrophils to the endothelial cells and lead to extravasation. At later stages other members of the selectin and Ig-superfamilies become involved which guarantees continued recruitment of leukocytes to the inflamed tissue.

Neural crest cell migration

Another thoroughly studied process is the migration of

neural crest cells in the embryo. When these cells dissociate from the dorsal part of the neural tube they lose expression of NCAM and N-cadherin[49]. At the same time they gain increased adhesivity to **fibronectin**[49], which is mediated by integrins. The dissociated cells migrate to new locations on tracks of fibronectin fibres. At their final locations they reaggregate, and a reexpression of both NCAM and N-cadherin has been observed[49]. During the continued differentiation also other adhesion molecules, e.g. Ng-CAM (a member of the Ig-superfamily), appear[49].

Neurite guidance

The outgrowth of axons during the embryonic development of the nervous system has received much attention and several different experimental systems are being explored. These processes are very complicated and no single chain of events applicable to all types of neurons has been found. However, in all the systems studied so far, Ig-superfamily molecules, cadherins, integrins and extracellular matrix molecules have been found to play crucial roles[37,48]. So far selectins have not been found to be involved. As in all processes involving cellular interactions during development and pathology, a highly dynamic pattern has been observed. Thus, various adhesion molecules appear and disappear in different combinations, which contribute to the correct cellular architectures of the mature tissues.

Tumour invasion

A characteristic feature of malignant tumours is that they invade surrounding tissues and form metastases. Tumour invasion can be regarded as an aberrant tissue formation due to failure of normal cell sorting mechanisms. Accordingly, it has been reasonable to suspect that alterations in cell adhesion mechanisms may contribute to invasion, in addition to other processes such as degradation of the extracellular matrix and changes in cellular motility. A loss of cadherin molecules from coherent epithelial tissues might for instance facilitate dissociation of tumour cells from the parent tumour. Several recent studies have indeed demonstrated a loss of E-cadherin expression in the invasive phenotype of tumour cells[50,51], and expression of anti-sense RNA for E-cadherin has been found to make cells more invasive[52]. It has also been demonstrated that transfection of E-cadherin cDNA blocked the invasive phenotype[51-53]. It remains to be seen if invasion-restriction is confined only to cadherins or if other cell adhesion molecules, e.g. of the Ig-superfamily might be important as well.

Also integrin receptors for ECM molecules could affect invasion. Overexpression of $\alpha_5\beta_1$ integrin increases adhesion to fibronectin and reduces tumorigenicity of virally transformed cells[54]. It is also possible that acquisition of new adhesion molecules might lead to invasion and metastasis. Thus, invasive growth of melanomas correlate with the expression of $\alpha V\beta_3$ integrin[55], which is a receptor for some ECM molecules.

REFERENCES

1.Wilson, H.V. (1907) J. Exp. Zool. 5, 245-258.
2.Holtfreter, J. (1939) Arch. Exp. Zellforschung 23, 169-209.
3.Beug, H., Katz, F.E. and Gerisch, G. (1973) J. Cell Biol. 56, 647-658.
4.Brackenbury, R., Thiery, J.P., Rutishauser, U. and Edelman, G.M. (1977) J. Biol. Chem. 252, 6835-6840.
5.Thiery, J.P., Brackenbury, R., Rutishauser, U. and Edelman, G.M. (1977) J. Biol. Chem. 252, 6841-6845.
6.Gallin, W.J., Edelman, G.M. and Cunningham, B.A. (1983) Proc. Natl. Acad. Sci. USA, 80, 1038-1042.
7.Peyriéras, N., Hyafil, F., Louvard, D., Ploegh, H.L. and Jacob, F. (1983) Proc. Natl. Acad. Sci. USA, 80, 6274-6277.
8.Damsky, C.H., Richa, J., Solter, D., Knudsen, K. and Buck, C.A. (1983) Cell 34, 455-466.
9.Ocklind, C. and Öbrink, B. (1982) J. Biol. Chem. 257, 6788-6795.
10.Öbrink, B. (1986) In:"Frontiers of Matrix Biology" (L. Robert, ed.) 11, 123-138. (Karger, Basel).
11.Edelman, G.M. (1988) Biochemistry 27, 3533-3543.
12.Takeichi, M. (1988) Development 102, 639-655.
13.Takeichi, M. (1991) Science 251, 1451-1455.
14.Öbrink, B. (1986) Exp. Cell Res. 163, 1-21.
15.Volk, T. and Geiger, B. (1984) EMBO J. 3, 2249-2260.
16.Giudice, G.J., Cohen, S.M., Patel, N.H. and Steinberg, M.S. (1984) J. Cell. Biochem. 26, 35-45.
17.Cowin, P., Mattey, D. and Garrod, D. (1984) J. Cell Sci. 70, 41-60.
18.Öbrink, B. (1991) Bioessays 13, 227-234.
19.Stoolman, L.M. (1989) Cell 56, 907-910.
20.Suzuki, S., Sano, K. and Tanihara, H. (1991) Cell Regulation 2, 261-270.
21.Williams, A.F. and Barclay, A.N. (1988) Ann. Rev. Immunol. 6, 381-405.
22.Mechanic, S., Raynor, K., Hill, J.E. and Cowin, P. (1991) Proc. Natl. Acad. Sci. USA 88, 4476-4480.
23.Collins, J.E., Legan, P.K., Kenny, T.P., Mac Garvie, J., Holton, J.L. and Garrod,
D.R. (1991) J. Cell Biol. 113, 381-391.
24.Koch, P.J., Walsh, M.J., Schmelz, M., Goldschmidt, M.D., Zimbelmann, R. and Franke, W.W. (1990) Eur. J. Cell Biol. 53, 1-12.
25.Goodwin, L., Hill, J.E., Raynor, K., Raszi, L., Manabe, M. and Cowin, P. (1990) Biochem. Biophys. Res. Commun. 173, 1224-1230.
26.Hynes, R.O. (1987) Cell 48, 549-554.
27.Springer, T.A. (1990) Nature 346, 425-434.
28.Burridge, K., Nuckolls, G., Otey, C., Pavalko, F., Simon, K. and Turner, C. (1990) Cell Differ. Dev. 32, 337-342.
29.Sonnenberg, A., Calafat, J., Janssen, H., Daams, H., van der Raaij-Helmer, L.M., Falcioni, R., Kennel, S.J., Aplin, J.D., Baker, J., Loizidou, M. and Garrod, D.J. (1991) J. Cell Biol. 113, 907-917.
30.Bevilacqua, M., Butcher, E., Furie, B., Gallatin, M., Gimbrone, M., Harlan, J., Kishimoto, K., Lasky, L., McEver, R., Paulson, J., Rosen, S., Seed, B., Siegelman, M., Springer, T., Stoolman, L., Tedder, T., Varki, A., Wagner, D., Weissman, I. and Zimmerman, G. (1991) Cell 67, 233.
31.Aruffo, A., Stamenkovic, I., Melnick, M., Underhill, C.B, and Seed, B. (1990) Cell 61, 1303-1313.
32.Elkins, T., Hortsch, M., Bieber, A.J., Snow, P.M. and Goodman, C.S. (1990) J. Cell Biol. 110, 1825-1832.
33.Gloor, S., Antonicek, H., Sweadner, K.J., Pagliusi, S., Frank, R., Moos, M. and Schachner, M. (1990) J. Cell Biol. 110, 165-174.
34.de la Escalera, S., Bockamp, E.O., Moya, F., Piovant, M. and Jimenez, F. (1990) EMBO J. 9, 3593-3601.

35.Barthalay, Y., Hipeau-Jacquotte, R., de la Escalera, S., Jimenez, F. and Piovant, M. (1990) EMBO J. 9, 3603-3609.
36.Butcher, E.C. (1991) Cell 67, 1033-1036.
37.Hynes, R.O. and Lander, A.D. (1992) Cell 68, 303-322.
38.Banerjee, U., Renfranz, P.J., Hinton, D.R., Rabin, B.A. and Benzer, S. (1987) Cell 51, 151-158.
39.Basler, K. and Hafen, E. (1988) Cell 54, 299-311.
40. Krämer, H., Cagan, R.L. and Zipurski, L. (1991) Nature 352, 207-212.
41.Fehon, R.G., Kooh, P.J., Rebay, I., Regan, C.L., Xu, T., Muskavitch, M.A.T. and Artavanis-Tsakonas, S. (1990) Cell 61, 523-534.
42.Figdor, C.G., van Kooyk, Y. and Keizer, G.D. (1990) Immunol. Today 11, 277-280.
43.Rudd, C.E., Barber, E.K., Burgess, K.E., Hahn, J.Y., Odysseos, A.D., Sy, M.S. and Schlossman, S.F. (1991) Adv. Exp. Med. Biol. 292, 85-96.
44.Dustin, M.L. and Springer, T.A. (1989) Nature 341, 619-624.
45.Schuch, U., Lohse, M.J. and Schachner, M. (1989) Neuron 3, 13-20.
46.Doherty, P., Ashton, S.V., Moore, S.E., and Walsh, F.S. (1991) Cell 67, 21-33.
47.Edelman, G.M. and Crossin, K.L. (1991) Ann. Rev. Biochem. 60, 155-190.
48.Doherty, P. and Walsh, F.S. (1989) Curr. Opin. Cell Biol. 1, 1102-1106.
49.Thiery, J.P., Boyer, B., Tucker, G., Gavrilovic, J. and Valles, A.-M. (1988) In "Metastasis" (Ciba Foundation Symposium 141) pp. 48-74, Wiley, Chichester.
50.Behrens, J., Mareel, M.M., Van Roy, F.M. and Birchmeier, W. (1989) J. Cell Biol. 108, 2435-2447.
51.Frixen, U.H., Behrens, J., Sachs, M., Eberle, G., Voss, B., Warda, A., Lochner, D. and Birchmeier, W. (1991) J. Cell Biol. 113, 173-185.
52.Vleminckx, K., Vakaet Jr., L., Mareel, M., Fiers, W. and Van Roy, F. (1991) Cell 66, 107-119.
53.Chen, W. and Öbrink, B. (1991) J. Cell Biol. 114, 319-327.
54.Giancotti, F.G. and Ruoslahti, E. (1990) Cell 60, 849-859.
55.Albelda, S.M., Mette, S.A., Elder, D.E., Stewart, R., Damjanovich, K., Herlyn, M. and Buck, C.A. (1990) Cancer Res. 50, 6757-6764.

■ *Björn Öbrink:*
Department of Medical Cell Biology,
Medical Nobel Institute, Karolinska Institute,
Stockholm, Sweden

AMOG

The adhesion molecule on glia (AMOG) is an integral membrane glycoprotein which shows a significant structural homology with, but is distinct from, the β–subunit of the Na^+/K^+-ATPase and is functionally associated with the catalytic α-subunit of this enzyme. AMOG is expressed by central nervous system glial cells and is involved in neuron-astrocyte adhesion and cerebellar granule cell migration.

AMOG is an integral membrane glycoprotein like the β-subunit of the Na^+/K^+-ATPase and has a molecular mass of ~45 kDa of which at least 30% is carbohydrate[1]. Sequence analysis of the macromolecule reveals 40% identity with the β-subunit of Na^+/K^+-ATPase at the amino acid level[2]. (AMOG has also been termed β2-subunit[3] by investigators in the Na^+/K^+-ATPase field). The α2, and possibly also α3, isoforms of the Na^+/K^+-ATPase copurify with AMOG in immunoaffinity isolation procedures from adult mouse brain[2]. A functional link between AMOG and Na^+/K^+-ATPase could be shown by triggering AMOG via monoclonal antibody: addition of AMOG antibodies to astrocytes in culture increased the ouabain inhibitable uptake of radioactive Rb^+, indicative of K^+ uptake[2]. The fact that AMOG-mediated adhesion occurs both at 4°C and in the presence of ouabain, an inhibitor of the catalytic activity of the enzyme, underscores that AMOG is indeed a cell recognition molecule[2]. Furthermore, the operational definition of an adhesion molecule is well satisfied by AMOG, as indicated by its ability to bind to particular types of neurons after purification and incorporation into liposomes[4]. AMOG is expressed predominantly in the central nervous system by glial cells and is involved in neuron-astrocyte, but not astrocyte-astrocyte adhesion, thus indicating a heterophilic adhesion mechanism[1]. AMOG is expressed during morphogenetically active periods, such as migration of cerebellar granule cell neurons along Bergmann glial cells. Fab-fragments of monoclonal anti-AMOG antibodies strongly inhibit this migration[1]. Whether this inhibition results from triggering of pump activity in Bergmann glial cells or by modification of the recognition process itself is presently unknown. After completion of granule cell migration, AMOG remains expressed by astrocytes predominantly in the cerebellum, but also in other brain regions, particularly in those where the packing density of neuronal cell bodies is high[5].

The link between a recognition molecule and pump activity is intriguing in that Na^+ and K^+ are known to determine the membrane potential, size of the extracellular space and cell volume. Thus, an influence on these basic cellular parameters by cell recognition processes may provide new venues of regulating fundamental cellular parameters not only during development, but also in the adult.

PURIFICATION

AMOG can be purified from adult mouse brain by immunoaffinity chromatography using a monoclonal antibody column[1].

ACTIVITIES

The biological activity of AMOG is best tested in *in vitro* assays by measuring the adhesive activity of AMOG containing liposomes to neurons[3] and by triggering Na^+/K^+-ATPase activity in cultured astrocytes by monoclonal antibodies[2].

ANTIBODIES

Polyclonal and monoclonal antibodies to AMOG from adult mouse brain have been described[1].

GENES

The full length cDNA sequence of AMOG from mouse has been described[2]. The complete sequence for the mouse AMOG gene is also available[6]. The gene is localized on mouse chromosome II and human chromosome 17[7]. EMBL Data Bank Accession number X16645.

REFERENCES

1. Antonicek, H., Persohn, E. and Schachner, M. (1987) J. Cell Biol. 104, 1587-1595.
2. Gloor, S., Antonicek, H., Sweadner, K., Pagliusi, S., Frank, R., Moos, M. and Schachner, M. (1990) J. Cell Biol. 110, 165-174.
3. Martin-Vasallo, P., Dackowski, W., Emanuel, J.R. and Levenson, R. (1989) J. Biol. Chem. 264, 4613-4618.
4. Antonicek, H. and Schachner, M. (1988) J. Neurosci. 8, 2961-2966.
5. Pagliusi, S.R., Schachner, M., Seeburg, P.H. and Shivers, B.D. (1990) Eur. J. Neurosci. 2, 471-480.
6. Magyar, J.P. and Schachner, M. (1990) Nucleic Acids Res. 18, 6695-6696.
7. Hsieh, C.-L., Cheng-Deutsch, A., Gloor, S., Schachner, M. and Francke, U. (1990) Somat. Cell Mol. Genet. 16, 401-405.

M. Schachner:
Department Neurobiology,
Swiss Federal Institute of Technology,
Hönggeberg, 8093 Zurich,
Switzerland

Cadherins

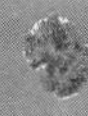

Cadherins[1,2] are a family of transmembrane glycoproteins that play a key role in Ca^{2+}-dependent cell-cell adhesion. They function not only for establishing tight cell-cell associations but also for defining adhesive specificities of cells. In development, the expression of each member of the family is spatio-temporally regulated so as to be correlated with morphogenetic events in which adhesion or dispersion of cells is involved.

Several members of the cadherin family have been characterized at the molecular level, including E-cadherin (identical to uvomorulin), N-cadherin, P-cadherin and L-CAM, and more members are being identified. These members are identical to each other in amino acid sequences (in the range of 43% to 58%), and have a common basic structure (120 to 130 kDa on SDS-PAGE). They are composed of an extracellular domain characterized by unique sequences repeated three to four times, a transmembrane domain and a cytoplasmic domain which is most highly conserved (Figure 2). Desmosomal intercellular proteins, **desmoglein** and **desmocollins**, are similar to cadherins, but their cytoplasmic domains are distinct from those of cadherins[3,4]. *Drosophila* has some molecules with cadherin-like sequences which are involved in tumour suppression, although their overall structure is different from that of the vertebrate cadherins so far identified[5]. Arc-1, cell-CAM 120/80, and rr-1, defined with antibodies, are probably homologous to E-cadherin. A-CAM and N-Cal-CAM are probably identical to N-cadherin[1,2].

Cadherins bind cells via homophilic interactions. In these interactions, they preferentially bind to the identical types; as a result, when they are mixed in culture, cells expressing distinct cadherins aggregate separately[2,6]. Thus, cadherins seem to be important for cell sorting mechanisms. The amino acid sequences responsible for the binding specificities of each member seem to be located within the N-terminal 113 amino acid region, some of which have been determined[7].

The cytoplasmic domain of cadherins play a crucial role in their cell-cell adhesive function; cells expressing mutated cadherins in which portions of the carboxy-half of the cytoplasmic domain were deleted cannot aggregate[2]. These cytoplasmic regions are associated with intracellular proteins, ***catenins***[8]. It is also known that cadherins are colocalized with cortical ***actin*** bundles at

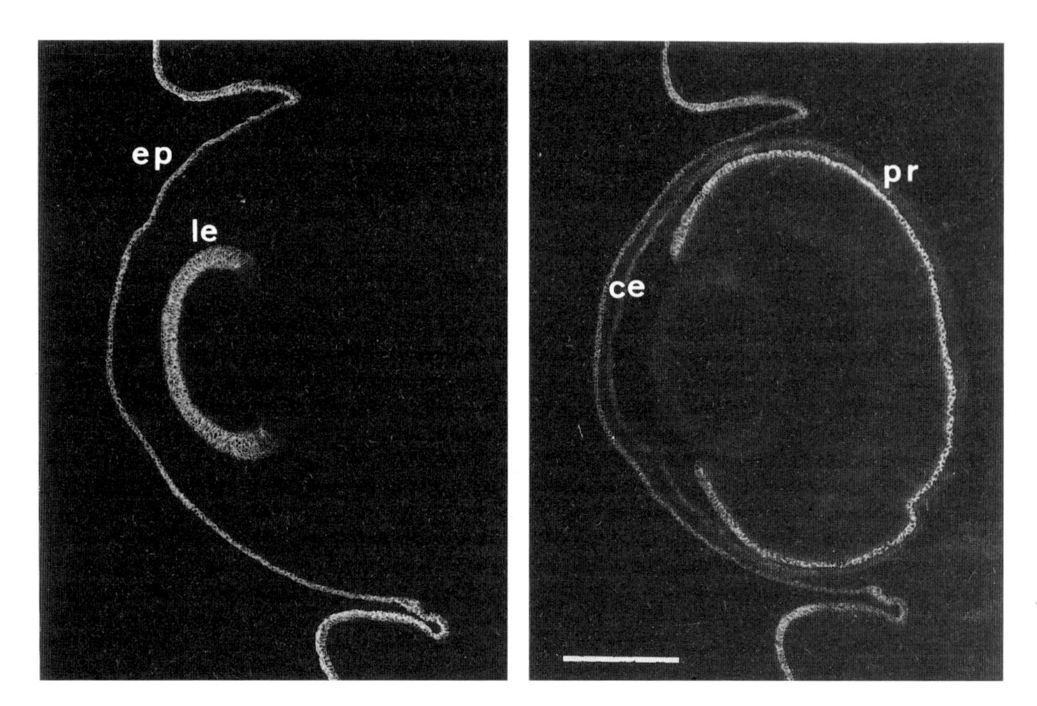

Figure 1. Double-immunofluorescent localization of E- (left) and P-cadherin (right) in the eye region of a foetal mouse. E-cadherin is expressed in the epidermis (ep) and the lens epithelium (le), while P-cadherin is expressed in the epidermis, the pigmented retina (pr) and the corneal endothelium (ce). Bar 250 μm. (Photographs provided by Dr. A. Nose).

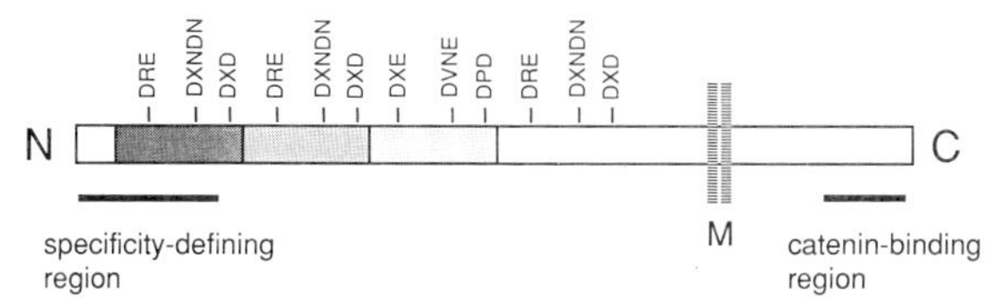

Figure 2. The primary structure of cadherins. Some of the sequences conserved in the cadherin family are shown. Stippled boxes represent repeated structures. N, N-terminus; M, transmembrane region; C, C-terminus.

cell-cell junctional sites and, at the ultrastructural level, they are concentrated at the *zonula adherens*. The association of cadherins with cytoskeletal components as well as catenins suggest that these interactions are essential for cadherins to operate as cell-cell adhesion receptors.

Each member of the cadherin family has a distinct tissue distribution[1] (Figure 2). In development, the expression of cadherins is spatio-temporally regulated so as to be associated with various morphogenetic events[1]; for example, when epithelia are transformed into mesenchymal cells, cadherin expression is down-regulated, and, when a cell layer is divided into distinct domains, they begin to express distinct cadherin types. Inhibition or ectopic expression of cadherins in tissues or whole embryos results in abnormal morphogenesis[1,2,9,10].

PURIFICATION

Fragments of the extracellular domain of cadherins, generated with proteolytic cleavage, can be purified by antibody affinity chromatography[11,12]. The native form, solubilized with detergents, is copurified with catenins[8]. E-cadherin adheres to ConA columns[1].

ACTIVITIES

Cadherins mediate cell-cell adhesion only in the presence of Ca^{2+} with concentrations of the order of 10^{-4}-10^{-3}M[1]. Cadherins on cell surfaces are highly sensitive to proteases, but protected by Ca^{2+} against proteolysis[1]. Some of the repeated sequences in the extracellular domain are assumed to be responsible for Ca^{2+}-binding[13]. Cadherin mediated adhesion is temperature dependent, not occurring at 4°C; but the adhesion once established is stable at low temperatures. Cadherins are required for compacted cell associations[14].

ANTIBODIES

Polyclonal and monoclonal antibodies specific to each type of cadherin are available, although many of them recognize only limited animal species. Many of these antibodies can disrupt cell-cell adhesion[1]. Antibodies against the conserved cytoplasmic regions of cadherins can react with multiple cadherin types. Monoclonal antibodies against N-cadherin (A-CAM) can be purchased from Sigma and BioMakor.

GENES

Complete or partial cDNA sequences are published for mouse[14,15] and human[16] E-cadherin, chicken[17], *Xenopus*[10], mouse[18], bovine[19] and human[20] N-cadherin, mouse[21], bovine[19] and human[22] P-cadherin, chicken L-CAM[23] and others[24-26]. The structures of the chicken L-CAM[27] gene and mouse P-cadherin genes[28] are also published. The mouse E-[29] and P-cadherin[28] loci are tightly linked on chromosome 8 and the mouse N-cadherin gene has been placed on chromosome 18 (Jenkins, Copeland and Takeichi, unpublished). The human E-cadherin gene[16] has been localized to chromosome 16, in the region of 16p11-16qter, and the human N-cadherin gene[20] mapped to chromosome 18. Many of these sequence data can be obtained using the key word "cadherin" from databases.

REFERENCES

1. Takeichi, M. (1988) Development 102, 639-655.
2. Takeichi, M. (1990) Ann. Rev. Biochem. 59, 237-252.
3. Koch, P.J., Walsh, M.J., Schmelz, M., Goldschmidt, M.D., Zimbelmann, R. and Franke, W.W. (1990) Euro. J. Cell Biol. 53, 1-12.
4. Mechanic, S., Raynor, K., Hill, J.E. and Cowin, P. (1991) Proc. Natl. Acad. Sci. (USA) 88, 4476-4480.
5. Mahoney, P.A., Weber, U., Onofrechuck, P., Biessmann, H., Bryant, P.J. and Goodman, C.S. (1991) Cell, 67, 853-868.
6. Mege, M., Cunningham, B.A. and Edelman, G.M. (1989) Proc. Natl. Acad. Sci. (USA) 86, 7043-7047.
7. Nose, A., Tsuji, K. and Takeichi, M. (1990) Cell 61, 147-155.
8. Ozawa, M., Baribault, J. and Kemler, R. (1989) EMBO J. 8, 1711-1717.
9. Fujimori, T., Miyatani, S. and Takeichi, M. (1990) Development 110, 97-104.
10. Detrick, R.J., Dickey, D. and Kintner, C.R. (1990) Neuron 4, 493-506.
11. Damsky, C.H., Richa, J., Solter, D., Knudsen, K. and Buck, C.A. (1983) Cell 34, 455-466.
12. Shirayoshi, Y., Hatta, K., Hosoda, M., Tsunasawa, S., Sakiyama, F. and Takeichi, M. (1986) EMBO J. 5, 2485-2488.
13. Ozawa, M., Engel, J. and Kemler, R. (1990) Cell 63, 1033-1038.
14. Nagafuchi, A., Shirayoshi, Y., Okazaki, K., Yasuda, K. and Takeichi, M. (1987) Nature 329, 341-343.
15. Ringwald, M., Schuh, R., Vestweber, D., Eistetter, H., Lottspeich, F., Engel, J., Dolz, R., Jahnig, F., Epplen, J., Mayer, S., Muller, C. and Kemler, R. (1987) EMBO J. 6, 3647-3653.
16. Mansouri, A., Spurr, N., Goodfellow, P.N. and Kemler, R. (1988) Differentiation 38, 67-71.
17. Hatta, K., Nose, A., Nagafuchi, A. and Takeichi, M. (1988) J. Cell Biol. 106, 873-881.
18. Miyatani, S., Shimamura, K., Hatta, M., Nagafuchi, A., Nose, A., Matsunaga, M., Hatta, K. and Takeichi, M. (1989) Science 245, 631-635.
19. Liaw, C.W., Cannon, C., Power, M.D., Kiboneka, P.K. and Rubin, L.L. (1990) EMBO J. 9, 2701-2708.
20. Walsh, F.S., Barton, H., Putt, W., Moore, S.E., Kelsell, D., Spurr, N. and Goodfellow, P.N. (1990) J. Neurochem. 55, 805-812.
21. Nose, A., Nagafuchi, A. and Takeichi, M. (1987) EMBO J. 6, 3655-3661.
22. Shimoyama, Y., Hirohashi, S., Hirano, S., Noguchi, M. and Shimosato, Y. (1989) Cancer Res. 49, 2128-2133.
23. Gallin, W.J., Sorkin, B.C., Edelman, G.M. and Cunningham, B.A. (1987) Proc. Natl. Acad. Sci. (USA) 84, 2808-2812.

24.Suzuki, S., Sano, K. and Tanihara, H. (1991) Cell Regul. 2, 261-270.
25.Ginsberg, D., DeSomone, D. and Geiger, B. (1991) Development 111, 315-325.
26.Herzberg, F., Wildermuth, V. and Wedlich, D. (1991) Mechan. Develop. 35, 33-42.
27.Sorkin, B.C., Hemperley, J.J., Edelman, G.M. and Cunningham, B.A. (1988) Proc. Natl. Acad. Sci. (USA) 85, 7617-7621.
28.Hatta, M., Miyatani, S., Copeland, N.G., Gilbert, D.J., Jenkins, N.A. and Takeichi, M. (1991) Nucl. Acid Res. 19, 4437-4441.
29.Eistetter, H.R., Adolph, S., Ringwald, M., Chazottes, D.S., Schuh, R., Guénet, J.-L. and Kemler, R. (1988) Proc. Natl. Acad. Sci. (USA) 85, 3489-3493.

Masatoshi Takeichi:
Department of Biophysics,
Kyoto University,
Kyoto, Japan

CD2

CD2[1-3] is a single chain transmembrane glycoprotein which is a member of the immunoglobulin (Ig) superfamily and it is predominantly expressed on thymocytes and mature T lymphocytes. CD2 binds to the LFA-3 cell surface glycoprotein which has a structure similar to CD2 and is expressed on a wide variety of cells. The CD2-LFA-3 interaction plays an important role in the adhesion of T lymphocytes to antigen presenting cells and crosslinking of CD2 by antibodies leads to T cell activation.

CD2 is a 50-55 kDa glycoprotein consisting of an extracellular region of 181-186 amino acids with several N-linked glycosylation sites, a single transmembrane region of 25-26 residues and a relatively long cytoplasmic domain of 115-116 amino acids depending on the species[4-7]. The extracellular region of CD2 is postulated to contain two Ig-related domains as shown in the Figure. These domains show particular structural similarities to the extracellular domains of two other cell surface molecules called CD58 (**LFA-3**) and CD48 (Blast-1, OX-45, BCM1)[8,9]. The solution structure for the first domain of CD2 has recently been determined by NMR spectroscopy and it confirmed that this domain is Ig-related[10]. The cytoplasmic domain of CD2 contains an unusually high proportion of prolines (about 22%) and five regularly spaced histidines[4-7]. Near the C-terminus, there is a patch of 24 amino acid residues which contains 23 identical residues in human, rat and mouse.

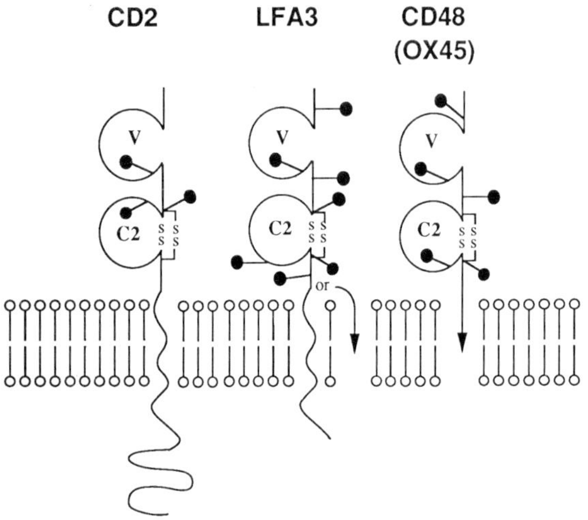

Figure. Models of human CD2 and LFA-3 and rat CD48. Potential N-linked glycosylation sites are indicated by the club shaped symbols, glycosylphosphatidylinositol anchors by arrows and postulated disulphide bonds by S symbols. The categorization of the domains as being of V or C-2 type is discussed[9].

The genes for CD2 and LFA-3 map to syntenic regions of human chromosome 1p and mouse chromosome 3[2]. The CD48 (BCM1) genes map to human chromosome 1q and mouse chromosome 1[11]. It has been postulated that the genetic regions containing CD48 and CD2/LFA-3 respectively evolved by gene duplication of a large chromosome segment[11]. The exon pattern of CD2 gene largely corresponds to the structural domains of the protein, i.e. the leader, each of the extracellular domains, the transmembrane domain and the cytoplasmic domain[2].

The tissue distribution of CD2 shows species heterogeneity. In the human, CD2 is expressed on thymocytes, mature T cells and NK cells[2,3]. In the rat, CD2 is similarly expressed, but also on splenic macrophages[1], and in the mouse, CD2 is expressed on T and B lineage cells[12].

CD2 mediates intercellular adhesion by binding to LFA-3[2,3,13,14]. The binding site for LFA-3 is located on the N-terminal domain of CD2 and has been mapped by mutagenesis studies[15]. The affinity of the interaction is about 10^6 M^{-1}[10,14]. Truncation of the cytoplasmic domain of CD2 does not affect the affinity for LFA-3[2,3]. Intercellular adhesion via the CD2-LFA-3 interaction may be enhanced after T cell activation due to an increase in the expression of CD2 by activated cells[2].

Crosslinking of CD2 by combinations of monoclonal antibodies (mAbs) leads to activation of the inositol phosphate pathway[16] and mitogenesis of human and rat T lymphocytes[1-3,17].The cytoplasmic domain of CD2 is required for the increase in cytosolic free Ca^{2+} concentration and IL-2 secretion mediated by anti-CD2 mAbs[1-3]. Activation of T cells via CD2 is weak in cell lines that lack the T cell receptor for antigen (TCR) indicating an involvement of the TCR in CD2 signalling[1,2]. Functional or physical interactions with other molecules may also occur and a physical association between CD2 and the transmembrane phosphotyrosine phosphatase CD45 (**LCA**) has been reported[18].

FUNCTIONAL STUDIES

Certain combinations of anti-CD2 mAbs induce the same intracellular events as triggering of the TCR namely the mobilization of inositol phosphates and diacylglycerol[16], an increase in cytosolic free Ca^{2+} concentration[1,2,16], activation of protein kinase C[19], phosphorylation of the ζ chain of the TCR complex[20], IL-2 secretion[2,3] and ultimately mitogenesis[1-3,17].

ANTIBODIES

A large number of anti-human CD2 mAbs have been described[21]. Early studies defined three functionally distinct epitopes on the extracellular region of CD2[17]. $T11_1$ was present on all T cells and was recognized by mAbs that inhibited rosetting with sheep red blood cells (SRBC). $T11_2$ was also present on all T cells but it was distinct form the SRBC binding site. $T11_3$ was only expressed on activated T cells and its expression could be induced by incubation with anti-$T11_2$ mAb. The induction of $T11_3$ epitope by conformational change is controversial and alternatively recognition of this epitope may reflect a requirement for high antigen density to allow bivalent binding of low affinity mAbs[1]. The anti-rat CD2 mAbs OX-34, OX-54 and OX-55 (of which the OX-54 plus OX-55 combination is mitogenic)[1] are available from Serotec, Kidlington, U.K. Recently anti-mouse CD2 mAbs have been described[12]. A rabbit polyclonal antiserum, raised against a synthetic peptide corresponding to a highly conserved region of the cytoplasmic domain of human CD2, crossreacts with CD2 of other mammals and chicken[22].

GENES

The GenBank accession numbers of the cDNA sequences of human[4,5], rat[6] and mouse[7] CD2 are M16445, X05111 and Y00023 respectively.

REFERENCES

1. Beyers, A.D., Barclay, A.N., Law,D.A., He, Q. and Williams, A.F. (1989) Immunol. Rev. 111, 59-77.
2. Moingeon, P., Chang, H.C., Sayre, P.H., Clayton, L.K., Alcover, A., Gardner, P. and Reinherz, E.L. (1989) Immunol. Rev. 111, 111-144.
3. Bierer, B.E. and Burakoff, S.J. (1989) Immunol. Rev. 111, 267-294.
4. Sewell, W.A., Brown, M.H., Dunne, J., Owen, M.J. and Crumpton, M.J. (1987) Proc. Natl. Acad. Sci. (USA) 84, 7256.
5. Seed, B. and Aruffo, A. (1987) Proc. Natl. Acad. Sci. (USA) 84, 3365-3369.
6. Williams, A.F., Barclay, A.N., Clark, S.J., Paterson, D.J. and Willis, A.C. (1987) J. Exp. Med. 165, 368-380.
7. Sewell, W.A., Brown, M.H., Owen, M.J., Fink, P.J., Kozak, C.A. and Crumpton, M.J. (1987) Eur. J. Immunol. 17, 1015-1020.
8. Williams, A.F., Davis, S.J., He, Q. and Barclay, A.N. (1989) Cold Spring Harb. Symp. Quant. Biol. 54,637-647.
9. Killeen, N., Moessner, R., Arvieux, J., Willis, A. and Williams, A.F. (1988) EMBO J. 7, 3087-3091.
10. Driscoll, P.C., Cyster, J.G., Campbell, I.D. and Williams, A.F. Nature 1991 353, 762-765.
11. Wong, Y.W., Williams, A.F., Kingsmore, S.F. and Seldin, M.F. (1990) J. Exp. Med. 171, 2115-2130.
12. Yagita, H., Nakamura, T., Karasuyama, H. and Okumura, K. (1989) Proc. Natl. Acad. Sci. (USA) 86, 645-649.
13. Hünig, T. (1985) J. Exp. Med. 162, 890-901.
14. Selvaraj, P., Plunkett, M.L., Dustin, M., Sanders, M.E., Shaw, S. and Springer, T.A. (1987) Nature 326, 400-403.
15. Peterson, A. and Seed, B. (1987) Nature 329, 842-846.
16. Pantaleo, G., Olive, D., Poggi, A., Kozumbo, W.J., Moretta, L. and Moretta, A. (1987) Eur. J. Immunol. 17, 55-60.
17. Meuer, S.C., Hussey, R.E., Fabbi, M., Fox, D., Acuto, O., Fitzgerald, K.A., Hodgdon, J.C., Protentis, J.P., Schlossman, S.F. and Reinherz, E.L. (1984) Cell 36, 897-906.
18. Altevogt, P., Schrek, J., Schraven, B., Meuer, S., Schirrmacher, V. and Mitsch, A. (1990) Internat. Immunol. 2, 353-360.
19. Alexander, D.R. and Cantrell, D.A. (1989) Immunol. Today 10, 200-205.
20. Monostori, E., Desai, D., Brown, M.H., Cantrell, D.A. and Crumpton, M.J. (1990) J. Immunol. 144, 1010-1014.
21. Knapp, W., Dörken, B., Gilks, W.R., Rieber, E.P., Schmidt, R.E., Stein, H. and von dem Borne, A.E.G.Kr. (1989) Leukocyte Typing IV, Oxford University Press.
22. Brown, M.H., Sewell, W.A., Mason, D.Y., Rothbard, J.B. and Crumpton, M.J. (1988) Eur. J. Immunol. 18, 1223-1227.

Alan F. Williams and Albertus D. Beyers:
MRC Cellular Immunology Unit,
University of Oxford,
UK

CD4

CD4 is a T cell surface glycoprotein that serves as a coreceptor, with the T cell antigen receptor, for transduction of developmental signals in thymocytes and activation signals in mature T cells. The CD4 molecule binds to class II MHC glycoproteins on antigen presenting cells and transmits signals through its association with a cytoplasmic tyrosine kinase, p56^{lck}. CD4 is also the cell surface receptor for the human and simian immunodeficiency viruses.

The CD4 glycoprotein is a 55-58 kDa monomer consisting of four distinct extracellular immunoglobulin (Ig) family domains, a hydrophobic transmembrane domain, and a highly conserved 38 amino acid cytoplasmic tail[1,2]. In the thymus, the molecule is expressed on immature double positive and mature single positive cells (Figure 1), while in the peripheral immune system it is expressed primarily on helper T cells. In human, CD4 is also expressed on other hematopoietic lineage cells.

The extracellular domains of CD4 bind to a presumed monomorphic region of class II MHC molecules on antigen presenting cells[3], while the cytoplasmic tail binds to a cytoplasmic tyrosine kinase, p56^{lck}[4-6] (Figure 2). Activation of T cell hybridomas bearing receptors specific for peptide antigen plus class II MHC molecules requires expression of CD4 and the ability of CD4 to interact with p56^{lck} [7]. Requirement for CD4 may not be so stringent if the T cell receptor has high affinity for antigen. In development,

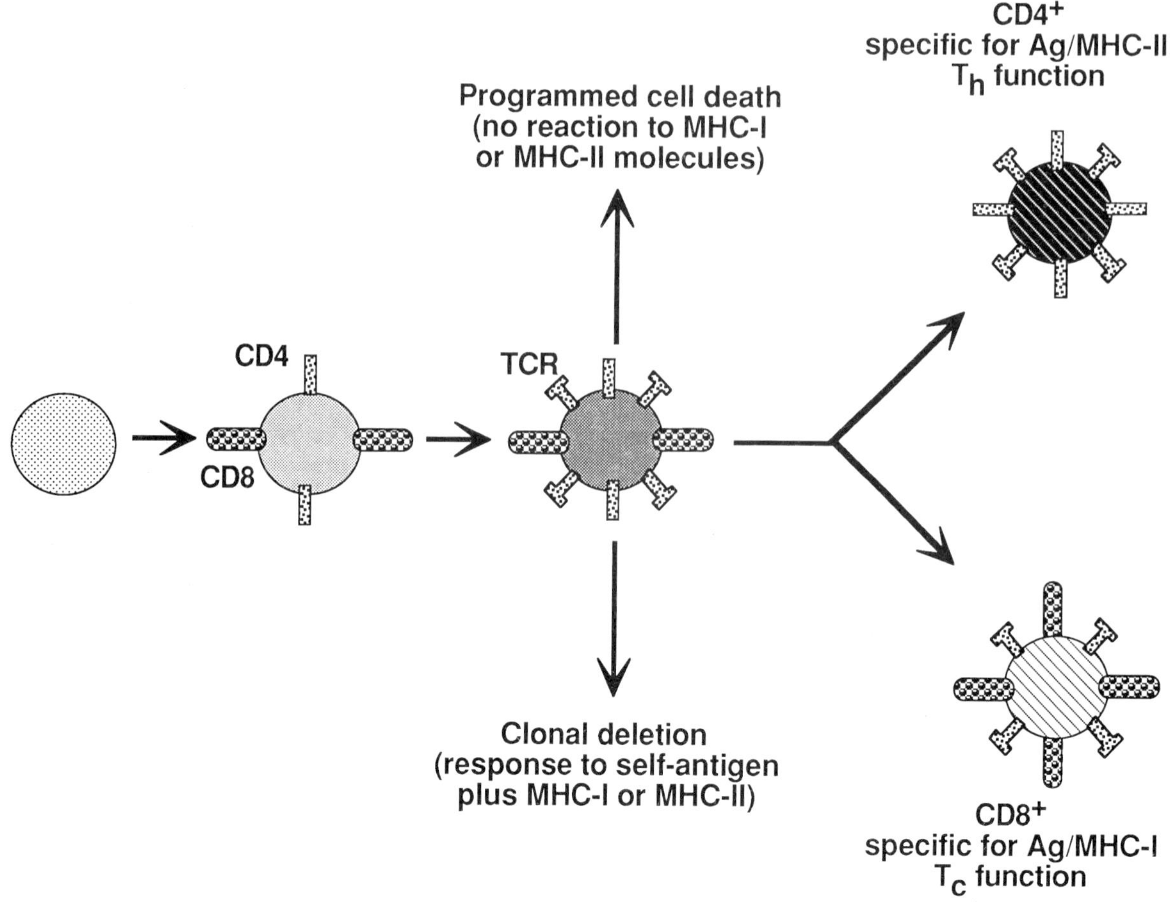

Figure 1. CD4 in T cell development. Expression of CD4 during T cell differentiation in the thymus. The earliest thymocyte precursors (0.05% of total thymocytes) express low levels of CD4 and give rise to double-negative cells shown at the left in the figure. CD4$^+$CD8$^+$ cells arise from double-negative cells and express clonotypic T cell receptors. These cells are then subjected to negative and positive selection, resulting in the appearance of single-positive mature T cells.

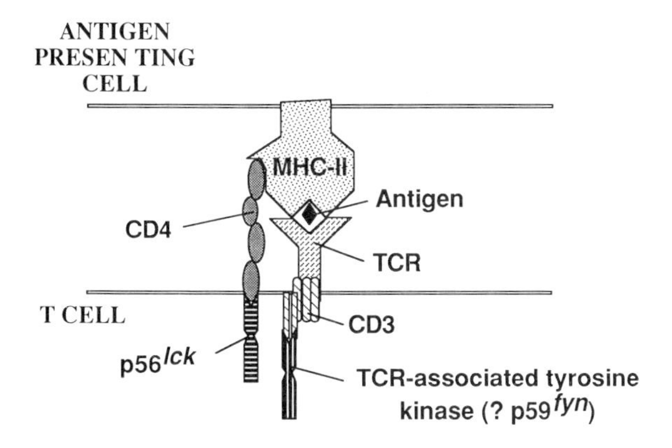

Figure 2. Co-receptor function of CD4. The figure depicts the coordinate recognition of MHC class II plus antigen by CD4 and the T cell receptor. Crosslinking of the p56[lck] tyrosine kinase with receptor-associated components (including a different tyrosine kinase) activates phospholipase-Cγ and results in lymphokine gene expression.

coreceptor function of CD4 is required for transmission of signals resulting in clonal deletion of self reactive double positive thymocytes and in positive selection of cells that have receptors that interact with self class II MHC plus foreign antigen[8]. The CD4 associated tyrosine kinase is thought to interact with substrate(s) associated with the T cell receptor complex, setting off a series of events that can result in production of lymphokines, in programmed cell death, or in progression to a mature differentiated T cell (Figures 1 and 2).

The CD4 molecule is the principal receptor for the human immunodeficiency virus (HIV)[9]. Binding of the HIV envelope glycoprotein, gp120, to CD4 on the cell surface is the first step in the HIV life cycle. This interaction requires the N-terminal domain of CD4, where the binding site for gp120 is located[9]. Chimeric molecules containing only the first two domains of CD4 are sufficient to permit HIV entry into cells[10]. Several soluble forms of CD4 have been designed for the purpose of blocking infection with HIV[9]. CD4$^+$ human cells infectable with HIV include T cells, monocytes, and macrophages such as brain microglial cells. Nonprimate cells transfected with human CD4 bind HIV but are not infected[11].

The crystal structure of the N-terminal two domains of CD4 has recently been reported[12,13]. Both domains have the configuration of Ig folds, even though the second domain is truncated, and the connection between the domains is quite stiff, resulting in an elongated molecule with extensive interdomain interaction.

PURIFICATION

CD4 can be readily purified by solubilization of cells with non-ionic detergents and subsequent affinity purification using any of a large variety of antibodies[14].

ACTIVITIES

Purified CD4 binds to the HIV envelope glycoprotein, gp120[9]. The detergent solubilized molecule also binds to the tyrosine kinase, p56[lck] [6]. Cells that overexpress CD4 bind to cells expressing class II MHC molecules[3].

ANTIBODIES

Many monoclonal antibodies against human[15], mouse[16], and rat[14] CD4 have been described and are commercially available.

GENES

Complete coding sequences are available for CD4 in human (GenBank HUMATCT4A)[17,18], rhesus (MACCD4A), chimpanzee (CHPCD4A), mouse (MUSCD4R)[16,19], and rat (RATCD4A)[14]. The human and murine genes have been localized to chromosomes 12 and 6, respectively[1,2].

REFERENCES

1. Littman, D.R. (1987) Ann. Rev. Immunol. 5, 561-584.
2. Parnes, J.R. (1989) Adv. Immunol. 44, 265-311.
3. Doyle, C. and Strominger, J.L. (1987) Nature 330, 256-259.
4. Rudd, C.E., Trevillyan, J.M., Dasgupta, J.D., Wong, L.L. and Schlossman, S.F. (1988) Proc. Natl. Acad. Sci. (USA) 85, 5190-5194.
5. Shaw, A.S., Amrein, K.E., Hammond, C., Stern, D.F., Sefton, B.M. and Rose, J.K. (1989) Cell 59, 627-636.
6. Turner, J.M., Brodsky, M.H., Irving, B.A., Levin, S.D., Perlmutter, R.M. and Littman, D.R. (1990) Cell 60, 755-765.
7. Glaichenhaus, N., Shastri, N., Littman, D.R. and Turner, J.M. (1991) Cell 65, 511-520.
8. Fowlkes, B.J. and Pardoll, D.M. (1989) Adv. Immunol. 44, 207-264.
9. Sattentau, Q. and Weiss, A. (1988) Cell 52, 631-633.
10. Bedinger, P., Moriarty, A., von Borstel, R., Donovan, N., Steimer, K. and Littman, D.R. (1988) Nature 334, 162-165.
11. Maddon, P.J., Dalgleish, A.G., McDougal, J.S., Clapham, P.R., Weiss, R.A. and Axel, R. (1986) Cell 47, 333-348.
12. Wang, J., Yan, Y., Garrett, T.P.J., Liu, J., Rodgers, D.W., Garlick, R.L., Tarr, G.E., Husain, Y., Reinherz, E.L. and Harrison, S.C. (1990) Nature 348, 411-418.
13. Ryu, S.-E., Kwong, P.D., Truneh, A., Porter, T.G., Arthos, J., Rosenberg, M., Dai, X., Xuong, N., Axel, R., Sweet, R.W. and Hendrickson, W.A. (1990) Nature 348, 419-425.
14. Clark, S.J., Jefferies, W.A., Barclay, A.M.N., Gagnon, J. and Williams, A.F. (1987) Proc. Natl. Acad. Sci. (USA) 84, 1649-1653.
15. McMichael, A.J., ed. (1987) Leukocyte Typing III, Oxford Univ. Press.
16. Littman, D.R. and Gettner, S. (1987) Nature 325, 453-455.
17. Maddon, P.J., Littman, D.R., Goddfrey, M., Maddon, D.E., Chess, L. and Axel, R. (1985) Cell 42, 93-104.
18. Littman, D.R., Maddon, P.J. and Axel, R. (1988) Cell 55, 541.
19. Tourvieille, B., Gorman, S.D., Field, E.H., Hunkapiller, T. and Parnes, J.R. (1986) Science 234, 610-614.

Dan R. Littman:
Dept. of Microbiology and Immunology
and the Howard Hughes Medical Institute,
UCSF, San Fransisco, CA, USA

CD8

CD8 is a homodimeric or heterodimeric glycoprotein expressed on the surface of thymocytes and mature cytotoxic T lymphocytes[1,2]. It binds to a nonpolymorphic domain of class I MHC molecules on antigen presenting cells and participates, with the T cell receptor (TCR), in signal transduction during thymic differentiation and T cell activation. The cytoplasmic domain of CD8 interacts with the cytoplasmic tyrosine kinase, $p56^{lck}$.

The CD8 glycoprotein exists either as an αα homodimer or as an αβ heterodimer[1,2]. The subunits of CD8 are members of the immunoglobulin (Ig) superfamily and have molecular weights of 32-37 kDa. Cell surface expression of the β-chain requires pairing with the α-chain. The two chains of CD8 have similar overall organization: a leader peptide is followed by a domain of approximately 100 amino acids that is related to Ig light chain variable regions; and a domain rich in serines, threonines, and prolines, containing O-linked glycosylation, links the Ig-like domain to a hydrophobic transmembrane domain and a short cytoplasmic tail[1,2]. In different species, alternative splicing results in multiple isoforms of both chains of CD8[3-5]; in addition, splicing out of the transmembrane domain results in secretion of the human CD8α homodimer[5]. CD8 is expressed on the surface of immature thymocytes (the early precursors express low levels of CD8 only, while later precursors that make up the largest fraction of thymocytes express both **CD4** and CD8), mature thymocytes that have TCR's interacting with class I MHC molecules, and cytotoxic T lymphocytes in the peripheral immune system[6]. In addition, CD8 αα homodimer is expressed on some natural killer cells and on some T cells bearing γδ TCR's.

The CD8α chain binds to the nonpolymorphic membrane proximal domain of class I MHC molecules (Figure)[7]. The role of CD8β in binding to MHC molecules is not known. The cytoplasmic tail of CD8α interacts directly with the N-terminal domain of the lymphocyte specific cytoplasmic tyrosine kinase, $p56^{lck}$[8,9]. Two cysteine residues, shared by the cytoplasmic domains of CD4 and CD8, are required for this interaction[8]. The cytoplasmic domain of CD8b does not bind $p56^{lck}$, and its function is not yet known[9].

CD8 serves as a coreceptor for the TCR, with which it functions to transduce transmembrane signals resulting in developmental decisions such as clonal deletion or positive selection (Figure) and in activation of cytotoxic functions[10]. T cell activation through the TCR results in increased avidity between CD8 on the T cell surface and class I MHC molecules on lipid bilayers[11].

Figure. Co-receptor function of CD8. The figure depicts the coordinate recognition of MHC class I pus antigen by CD8 and the T-cell receptor. Crosslinking of the $p56^{lck}$ tyrosine kinase with receptor-associated components (including a specific receptor-associated tyrosine kinase) activates phospholipase Cγ and results in lymphokine gene expression.

PURIFICATION

CD8 heterodimers and homodimers are routinely purified from nonionic detergent extracts of cells by immunoaffinity chromatography[12,13].

ACTIVITIES

CD8α homodimers, expressed at high levels on the surface of transfected cells, mediate cell-cell adhesion to B cells that express class I MHC molecules. Transfection of CD8α into a $CD8^-$ T cell hybridoma restores its secretion of IL-2 in response to a specific class I MHC alloantigen[13].

ANTIBODIES

Numerous monoclonal antibodies against human[14], murine[15], and rat[12,13] CD8 have been described and are commercially available. Most antibodies react with CD8α, but a few react with CD8β or with heterodimer only[4,14].

GENES

Complete coding sequences are available for CD8α in human (GenBank HUMANTCD8)[16], mouse (MUSLYT2A, MUSLYT2AP)[3], and rat (RATATCOX8)[12] and for CD8β in human (HUMLYT3, X13444, X13445, X13446)[4], mouse (MUSLY3A)[17], and rat (RATCD837K)[13]. Complete sequence of the human CD8α gene is also available (HUMMHCD8A). The murine CD8 genes are separated by 36 kB on chromosome 6, while the human CD8α gene has been mapped to chromosome 2p12[2].

■ REFERENCES

1. Littman, D.R. (1987) Ann. Rev. Immunol. 5, 561-584.
2. Parnes, J.R. (1989) Adv. Immunol. 44, 265-311.
3. Zamoyska, R., Vollmer, A.C., Sizer, K.C., Liaw, C.W. and Parnes, J.R. (1985) Cell 43, 153-163.
4. Norment, A. and Littman, D.R. (1988) EMBO J. 7, 3433-3439.
5. Norment, A.M., Lonberg, N., Lacy, E. and Littman, D.R. (1989) J. Immunol. 142, 3312-3319.
6. Fowlkes, B.J. and Pardoll, D.M. (1989) Adv. Immunol. 44, 207-264.
7. Salter, R.D., Benjamin, R.J., Wesley, P.K., Buxton, S.E., Garrett, T.P.J., Clayberger, C., Krensky, A.M., Norment, A.M., Littman, D.R. and Parham, P. (1990) Nature 345, 41-46.
8. Turner, J.M., Brodsky, M.H., Irving, B.A., Levin, S.D., Perlmutter, R.M. and Littman, D.R. (1990) Cell 60, 755-765.
9. Zamoyska, R., Derham, P., Gorman, S.D., von Hogen, P., Bolen, J.B., Vaillette, A. and Parnes, J.R. (1989) Nature 342, 278-281.
10. Littman, D.R. (1989) Curr. Opin. Cell Biol. 1, 920-928.
11. O'Rourke, A.M., Rogers, J. and Mescher, M.F. (1990) Nature 346, 187-189.
12. Johnson, P., Gagnon, J., Barclay, A.N. and Williams, A.F. (1985) EMBO J. 4, 2539- 2545.
13. Johnson, P. and Williams, A.F. (1986) Nature 323, 74-76.
14. McMichael, A.J., ed. (1987) Leukocyte Typing III, Oxford Univ. Press.
15. Swain, S.L. (1983) Immunol. Rev. 74, 129-142.
16. Littman, D.R., Thomas, Y., Maddon, P., Chess, L. and Axel, R. (1985) Cell 40, 237-246.
17. Nakauchi, H., Shinkai, Y.-I. and Okumura, K. (1987) Proc. Natl. Acad. Sci. (USA) 84, 4210-4214.

■ *Dan R. Littman:*
Dept. of Microbiology and Immunology
and the Howard Hughes Medical Institute,
UCSF, San Francisco, CA, USA

C-CAM (CELL-CAM 105)

C-CAM[1,2] is an integral membrane glycoprotein that mediates cell-cell adhesion of isolated rat hepatocytes by Ca^{2+}-independent homophilic binding. It is a member of the immunoglobulin (Ig) superfamily and appears in several isoforms. C-CAM is present in epithelia (including hepatocytes), vessel endothelia, megakaryocytes, platelets, polymorphonuclear leukocytes, and a subset of mononuclear leukocytes.

C-CAM isolated from rat liver consists of two highly N-glycosylated peptide chains with apparent molecular weights of 105,000 and 110,000 respectively[3]. The two chains give similar peptide maps and represent different isoforms of C-CAM. Other isoforms differing in apparent molecular weights from 90,000 to 130,000 are present in small intestine, kidney, salivary glands, spleen and platelets[4]. One of the liver isoforms is identical to rat liver ecto-ATPase[5], which has been cloned and sequenced[6]. The deduced amino acid sequence suggests that rat liver C-CAM is a single-spanning transmembrane protein with four Ig like domains in the extracellular, N-terminal part[2,6] (Figure 1). Sequence comparison shows that C-CAM is a member of the carcinoembryonic antigen (CEA) gene family, which contains several different proteins with varying numbers of Ig like domains[2]. The human homologue of rat liver C-CAM is known as biliary glycoprotein 1 (BGP 1)[7]. Rat liver C-CAM has a cAMP dependent phosphorylation sequence in the cytoplasmic C-terminal part[6] and can be phosphorylated both on serine[3] and on tyrosine[8] residues *in vivo*. Calmodulin can bind to C-CAM in a specific Ca^{2+}-dependent manner[9].

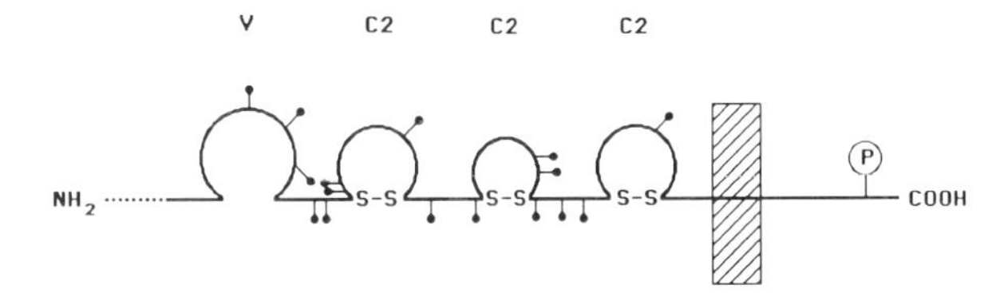

Figure 1. Linear domain structure of C-CAM in rat liver. The protein has a hydrophobic putative transmembrane domain (▨). The extracellular portion contains one V-like and three C2-like Ig-domains, and has 16 potential N-glycosylation sites (⫯). The cytoplasmic portion has several threonine and serine residues, one of which is part of a cAMP dependent phosphorylation consensus sequence (Ⓟ).

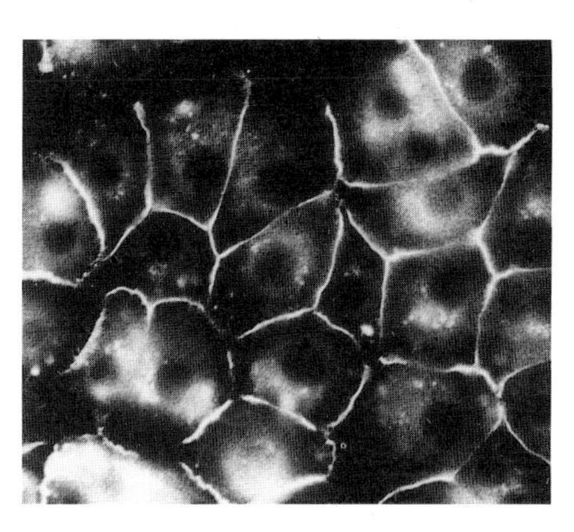

Figure 2. Immunofluorescence of primary rat hepatocytes in culture. Antibodies against C-CAM specifically stain the cell-cell borders of hepatocytes in contact with each other. Free cell surfaces are not stained.

The expression of C-CAM is developmentally regulated. It appears rather late in the foetal development of the

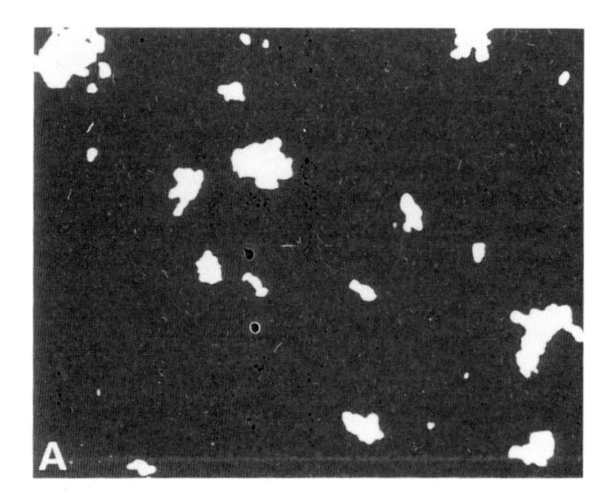

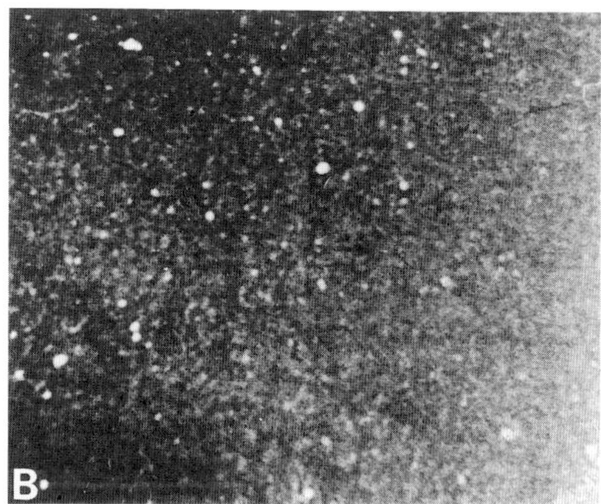

Figure 3. Self-aggregation of C-CAM containing liposomes. Carboxy-fluorescein containing liposomes reconstituted with C-CAM from rat liver form large aggregates (A). Incubation in the presence of Fab-fragments of antibodies directed against C-CAM dissociates the aggregated liposomes (B).

liver, and undergoes a transient decrease during liver regeneration after partial hepatectomy[10]. In both primary hepatocellular carcinomas and transplantable hepatocellular carcinomas C-CAM is either missing or occurs in reduced amounts as a chemically altered form[11,12].

The subcellular location of C-CAM differs in different cell types[2,4]. It is highly concentrated in the microvillar membranes of brush borders of small intestinal epithelial cells and renal proximal tubular cells, and of bile canaliculi in mature liver. It is also found in the lateral membranes of the pericanalicular domain in mature liver[13]. In stratified epithelia and in some epithelial cells in culture C-CAM occurs predominantly in the cell-cell contact areas (Figure 2). C-CAM has a cytoplasmic localization in unactivated granulocytes and platelets.

PURIFICATION

C-CAM can be purified from rat liver by immunoaffinity chromatography on polyclonal or monoclonal antibodies, followed by gel permeation chromatography and ion exchange chromatography[3].

ACTIVITIES

Purified liver C-CAM binds to itself in a Ca^{2+}-independent manner both in dot blotting assays and when incorporated into liposomes[14]. Cells transfected with full length cDNA exhibit increased levels of ecto-ATPase activity[6].

ANTIBODIES

Polyclonal antibodies against C-CAM have been described[3,4,12]. They react with the various isoforms in rat tissues, but crossreact poorly with other species. Monoclonal antibodies against C-CAM (MAb 362.50[12]; MAb HA-4[15]; MAb Be 9.2[16]; MAb F7[17]) have been published. The monoclonal antibodies have limited reactivities against the various isoforms of C-CAM in the rat. Monovalent antibodies (Fab) against C-CAM inhibit aggregation of isolated rat hepatocytes[1] (Figure 3).

GENES

Liver ecto-ATPase has been cloned and sequenced from rat[6] (GenBank/EMBL J04963). The sequence of human BGP 1 is also available[7] (GenBank/EMBL J03858).

REFERENCES

1. Ocklind, C. and Öbrink, B. (1982) J. Biol. Chem. 257, 6788-6795.
2. Öbrink, B. (1991) BioEssays 13, 227-234.
3. Odin, P., Tingström, A. and Öbrink, B. (1986) Biochem. J. 236, 559-568.
4. Odin, P., Asplund, M., Busch, C. and Öbrink, B. (1988) J. Histochem. Cytochem. 36, 729-739.
5. Aurivillius, M., Hansen, O.C., Lazrek, M.B.S., Bock, E. and Öbrink, B. (1990) FEBS Lett. 264, 267-269.
6. Lin, S.H. and Guidotti, G. (1989) J. Biol. Chem. 264, 14408-14414.
7. Hinoda, Y., Neumaier, M., Hefta, S.A., Drzeniek, Z., Wagener, C., Shively, L., Hefta, L.J.F., Shively, J.E. and Paxton, R.J. (1988) Proc. Natl. Acad. Sci. (USA) 85, 6959-6963.
8. Margolis, R.N., Schell, M.J., Taylor, S.I. and Hubbard, A.L. (1990) Biochem. Biophys. Res. Commun. 166, 562-566.
9. Öbrink, B., Blikstad, I., Hansson, M., Odin, P., Svalander, P. and Tingström, A. (1988) Protoplasma 145, 182-187.
10. Odin, P. and Öbrink, B. (1986) Exp. Cell Res. 164, 103-114.
11. Hixson, D.C., McEntire, K.D. and Öbrink, B. (1985) Cancer Res. 45, 3742-3749.
12. Hixson, D.C. and McEntire, K.D. (1989) Cancer Res. 49, 6788-6794.
13. Mowery, J. and Hixson, D.C. (1991) Hepatology 13, 47-56.
14. Tingström, A., Blikstad, I., Aurivillius, M. and Öbrink, B. (1990) J. Cell Sci. 96, 1725.
15. Hubbard, A., Bartles, J.R. and Braiterman, L.T. (1985) J. Cell Biol. 100, 1115-1125.
16. Becker, A., Neumaier, R., Park, C.S., Gossrau, R. and Reutter, W. (1985) Eur. J. Cell Biol. 39, 417-423.
17. Aurivillius, M. (1990) Thesis, Karolinska Institutet, Stockholm, Sweden.

Björn Öbrink:
Department of Medical Cell Biology, Medical Nobel Institute, Karolinska Institute, Stockholm, Sweden

Cell Surface Galactosyltransferase

β1,4 Galactosyltransferase (GalTase) is one member of the glycosyltransferase family of enzymes responsible for oligosaccharide chain biosynthesis. GalTase has two distinct subcellular distributions resulting in two different biological functions. In the Golgi, GalTase catalyzes the transfer of galactose from UDP-galactose to terminal N-acetylglucosamine residues on oligosaccharides of membrane bound and secretory glycoproteins. GalTase is also found on the plasma membrane of many cells, where it mediates a variety of cell-cell and cell-matrix interactions by binding to appropriate glycoside substrates on adjacent cell surfaces or in the extracellular matrix.

GalTase is a 54-60 kDa type II integral membrane glycoprotein with a C-terminal catalytic domain oriented lumenally in the Golgi compartment and extracellularly on the plasma membrane[1,2]. Most cells contain two size classes of GalTase mRNA that are produced by differential transcription initiation from a single GalTase gene[1,2]. The two classes of GalTase mRNA encode identical proteins except that one has an additional 13 amino acid residue in its N-terminal cytoplasmic domain. Preliminary evidence suggests that this unique 13 amino acid sequence directs the long form of GalTase to the plasma membrane, whereas the shorter GalTase protein remains in the Golgi[3]. A spermatid specific transcript that encodes the long form of the GalTase protein has also been described and results from alternative polyadenylation signals[4]. A catalytically active, proteolytic fragment of GalTase is found in most body fluids including milk, where it interacts with α-lactalbumin to form the lactose synthetase complex[5]. Other than in milk, the function of soluble GalTase is unknown.

GalTase has been detected on the surface of virtually all cells examined, although its function has been studied in most detail on sperm, embryonic cells, mesenchymal cells and neuronal cells[6]. Its general function appears to be as a cell adhesion molecule, whereby it mediates a variety of cell-cell and cell-matrix interactions (Figure 1). It is unknown whether GalTase functions catalytically on the cell surface or acts in a nonenzymatic, lectin like capacity. GalTase expression is restricted to specific plasma membrane domains appropriate for its adhesive function (Figure 2). In this regard, GalTase is localized to broad areas of intercellular contact during embryonal carcinoma cell adhesion[7], but is restricted to the dorsal aspect of the anterior mouse sperm head, where it binds ZP3 oligosaccharides during fertilization[8,9]. GalTase is preferentially localized to newly formed lamellipodia and neurites on mesenchymal cells and neurons, respectively, where it functions during cell spreading[10], migration[11] and neurite outgrowth[12] by binding to *N*-linked oligosaccharides within the E8 domain of **laminin**[13]. During cell migration, laminin induces GalTase expression on the cell surface, at which time it becomes associated with the cytoskeleton[11,14]. Perturbing surface GalTase activity in avian embryos inhibits cell migration and leads to teratological abnormalities, suggesting that surface GalTase functions during embryonic development *in vivo*[6,16]. Surface GalTase activity is elevated on metastatic cells[15] and on mouse sperm and embryos that carry mutant alleles of the T/t complex[6], although the physiological significance of elevated GalTase activity is unknown. There is evidence that surface GalTase plays some undefined role in growth control, since cell proliferation can be inhibited by selectively perturbing surface GalTase activity[17].

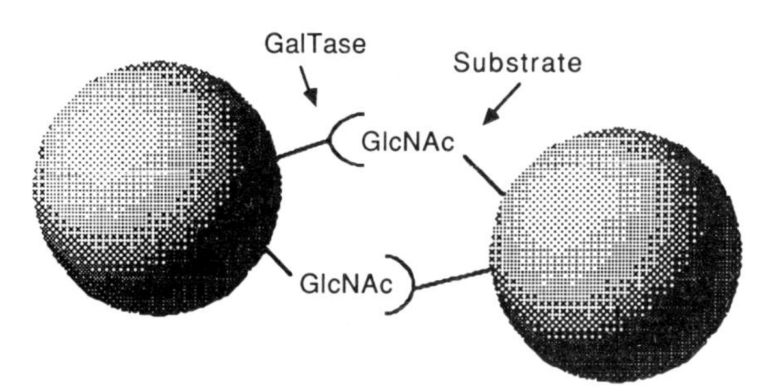

Figure 1. Surface GalTase is thought to participate in intercellular adhesion by binding to its glycoside substrate on adjacent cell surfaces an/or in the extracellular matrix.

PURIFICATION

GalTase is most often purified by affinity chromatography using α-lactalbumin conjugated columns[18,19]. Additional purification is achieved by application on *N*-acetylglucosamine and/or UDP substrate affinity columns. There are many reports of GalTase purified from cell lysates and one report of GalTase purified specifically from the cell surface[20].

ACTIVITIES

GalTase is assayed by the incorporation of radiolabeled galactose, donated from UDP-galactose, to glycoside acceptor substrates terminating in *N*-acetylglucosamine. The radiolabeled galactosylated product can be isolated by high voltage paper electrophoresis, acid precipitation, ion exchange or thin layer chromatography, or can be visualized by autoradiography[6,20]. Its cell adhesion activity can be assessed by reagents that perturb enzyme activity and/or substrate binding, including appropriate antibodies, competitive substrates, α-lactalbumin, hexosaminidase digestion, and UDP-galactose[6-8,10-12].

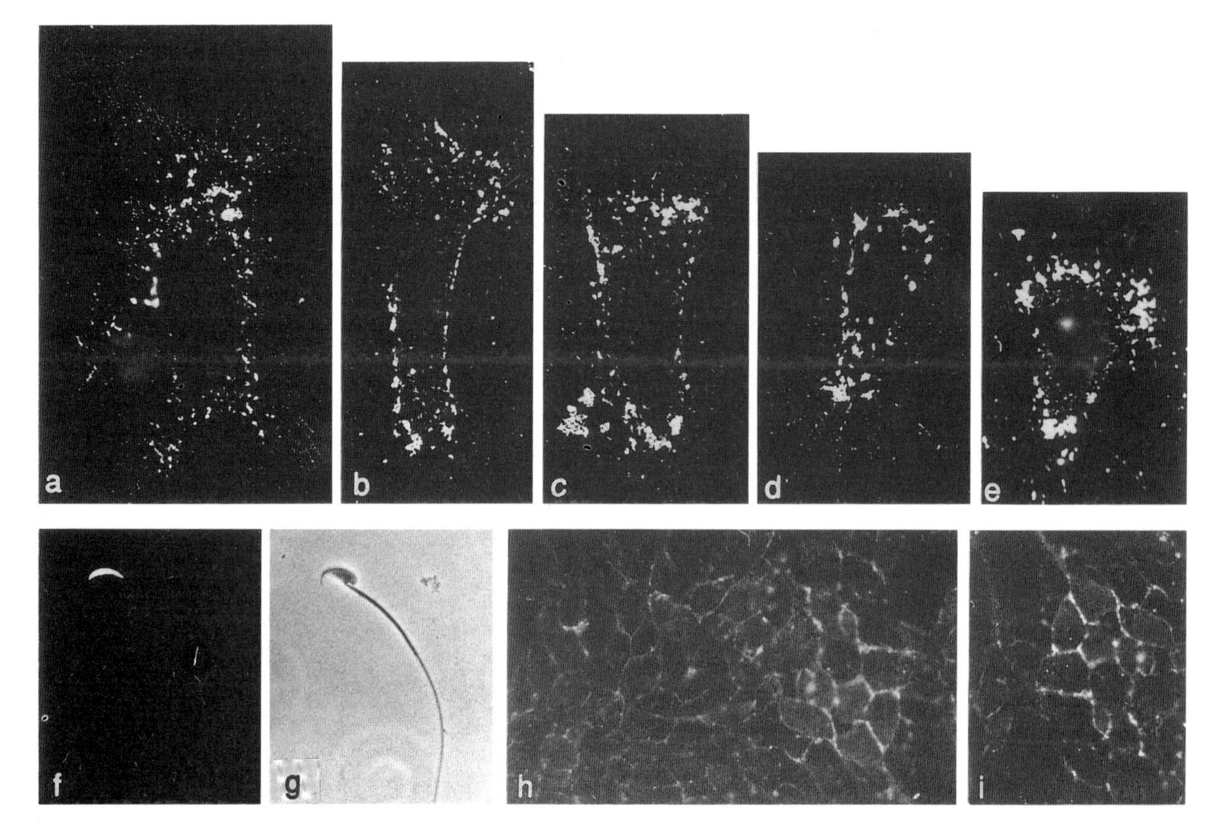

Figure 2. Surface GalTase is restricted to plasma membrane domains appropriate for its adhesive function. As mesenchymal cells polarize to initiate migration, surface GalTase becomes concentrated to the leading and trailing edges of the migrating cell (a-e). GalTase is restricted to the dorsal aspect of the anterior mouse sperm head where it mediates fertilization (f,g), and is localized to areas of intercellular contact during embryonal carcinoma cell adhesion (h,i).

ANTIBODIES

Most investigators have raised antibodies against GalTase using affinity purified bovine milk GalTase as immunogen[7]. The resulting anti-GalTase antiserum has been made monospecific by application to a GalTase affinity column[7]. Polyclonal anti-bovine milk GalTase IgG crossreacts with other mammalian GalTases to varying degrees, and has been useful for immunofluorescence and immunoprecipitation analysis, as well as for inhibiting GalTase-mediated adhesions[6-8,11]. Monoclonal antibodies prepared against bovine GalTase do not crossreact with other species[21]. Antibodies have also been prepared against chicken serum[19] and human milk[22] GalTases. However, whether anti-human milk GalTase antibodies recognize surface GalTase is questionable, since these antibodies recognize carbohydrate epitopes shared with other glycoproteins.

GENES

Full length cDNAs for both forms of GalTase have been published for mouse (J03880)[1], cow (J05217, M25398, X14558 GALT)[2,23,24] and human[25]. The gene for GalTase has been characterized in mouse[26] and is located on mouse chromosome 4[27] and human chromosome 9[28].

REFERENCES

1. Shaper, N.L., Hollis, G.F., Douglas, J.G., Kirsch, I.R. and Shaper, J.H. (1988) J. Biol. Chem. 263, 10420-10428.
2. Russo, R.N., Shaper, N.L. and Shaper, J.H. (1990) J. Biol. Chem. 265, 3324-3331.
3. Lopez, L.C., Youakim, A., Evans, S.C. and Shur, B.D. (1991) J. Biol. Chem. 266, 15984-15991.
4. Shaper, N.L., Wright, W.W. and Shaper, J.H. (1990) Proc. Natl. Acad. Sci. (USA) 87, 791-795.
5. Brodbeck, U., Denton, W.L., Tanahashi, N. and Ebner, K.E. (1967) J. Biol. Chem. 242, 1391-1397.
6. Shur, B.D. (1989) Biochim. Biophys. Acta Rev. Biomembranes 988, 389-409.
7. Bayna, E.M., Shaper, J.H. and Shur, B.D. (1988) Cell 53, 145-157.
8. Lopez, L.C., Bayna, E.M., Litoff, D., Shaper, N.L., Shaper, J.H. and Shur, B.D. (1985) J. Cell Biol. 101, 1501-1510.
9. Miller, D.J., Macek, M.B. and Shur, B.D. (1992) Nature 357, 589-593.
10. Runyan, R.B., Versalovic, J. and Shur, B.D. (1988) J. Cell Biol. 107, 1863-1871.

11. Eckstein, D.J. and Shur, B.D. (1989) J. Cell Biol. 108, 2507-2517.
12. Begovac, P.C. and Shur, B.D. (1990) J. Cell Biol. 110, 461-470.
13. Begovac, P.C., Hall, D.E. and Shur, B.D. (1991) J. Cell Biol. 113, 637-644.
14. Eckstein, D.J. and Shur, B.D. (1990) J. Cell Biol. 111, 423a.
15. Hathaway, H.J. and Shur, B.D. (1992) J. Cell Biol. 117, 369-382.
16. Passaniti, A. and Hart, G.W. (1990) Cancer Res. 50, 7261-7271.
17. Humphreys-Beher, M.G., Schneyer, C.A., Kidd, V.J. and Marchase, R.B. (1987) J. Biol. Chem. 262, 11706-11713.
18. Hill, R.L. and Brew, K. (1975) Adv. Enzymol. 43, 411-490.
19. Hathaway, H.J., Runyan, R.B., Khounlo, S. and Shur, B.D. (1991) Glycobiology 1, 211-221,
20. Shur, B.D. and Neely, C.A. (1988) J. Biol. Chem. 263, 17706-17714.
21. Ulrich, J.T., Schenck, J.R., Rittenhouse, H.G., Shaper, N.L. and Shaper, J.H. (1986) J. Biol. Chem. 261, 7975-7981.
22. Roth, J., Lentze, M.J. and Berger, E.G. (1985) J. Cell Biol. 100, 118-125.
23. Masibay, A.S. and Qasba, P.K. (1989) Proc. Natl. Acad. Sci. (USA) 86, 5733-5737.
24. D'Agostaro, G., Bendiak, B. and Tropak, M. (1989) Eur. J. Biochem. 183, 211-217.
25. Masri, K.A., Appert, H.E. and Fukuda, M.N. (1988) Biochem. Biophys. Res. Commun. 157, 657-663.
26. Hollis, G.F., Douglas, J.G., Shaper, N.L., Shaper, J.H., Stafford-Hollis, J.M., Evans, R.J. and Kirsch, I.R. (1989) Biochem. Biophys. Res. Commun. 162, 1069-1075.
27. Shaper, N.L., Shaper, J.H., Hollis, G.F., Chang, H., Kirsch, I.R. and Zozak, C.A. (1987) Cytogenetics and Cell Genetics 44, 18-21.
28. Duncan, A.M.V., McCorquodale, M.M., Morgan, C., Rutherford, T.J., Appert, H.E. and McCorquodale, D.J. (1986) Biochem. Biophys. Res. Commun. 141, 1185-1188.

Barry D. Shur:
Department of Biochemistry and Molecular Biology,
The University of Texas,
M.D. Anderson Cancer Center,
Houston, TX, USA

Connexins

The connexins comprise a family of gap junction structural proteins. Hexameric assemblies of these proteins in the plasma membranes of adjacent cells interact to form intercellular channels. These channels are permeable to a variety of molecules smaller than 1200-2000 Daltons. They play an essential role in the coordination of contraction in smooth and cardiac muscle and compose electrical synapses which couple some neurons. In addition, it is likely that communication through gap junctions may regulate aspects of cell proliferation and differentiation.

Structural models based on X-ray diffraction (Figure 1) show that gap junctions are aggregations of intercellular channels which directly connect the cytoplasms of adjacent cells. Unlike other membrane channels, intercellular channels span two plasma membranes and require the contribution of hemi-channels, called connexons, from both participating cells. Two connexons interact in the extracellular space to form the complete intercellular channel. Each connexon is composed of six similar or identical proteins, which have been termed connexins.

A variety of studies indicate that the connexins are a multi-gene family of highly related proteins. DNAs encoding nine rat connexins have been reported[1-6]. Homologs of some of these, in addition to several possibly unique connexins, have been isolated from human[7,8], bovine[9], chicken[10,11], *Xenopus*[12,13] and murine[14] DNAs. To distinguish individual connexins, a system of nomenclature based on species of origin and molecular mass predicted by cDNA analysis has been suggested[3]. For example, the connexins expressed in rat heptocytes were named rat connexin (Cx) 26 and Cx32. An alternative nomenclature termed these proteins β_2 and β_1[15], respectively.

Immunocytochemistry and functional expression have in several cases confirmed that connexins are gap junctional structural proteins (Figure 2, Table 1).

The suggested topological orientation of connexin proteins in the plasma membrane is depicted in Figure 3 and is supported experimentally by hydropathy analysis[2,3], immunocytochemical localization of defined epitopes[16,17] and controlled proteolysis/Edman sequence analysis[18-20]. The four transmembrane, two extracellular and short N-terminal cytoplasmic domains are well conserved while the central (Figure 3 A) and C-terminal cytoplasmic (Figure 3 B) domains are highly variable in sequence and in size. The high degree of conservation in extracellular domains may permit different connexins in adjacent cells to functionally interact. This has been demonstrated experimentally for rat Cx32 and 43[21]. Conversely, the lack of similarity in the major cytoplasmic regions may underlie the differences in modulation that junctions in different tissues display. For example, **pp60**$^{v\text{-}src}$ inhibits communication resulting from rat Cx43 expression by inducing tyrosine phosphorylation at residue 265. In contrast, neither tyrosine phosphorylation nor inhibition of communication occurs with rat Cx32[22].

The connexins are expressed in complicated, overlapping patterns, which are summarized in table 1[23]. Specific connexins may exhibit relatively broad (e.g. rat Cx32 and Cx43) or highly restricted distributions (e.g. rat Cx46 and Cx33). One cell may express two connexins, which are both detected in the same junctional maculae (e.g. Cx26/32 in hepatocytes; Cx26/43 in leptomeningeal cells;

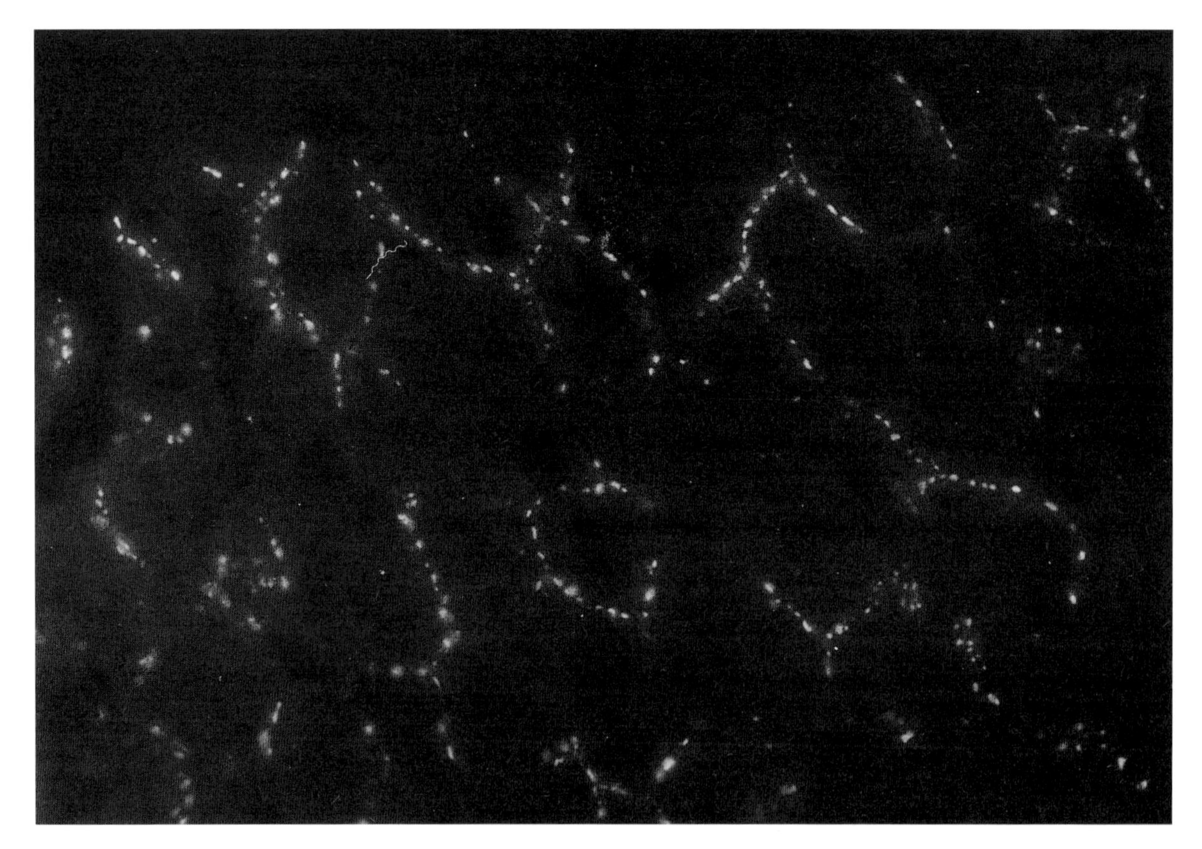

Figure 1. Immunofluorescence localization of Cx32 in a frozen section of rat liver.

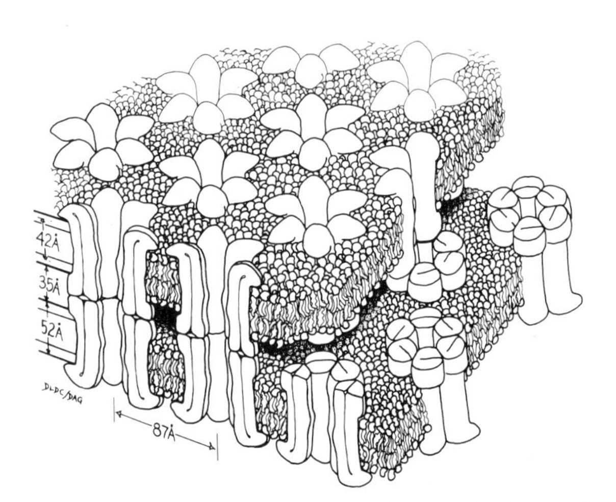

Figure 2. Low resolution diagram of one of the forms of isolated gap junctions from mouse liver, based on data from X-ray diffraction and electron microscopy. The protein subunits are arrayed in hexamers, or connexons, in each of the paired junctional membranes. (Reprinted with permission from the Jounal of Cell Biology, 1977, Vol. 74, p. 643).

Cx46/MP70 (Cx50) in lens fibres). However, it is not yet clear if different connexins are present in the same connexon. Currently, the biological significance of these patterns of distribution is not clear.

There are significant differences in the functional properties of the channels composed of different connexins. Connexin diversity may in part account for differences in the ability of cyclic nucleotides, pH and voltage gradients to modulate intercellular communication[23]. Differences in single intercellular channel conductances have also been observed[23]. Although not directly demonstrated, this may indicate differences in permeability to biologically significant molecules.

PURIFICATION

Methods for bulk purification of gap junction maculae have been reported using rat hepatocytes[23,24], rat myocardium[25] and bovine lens fibres[26].

ANTIBODIES

A variety of polyclonal and monoclonal antibodies which are specific for individual connexins have been reported; rat Cx26[1]; rat Cx32[16,17]; rat Cx43[20,27]; rat Cx46[4]; sheep MP70[28] (rat Cx50).

Table

CONNEXIN	PROOF OF CELL/ORGAN	LOCATION IDENTITY
Rat Cx26 (β_2)	Immunocytochemistry, functional expression	Hepatocytes, mammary epithelium, leptomeninges, pinealocytes, endometrium
Rat Cx31	Homology to connexins	Skin, placenta, Harderian gland
Rat Cx31.1	Homology to connexins	Skin, esophagus
Rat Cx32 (β_1)	Immunocytochemistry,	Hepatocytes, various secretory and functional expression absorptive epithelia, retinal horizontal cells, oligodendrocytes
Rat Cx33	Homology to connexins	Sertoli cells
Rat Cx37	Functional expression	?
Rat Cx40	Homology to connexins	Vascular endothelium
Rat Cx43 (α_1)	Immunocytochemistry functional expression	Smooth and cardiac muscle, fibroblasts, ovarian granulosa, lens and corneal epithelia, pancreatic β-cells, distal convoluted tubules, astrocytes, leptomeninges
Rat Cx46	Immunocytochemistry	Lens fibres, myocardium, kidney
Xen Cx38	Functional expression up to stage 11	*Xenopus* oocytes and embryos
Chick Cx42	Homology to connexins	Myocardium, liver, stomach, pectoral muscle
Chick Cx45	Homology to connexins	Embryonic myocardium, stomach
Sheep MP70 (Rat Cx50)	Imunocytochemistry, Lens fibres homology to connexins	

Table 1. Expression pattern of connexins[25] (unpublished observations).

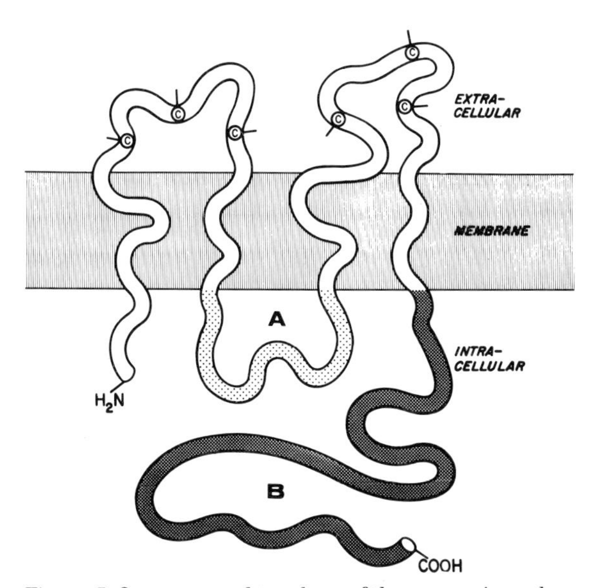

Figure 3. Structure and topology of the connexins relative to the junctional plasma membrane. Unshaded portions represent regions of connexins which are relatively conserved among connexins. Two extracellular domains each contain three invariant cysteines. Cytoplasmic domains A and B differ among connexins in both sequence and length.

GENES

cDNA sequences are available for rat Cx26[1]; rat and human Cx32[2,7] (X04070, X04325) and its *Xenopus* homolog Cx30[13] (Y00791); rat Cx31[6] (M59936); rat Cx31.1[6] (M76533); rat Cx33[6] (M76534); rat[6] and mouse[14] Cx37 (M76532, X57971); rat Cx40 (M57635); rat[3] (M19317), human[8], bovine[9], chicken[11] (M29003) and *Xenopus*[13] (X17243) Cx43; rat Cx46[4] (X57970); *Xenopus* Cx38[12] (J03091); chicken Cx42 and 45[10] (M35043, M35044). Analysis of genomic clones of rat Cx32[29] and Cx43[8] indicate that the coding regions do not contain introns. Southern analysis of the other connexins suggests that this feature is generally shared.

REFERENCES

1.Zhang, J.-T. and Nicholson, B.J. (1989) J. Cell Biol. 109, 3391-3401.
2.Paul, D.L. (1986) J. Cell Biol. 103, 123-134.
3.Beyer, E.C., Paul, D.L. and Goodenough, D.A. (1987) J. Cell Biol. 105, 2621-2629.
4.Paul, D.L., Swenson, K.I., Takemoto, L.J. and Goodenough, D.A. (1991) J.Cell Biol. 115, 1077-1089.
5.Hoh, J.H., John, S.A. and Revel, J.-P. (1991) J. Biol. Chem. 266, 6524-6531.
6.Haefliger, J.-A., Bruzzone, R., Jenkins, N.A., Gilbert, D.J., Copeland, N.G. and Paul, D.L. (1992) J. Biol. Chem. 267, 2057-2064.
7.Kumar, N. and Gilula, N.B. (1986) J. Cell Biol. 103, 767-776.

8.Fishman, G.I., Spray, D.C. and Leinwand, L.A. (1990) J. Cell Biol. 111, 589-598.
9.Lash, J.A., Critser, E.S. and Pressler, M.L. (1990) J. Biol. Chem. 265, 13113-13117.
10.Beyer, E.C. (1990) J. Biol. Chem. 265, 14439-14443.
11.Musil, L.S., Beyer, E.C. and Goodenough, D.A. (1990) J. Membr. Biol. 116, 163-175.
12.Ebihara, L., Beyer, E.C., Swenson, K.I., Paul, D.L. and Goodenough, D.A. (1988) Science 243, 1194-1195.
13.Gimlich, R.L., Kumar, N.M. and Gilula, N.B. (1990) J. Cell Biol. 110, 597-605.
14.Willecke, K., Heykes, R., Dahl, E., Stutenkemper, R., Hennemann, H., Jungbluth, S., Suchyna, T. and Nicholson, B.J. (1991) J. Cell Biol. 114, 1049-1058.
15.Risek, B., Guthrie, S., Kumar, N. and Gilula, N.B. (1990) J. Cell Biol. 110, 269-282.
16.Goodenough, D.A., Paul, D.L. and Jesaitis, L. (1988) J. Cell Biol. 107, 1817-1824.
17.Milks, L.C., Kumar, N.M., Houghten, R., Unwin, N. and Gilula, N.B. (1988) EMBO J. 7, 2967-2975.
18.Zimmer, D.B., Green, C.R., Evans, W.H. and Gilula, N.B. (1987) J. Biol. Chem. 262, 7751-171763.
19.Hertzberg, E.L., Disher, R.M., Tiller, A.A., Zhou, Y. and Cook, R.G. (1988) J. Biol. Chem. 263, 19105-19111.
20.Yancey, S.B., John, S.A., Ratneshwar, L., Austin, B.J. and Revel, J.-P. (1989) J. Cell Biol. 108, 2241-2254.
21.Swenson, K.I., Jordan, J.R., Beyer, E.C. and Paul, D.L. (1989) Cell 57, 145-155.
22.Swenson, K.I., Piwnica-Worms, H., McNamee, H. and Paul, D.L. (1990) Cell Regulation 13, 989-1002.
23.Bennett, M.V.L., Barrio, L., Bargiello, T.A., Spray, D.C., Hertzberg, E.L. and Saez, J. (1991) Neuron 6, 305-320.
24.Hertzberg, E.L. (1984) J. Biol. Chem. 259, 9936-9943.
25.Manjunath, C.K., Nicholson, B.J., Teplow, D., Hood, L., Page, E. and Revel, J.-P. (1987) Biochem. Biophys. Res. Comm. 142, 228-234.
26.Paul, D.L. and Goodenough, D.A. (1983) J. Cell Biol. 96, 625-632.
27.Beyer, E.C., Kistler, J., Paul, D.L. and Goodenough, D.A. (1988) J. Cell Biol. 108, 595-605.
28.Kistler, J., Christie, D. and Bullivant, S. (1988) Nature 331, 721-723.
29.Miller, T., Dahl, G. and Werner, R. (1988) Biosci. Rep. 8, 445-464.

■ *David L. Paul:*
Dept. of Anatomy and Cellular Biology,
Harvard Medical School,
Boston, MA, USA

Desmocollins

Desmocollins are a related pair of transmembrane glycoproteins which together with desmoglein constitute the major glycoproteins of desmosomes. They are involved in desmosomal adhesion and belong to the cadherin family of Ca^{2+}-dependent cell adhesion molecules. Their extracellular domains are located in the desmosomal intercellular material or desmoglea and their cytoplasmic domains in the desmosomal plaque.

Desmocollins are major glycoprotein components of desmosomes. These adhesive intercellular junctions are found in epithelia, cardiac muscle, follicular dendritic cells and the pia and arachnoid of the meninges. The major desmosomal glycoproteins were first clearly defined in bovine nasal epidermis[1,2] (Figure 1). Components immunologically related to desmocollins have been demonstrated in all desmosome bearing tissues examined from mammalian, avain, reptilian and anuran amphibian species[3-5]. The relative molecular weights of desmocollins from various sources and the number of isoforms are somewhat variable: immunoblotting of bovine nasal epidermal desmosomes and MDCK cells usually reveals two polypeptide of 115 and 107 kDa, desmocollins 1 and 2, respectively. The close relationship between them has been established by crossreactivity of monoclonal and polyclonal antibodies, peptide mapping and amino acid analysis[3,6-8]. Desmocollins have been localised within the desmosome by immuno electronmicroscopy[9,10] (Figure 2). Their transmembrane organization has been confirmed by whole cell trypsinisation and phosphorylation combined with immunoprecipitation, and a cytoplasmically reacting monoclonal antibody[11]. In MDCK cells, desmocollin 1 becomes serine phosphorylated in its cytoplasmic domain but the smaller cytoplasmic domain of desmocollin 2 does not[11]. The desmocollins have been shown to bind calcium[10,12].

N-terminal amino acid sequencing demonstrated a relationship between the desmocollins and the **cadherin** family of Ca^{2+}-dependent cell adhesion molecules[8] (Figure 3). Immuno electronmicroscopy with anti-peptide antibodies located the N-terminus extracellularly, in the desmosomal intercellular material or desmoglea[8]. The amino acid sequences of desmocollins deduced from cDNA clones of the bovine nasal proteins reveal transmembrane proteins with single membrane spanning domains[13-15]. The extracellular domains show the same organization of five repeats as found in the cadherins, and have 39.4% sequence identity with bovine N-cadherin. Putative Ca^{2+}-binding sites are conserved, there are three N-glycosylation sites and the His-Ala-Val putative adhesion site of cadherins is replaced by Tyr-Ala-Thr. The homology with cadherin is greatly reduced in the cytoplasmic domain (23%). The sequence begins with a signal peptide of 28 amino acids and a precursor sequence of 104 amino acids. The calculated molecular weights of the mature polypeptides are 85,200 and 79,081. The larger

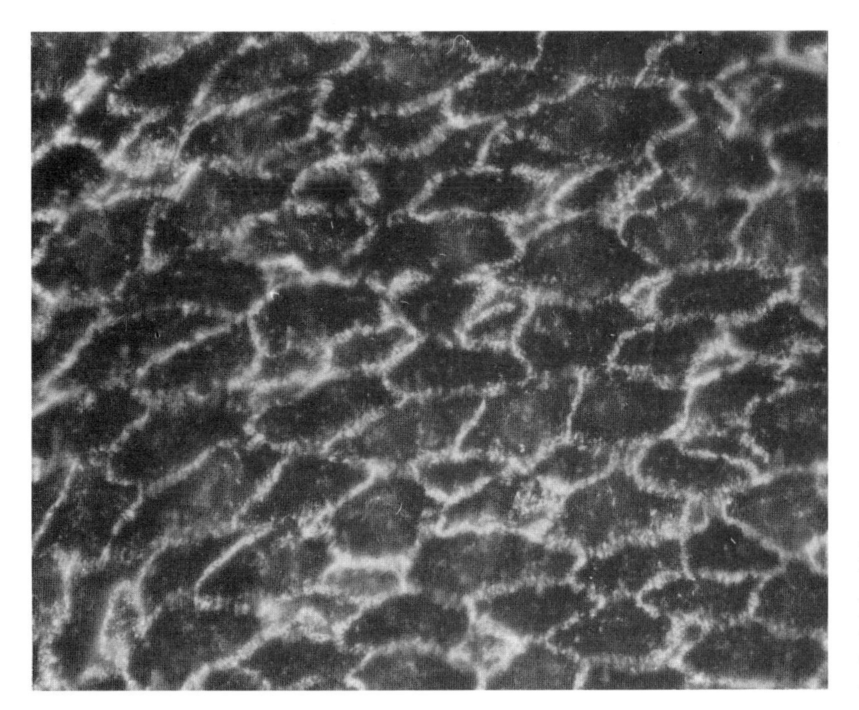

Figure 1. Frozen section (5 µm) of bovine nasal epidermis showing pattern of staining with monoclonal antibody, 52-3D, to the cytoplasmic domains of desmocollins 1 and 2 (x620).

Figure 2. Ultrathin frozen section of bovine nasal epidermis showing immunogold localization of polyclonal antiserum to desmocollins 1 and 2. The majority of gold particles are situated over the desmoglea with a few over the plasma membrane and desmosomal plaque (x120,000).

cytoplasmic domain of desmocollin 1 is generated by splicing out of a 46bp sequence coded by a single exon[13,16]. This removes an in frame stop codon, extending the open reading frame to encode a cytoplasmic domain larger by 54 amino acids, including a consensus serine phosphorylation site. Thus the desmocollins appear as hybrid molecules with extracellular domains specialized for Ca^{2+}-dependent adhesion and cytoplasmic domains for desmosomal plaque formation.

Two different subtypes of desmocollins, the products of different genes, have been found. Those described above (Type 1) are expressed in tongue mucosa and subrabasally

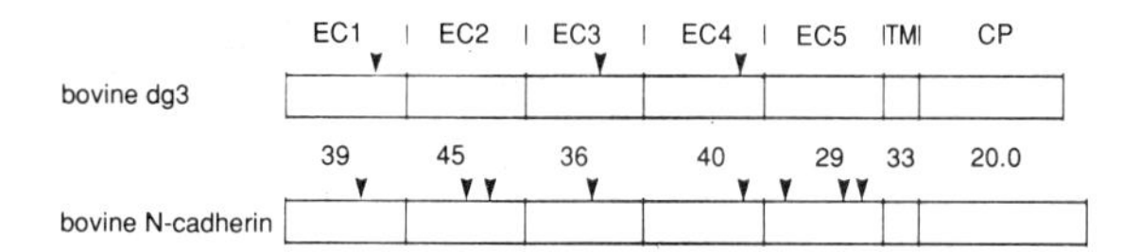

Figure 3. Diagram showing per cent identity of amino acid sequence between desmocollin 2 (dg3) and bovine N-cadherin. Both molecules are represented as five extracellular repeats (EC), a transmembrane domain (TM) and cytoplasmic domain (CP). Per cent identities between the corresponding regions are indicated by the numbers. Arrowheads indicate the consensus N-linked glycosylation sites.

in epidermis[14,17]. A subtype (Type 2) showing ~50% identity to Type 1 is expressed in basal keratinocytes and some simple epithelial cell lines[16,18].

Desmocollins 1 and 2 are also known as desmoglein IIa and IIb, desmosomal glycoproteins 2 and 3 and bands 4a and 4b.

PURIFICATION

Bovine nasal epidermis and tongue epithelium provide rich sources of desmosomes, from which they are readily extracted with nonionic detergent in citrate buffer[1,2]. Further solubilization of desmosomal glycoproteins is difficult. Glycoprotein enriched fractions may be obtained by treatment with metrizamide[2] or 9 M urea[19,20]. Individual glycoproteins may then be purified by preparative SDS-PAGE[3]. A method for purification of desmocollins by isolation of desmosomes in 6 M guanidine HCl followed by hydroxylapatite chromatography in SDS has been described[21].

ACTIVITIES

Desmocollins are believed to be directly involved in desmosomal adhesion[2,22].

ANTIBODIES

Polyclonal sera showing tissue and species reactivity are available[3,5,11,23]. Monoclonal antibodies to bovine proteins have been described[6]. A monoclonal antibody to the cytoplasmic domains and showing species crossreactivity is available[11]. N-terminal anti-peptide antisera showing tissue and species crossreactivity have been described[8].

GENES

Bovine nasal epidermis (Type 1), GenBank Data Base accession Nos. X56966/7/8; M61750. Bovine nasal epidermis (Type 2), GenBank Data Base accession No. M81190. Human keratinocyte (Type 2), X56807.

REFERENCES

1. Skerrow, C.J. and Matoltsy, A.G. (1974) J. Cell Biol. 63, 515-523 and 524-530.
2. Gorbsky, G. and Steinberg, M.S. (1981) J. Cell Biol.90, 243-248.
3. Cowin, P. and Garrod, D.R. (1983) Nature 302, 148-150.
4. Cowin, P., Mattey, D.L. and Garrod, D.R. (1984) J. Cell Sci. 66, 119-132.
5. Suhrbier, A. and Garrod, D.R. (1986) J. Cell Sci. 81,223-242.
6. Cohen, S.M., Gorbsky, G. and Steinberg, M.S. (1983) J. Biol. Chem. 252, 1102-1106.
7. Kapprell, H.P., Cowin, P., Franke, W.W., Postingl, H. and Opferkuch, H.J. (1985) Eur. J. Cell Biol. 36, 217-229.
8. Holton, J.L., Kenny, T.P., Legan, P.K., Collins, J.E., Keen, J.N., Sharma, R. and Garrod, D.R. (1990) J. Cell Sci. 97, 239-246.
9. Miller, K., Mattey, D., Measures, C., Hopkins, C. and Garrod, D.R. (1987) EMBO J. 6, 3655-3661.
10. Steinberg, M.S., Shida, H., Guidice, G.J., Patel, N.H. and Blaschuk, O. (1987) In: Junctional Complexes of Epithelial Cells (ed. G. Bock and S.Clarke) pp 3-25. Ciba Foundation Symp. 125, John Wiley and Sons, Chichester.
11. Parrish, E.P., Marston, J.E., Mattey, D.L., Measures, H.R., Venning, R. and Garrod, D.R. (1990) J. Cell Sci. 96, 239-248.
12. Mattey, D.L., Suhrbier, A., Parrish, E.P. and Garrod, D.R. (1987) In: Junctional Complexes of Epithelial Cells (ed. G. Bock and S. Clarke) pp 49-65. Ciba Foundation Symp. 125, John Wiley and Sons, Chichester.
13. Collins, J.E., Legan, P.K., Kenny, T.P., MacGarvie, J.E., Holton, J.L. and Garrod, D.R. (1991) J: Cell Biol. 113, 381-391.
14. Koch, P.J., Goldschmidt, M.D., Walsh, M.J., Zimbelmann, R., Schmelz, M. and Franke, W.W. (1991) Differentiation 47, 29-36.
15. Mechanic, S., Raynor, K., Hill, J.E. and Cowin, P. (1991) Proc. Natl. Acad. Sci. (USA) 88, 4476-4480.
16. Parker, A.E., Wheeler, G.N., Arnemann, J., Pidsley, S.C., Rutman, A.J., Thomas, C.L., Ataliotis, P., Rees, D.A., Magee, A.I. and Buxton, R.S. (1991) J. Biol. Chem. 266, 10438-10445.
17. King, I.A., Magee, A.I., Rees, D.A. and Buxton, R.S. (1991) FEBS Lett. 286, 9-12.
18. Koch, P.J., Goldschmidt, M.D., Zimbelmann, R. and Troyanovsky, R. (1992) Proc. Natl. Acad. Sci. (USA) 89, 353-357.
19. Franke, W.W., Kapprell, H.-P. and Mueller, H. (1983) Eur. J. Cell Biol. 32, 117-130.
20. Skerrow, C.J., Hunter, I. and Skerrow, D. (1987) J. Cell Sci. 87, 411-421.
21. Blaschuk, O.W., Manteuffel, R.L. and Steinberg, M.S. (1986) Biochem. Biophys. Acta 883, 426-431.
22. Cowin, P., Mattey, D.L. and Garrod, D.R. (1984) J. Cell Sci. 70, 41-60.
23. Penn, E.J., Burdett, I.D.J., Hobson, C., Magee, A.I. and Rees, D.A. (1987) J. Cell Biol. 105, 2327-2334.

David R. Garrod:
Cancer Research Campaign,
Epithelial Morphogenesis Research Group,
Department of Cell and Structural Biology,
University of Manchester,
Manchester, M13 9PT, UK

Desmoglein

Desmoglein is a hallmark glycoprotein of desmosomes (maculae adhaerentes) and belongs, together with the desmocollins, to the larger family of Ca^{2+}-dependent cell adhesion molecules, the cadherins. It is a transmembrane glycoprotein with an outer domain, contributing to the formation and maintenance of specific intercellular contacts (junctions of the desmosomal category), and with an extensive cytoplasmic domain that is an integral part of the desmosomal plaque structure.

Desmoglein (synonyms: band 3 polypeptide, dg 1) is a glycosylated polypeptide of 165 kDa (Table), showing in some SDS-PAGE systems slight differences of electrophoretic mobility in desmosomes from different tissues and cell lines[1-3]. Upon treatment of cells with tunicamycin, to inhibit N-glycosylation, it has a lower apparent molecular mass of ~150 kDa, and upon translation of mRNAs *in vitro* a similar molecular mass is observed[4,5] (Table). The carbohydrate containing portion of this molecule contributes to the intercellular "cement" ("glue", desmoglea), including the "midline" region of the desmosome[6-9], and the carbohydrate composition has been determined for bovine snout epithelial desmosomes[7]. The portion of the molecule which is located on the cytoplasmic side spans the desmosomal plaque structure as it is partially exposed on the plaque's cytoplasmic surface[8-11].

By immunofluorescence microscopy, desmoglein can be identified in epithelial as well as in certain nonepithelial cells[10-12], in punctate patterns (Figure 1a-c), mostly at cell-cell boundaries, representing desmosomes. Cytoplasmically located fluorescent "dots" may indicate desmosomal domains internalized in vesicles[13,14].

Desmoglein has been identified as a constitutive component of desmosomes in a variety of tissues, including nonstratified (simple), complex and stratified epithelia, urothelium, myocardial and Purkinje fibre cells of the heart, arachnoidal cells of meninges, and optic nerve astrocytes of some lower vertebrates[15].

The amino acid sequence of desmoglein and its precursor has been deduced from the sequence of cDNA clones encoding a bovine muzzle desmoglein (Table). Comparison with other protein sequences showed that desmoglein is a member of the **cadherin** family of cell adhesion glycoproteins; its cytoplasmic domain, however, is much longer and exceeds that of the other cadherins by more than 270 amino acids, revealing some sequence features which are exclusive for desmoglein (Figure 2)[8,16,17].

Desmoglein has also been identified as the major antigen recognized by autoantibodies from patients suffering from *pemphigus foliaceus*[18].

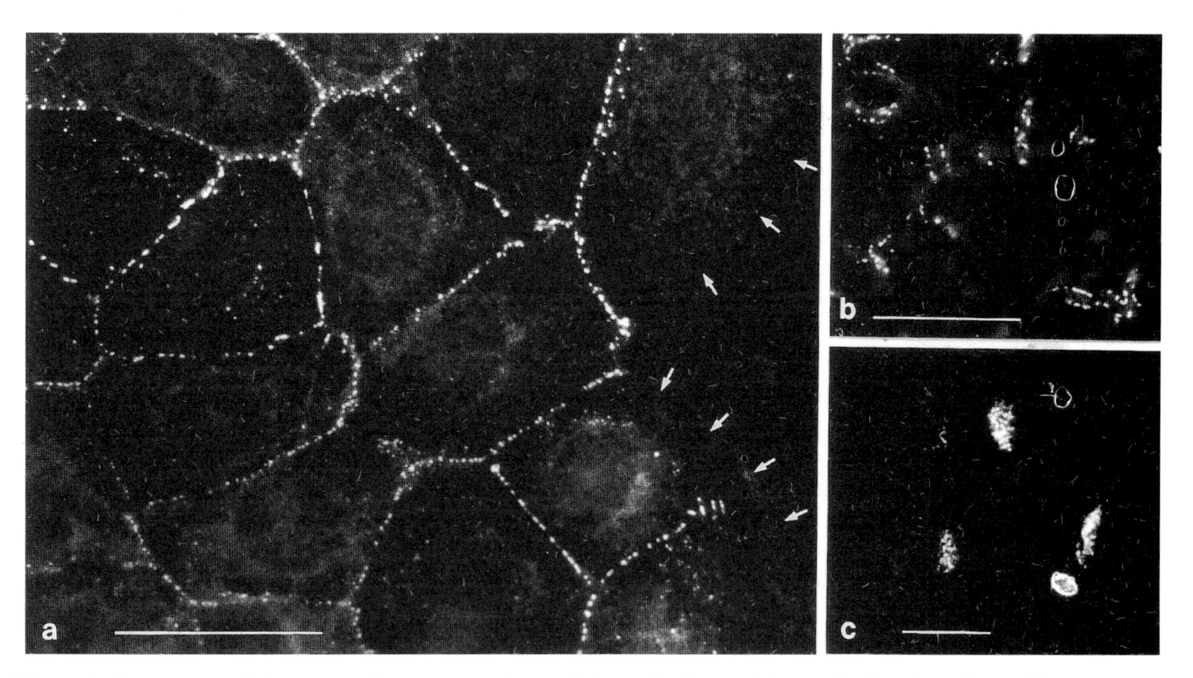

Figure 1. Appearance of desmosomal structures after staining with desmoglein antibodies in various cells or tissues: (a) cultured bovine kidney epithelial cells of line MDBK; (b) along bile canaliculi of rat liver; (c) intercalated disks of bovine myocardial cells. Arrows demarcate free cell boundaries. Bar 25 µm.

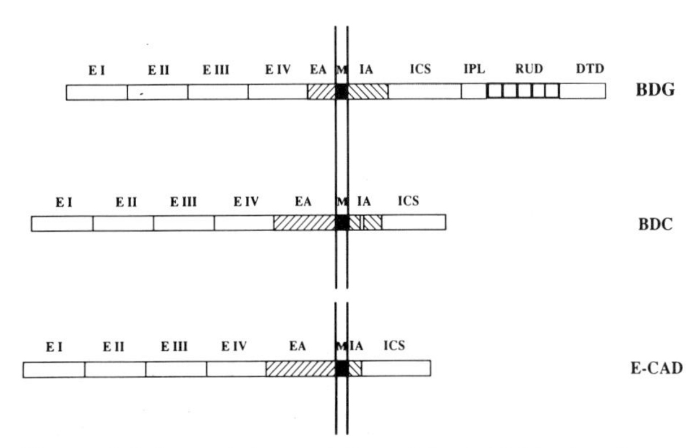

Figure 2. Schematic presentation of the individual molecular domains in a bovine epidermal desmoglein (BDG), compared with a **desmocollin** (BDC) and murine E-cadherin (E-CAD): EI-EIV, extracellular repeating elements; EA, extracellular anchoring domain; M, membrane-spanning domain; IA intracellular anchoring domain (the IA segment of BDC contains a special insertion); ICS, intracellular cadherin-type segment; IPL; "intracellular proline-rich linker" region; RUD, domain containing five repeating units of ~29 amino acids each; DTD, desmoglein-specific terminal domain (for amino acid sequences see references **8,16,17,25**).

■ PURIFICATION

Desmoglein can be purified by elution from electrophoretically separated polypeptides of isolated desmosomes. The most common source of desmosomes for preparative purposes is bovine muzzle epidermis[6,9,19].

■ ACTIVITIES

Desmoglein is believed to promote intercellular adhesion[5,15,20] in a Ca^{2+}-dependent manner.

TABLE

MOLECULAR DATA OF DESMOGLEIN (DG) AND DESMOCOLLIN (DC) IN COMPARISON WITH E-CADHERIN (E-CAD, UVOMORULIN)

	Precursor Protein	Mature, Processed Protein	
	amino / mol. wt. acids	amino / mol. wt. acids	App. M_r value $(10^3)^+$
DG-M (bovine)[1-3]	1043* / 112,242*	994 / 106,405	165
DC-M (bovine)[2]	n. d. / n. d.	761 / 85,006	130/115
E-CAD (murine)[4]	884* / 98,255*	728 / 80,350	120

Abbreviations: M, as isolated from muzzle epithelium; mol. wt., molecular weight (calculated), n.d., not determined.
* Including the terminal methionine
+ Estimated from SDS-PAGE
1 Koch et al.: Eur. J. Cell Biol. 53:1-12 (1990)
2 Koch et al.: (1991) Differentiation 47, 29-36
3 Koch et al.: (1991) Eur. J. Cell Biol., 55, 200-208
4 Ringwald et al.: EMBO J. 6:3647-3653 (1987)

■ ANTIBODIES

Characterized murine monoclonal antibodies with a wide range of crossspecies reactivity as well as more restricted ones, recognizing only stratified epithelial cells of certain species, have been published, in addition to specific sera[10,11,21] (some are commercially available from Progen Biotechnics, Heidelberg, and Boehringer Mannheim).

■ GENES

Full length cDNAs for bovine epidermal desmoglein and its precursor molecule have been published[17] (GenBank X57784). The corresponding human sequence has also been published[22] as well as two other human desmoglein types[17,23,24]. One of these has been reported to be the major antigen of the human skin disease, *Pemphigus vulgaris*.

■ REFERENCES

1. Franke, W.W., Schmid, E., Grund, C., Müller, H., Engelbrecht, I., Moll, R., Stadler, J. and Jarasch, E.-D. (1981) Differentiation 20, 217-241.
2. Giudice, G.J., Cohen, S.M., Patel, N.H. and Steinberg, M.S. (1984) J. Cell Biochem. 25, 35-45.
3. Suhrbier, A. and Garrod, D. (1986) J. Cell Sci. 81, 223-242.
4. Kapprell, H.-P., Duden, R., Owaribe, K., Schmelz, M. and Franke, W.W. (1990) In Morphoregulatory Molecules (G.M. Edelman et al. eds.) Wiley, New York, 285-314.
5. Penn, E.J., Hobson, C., Rees, D.A. and Magee, A.I. (1987) J. Cell Biol. 105, 57-68.
6. Gorbsky, G. and Steinberg, M.S. (1981) J. Cell Biol. 90, 243-248.
7. Kapprell, H.-P., Cowin, P., Franke, W.W., Ponstingl, H. and Opferkuch, H.J. (1985) Eur. J. Cell Biol. 36, 217-229.
8. Koch, P.J., Walsh, M.J., Schmelz, M., Goldschmidt, M.D., Zimbelmann, R. and Franke, W.W. (1990) Eur. J. Cell Biol. 53, 1-12.
9. Steinberg, M.S., Shida, H., Giudice, G.J., Shida, M., Patel, N.H. and Blaschuk, O.W. (1987) Ciba Found. Symp. 125, 3-25.
10. Schmelz, M., Duden, R., Cowin, P. and Franke, W.W. (1986) Eur. J. Cell Biol. 42, 177-183.
11. Schmelz, M., Duden, R., Cowin, P. and Franke, W.W. (1986) Eur. J. Cell Biol. 42, 184-199.
12. Cowin, P. and Garrod, D.R. (1983) Nature 302, 148-150.
13. Duden, R. and Franke, W.W. (1988) J. Cell Biol. 107, 1049-1063.
14. Kartenbeck, J., Schmelz, M., Franke, W.W. and Geiger, B. (1991) J. Cell Biol. 113, 881-892.
15. Schwarz, M.A., Owaribe, K., Kartenbeck, J. and Franke, W.W. (1990) Annu. Rev. Cell Biol. 6, 461-491.
16. Koch, P.J., Goldschmidt, M.D., Walsh, M.J., Zimbelmann, R., Schmelz, M. and Franke, W.W. (1991) Differentiation 47, 29-36.
17. Koch, P.J., Goldschmidt, M.D., Walsh, M.J., Zimbelmann, R. and Franke, W.W. (1991) Eur. J. Cell Biol. 55, 200-208.
18. Korman, N.J., Eyre, R.W., Klaus-Kovtun, V. and Stanley, J.R. (1989) N. Engl. J. Med. 32, 631-635.
19. Skerrow, C.J. and Matoltsy, A.G. (1974) J. Cell Biol. 63, 524-531.
20. Garrod, D.R., Parrish, E.P., Mattey, D.L., Marsteon, J.E., Measures, H.R. and Vilela, M.J. (1990) In: Morphoregulatory Molecules (G.M. Edelman et al. eds.) Wiley, New York, 315-339.

21.Vilela, M.J., Parrish, E.P., Wright, D.H. and Garrod, D.R. (1987) J. Pathol. 153, 365-375.
22.Parker, A.E., Wheeler, G.N., Arnemann, J., Pidsley, S.C., Ataliotis, P., Thomas, C.L., Rees, D.A., Magee, A.I. and Buxton, R.S. (1991) J.Biol. Chem. 266, 10438-10445.
23.Amagai, M., Klaus-Kovtun, V. and Stanley, J.R. (1991) Cell 67, 869-877.
24.Koch, P.J., Goldschmidt, M.D., Zimbelmann, R., Troyanovsky, R. and Franke, W.W. (1992) Proc. Natl. Acad. Sci. (USA) 89, 353-357.
25.Ringwald, M., Schuh, R., Vestweber, D., Eistetter, H., Lottspeich, F., Engel, J., Dölz, R., Jähnig, F., Epplen, J., Mayer, S., Müller, C. and Kemler, R. (1987) EMBO J. 6, 3647-3653.

J. Kartenbeck, P.J. Koch and W.W. Franke:
Institute of Cell and Tumour Biology,
German Cancer Research Center,
D-6900 Heidelberg, Germany

Fasciclin I

Fasciclin I[1,2] is a homophilic, Ca^{2+}-independent cell adhesion[3] molecule in grasshopper and Drosophila that is expressed on a subset of growth cones, axon pathways, and glia during embryonic development[1,2]; it is also expressed at other times and places outside of the developing nervous system[1,2]. During neuronal development, fasciclin I appears to play a role in growth cone guidance and axon fasciculation[4,5].

The fasciclin I protein was first identified in the grasshopper embryo as a result of a monoclonal antibody (mAb) screen[1]; it is a membrane associated glycoprotein with an apparent molecular weight of 70,000 that is expressed on the surface of a specific subset of axon fascicles in both the CNS and the PNS during embryogenesis[1,6]. It is expressed in a similar pattern during *Drosophila* embryogenesis, including on many nonneuronal tissues such as the hindgut, the salivary glands, and the ovaries[2].

Fasciclin I cDNAs were initially isolated in grasshopper using oligonucleotides[6] and subsequently in *Drosophila* using a low stringency screen[2]. The cDNA sequences of grasshopper and *Drosophila* fasciclin I encode for polypeptides of 662 and 652 amino acids, respectively, that have no sequence homology to any known polypeptide[2]. Both proteins are comprised of four tandem domains of 150 amino acids each, identified by virtue of their weak homology to each other and by the presence of more highly conserved ~40 amino acid repeats at the end of the second, third, and fourth domains. The last C-terminal 30 amino acids of the fasciclin I protein are mainly uncharged and provide the signal for the attachment of a glycosyl-phosphatidylinositol (GPI) lipid membrane anchor[7]. In *Drosophila* embryos this GPI linkage is developmentally regulated and a large fraction of the protein appears to lose its membrane anchor during embryonic development.

In both grasshopper and *Drosophila*, the *fasciclin I* gene gives rise to multiple transcripts which show different temporal and spatial regulation. The *Drosophila* transcripts differ in their 3′ untranslated regions and by the alternative splicing of two nine nucleotide (3 amino acid) micro-exons at the end of the conserved domain two repeat, giving rise to three different forms of the protein (the 0, +3, and +6 forms) (McAllister et al., unpublished).

Genetic analysis has shown that fasciclin I is involved in growth cone guidance. Protein null mutations in the *fas I*

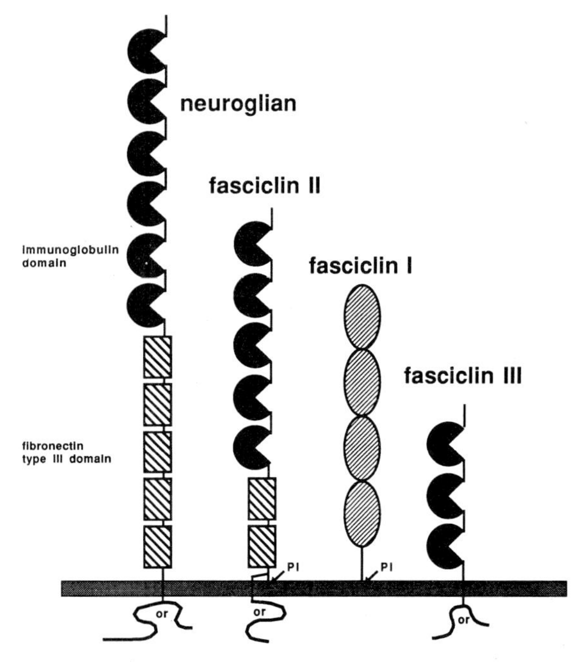

Figure. Schematic domain structure of the **neuroglian**, **fasciclin II**, fasciclin I, and **fasciclin III** glycoproteins. Three of the four glycoproteins are members of the immunoglobulin superfamily. Two of these proteins, neuroglian and fasciclin II, have multiple Ig domains followed by multiple **fibronectin** type III domains; the third, fasciclin III, has three more divergent Ig domains. Fasciclin I has a unique structure made up of four tandem domains and is anchored in the membrane via a GPI lipid membrane anchor. Neuroglian and fasciclin III have alternative cytoplasmic domains; fasciclin II also comes in different forms including one with a GPI membrane anchor and at least two different transmembrane forms with distinct cytoplasmic domains.

gene are homozygous viable, display strong behavioral defects (uncoordinated motor activity), but show no gross anatomical defects in CNS morphogenesis. Embryos doubly mutant for *fas I* and *abl*, the *Drosophila* Abelson protooncogene homologue, which encodes a cytoplasmic tyrosine kinase[8], display major defects in CNS axon pathways, particularly in the commissural tracts where the expression of these two proteins normally overlaps[4]. An anti-fasciclin I mAb was used to perturb fasciclin I protein on the surface of the Ti1 pioneer neurons in the developing grasshopper limb bud by chromophore assisted laser inactivation (CALI)[5]. CALI disrupted the fasciculation of these axons without affecting their growth or guidance.

PURIFICATION

Fasciclin I protein was isolated from grasshopper embryonic lysates by affinity chromatography on 3B11 mAb columns[1,6].

ACTIVITIES

Transfection of *Drosophila* S2 cells, and *in vitro* cell aggregation experiments show that fasciclin I can mediate homophilic, Ca^{2+}-independent cell adhesion[3].

ANTIBODIES

The generation of monoclonal and polyclonal antibodies in mice and rats against grasshopper fasciclin I has been reported[1,6]. Rat antisera and mouse mAbs have also been generated using different *Drosophila* fasciclin I fusion proteins[2,7].

GENES AND MUTATIONS

Full cDNA sequences for grasshopper (GenBank M20544) and *Drosophila* fasciclin I (GenBank M20545, genomic M32311) have been published[2]. The *Drosophila fasciclin I* gene maps to the third chromosome at position 89 D. Mutations have been identified and generated in the *fasciclin I* gene[4]. A large (~100 kB) transposable element, called a TE element, in one stock, TE77[9], was found to be located in a large intron in the *fasciclin I* gene[4]. The TE77 insert leads to a mRNA and protein null mutation in the *fasciclin I* gene. Additional *fas I* alleles were generated using X-rays by screening for *w*-revertants (having a white mutant eye color), since the large TE insert contains a copy of the *white* (*w*) gene. One class of *w*-revertants, represented by at least three alleles, produces apparently normal levels of fasciclin I protein in the embryo, whereas another class of *w*-revertants, represented by at least two alleles (e.g. *R8* and *R40.1*), appears to represent complete protein null mutations[4].

REFERENCES

1. Bastiani, M.J., Harrelson, A.L., Snow, P.M. and Goodman, C.S. (1987) Cell 48, 745-755.
2. Zinn, K., McAllister, L. and Goodman, C.S. (1988) Cell 53, 577-587.
3. Elkins, T., Hortsch, M., Bieber, A.J., Snow, P.M. and Goodman, C.S. (1990) J. Cell Biol. 110, 1825-1832.
4. Elkins, T., Zinn, K., McAllister, L., Hoffmann, F.M. and Goodman, C.S. (1990) Cell 60, 565-575.
5. Jay, D.G. and Keshishian, H. (1990) Nature 348, 548-550.
6. Snow, P.M., Zinn, K., Harrelson, A.L., McAllister, L., Schilling, J., Bastiani, M.J., Makk, G. and Goodman, C.S. (1988) Proc. Natl. Acad. Sci. (USA) 85, 5291-5295.
7. Hortsch, M. and Goodman, C.S. (1990) J. Biol. Chem. 265, 15104-15109.
8. Henkemeyer, M.J., Gertler, F.B., Goodman, W. and Hoffmann, F.M. (1988) Cell 51, 821-828.
9. Paro, R., Goldberg, L.G. and Gehring, W.J. (1983) EMBO J. 2, 853-860.

Michael Hortsch and Corey S. Goodman:*
Howard Hughes Medical Institute,
Department of Molecular and Cell Biology,
University of California, Berkeley, CA, USA
**Present address:*
Department of Anatomy and Cell Biology,
University of Michigan, Ann Arbor,
Michigan, USA

Fasciclin II

Fasciclin II[1-4] is a homophilic, Ca^{2+}-independent cell adhesion molecule[4] in insects that is a member of the immunoglobulin gene superfamily[3,4]; it shares a common ancestor with vertebrate NCAM[5,6]. It is expressed on a subset of longitudinal axon fascicles and other axon pathways in grasshopper and Drosophila embryos[1,3,4]. The protein comes in at least two different forms generated by alternative splicing[4,7]. During neuronal development, fasciclin II appears to play a role in specific growth cone guidance and selective fasciculation[3,7].

The fasciclin II protein was first identified in the grasshopper embryo as a result of a monoclonal antibody (MAb) screen[1]; it is a membrane glycoprotein with an apparent molecular weight of a 95 kDa that is expressed on the surface of a specific subset of axon fascicles in both the CNS and PNS during embryogenesis[1-3]. Fasciclin II is primarily expressed on a subset of longitudinal axon pathways, the intersegmental nerve root, and transiently on some commissural axon bundles[1,3]. The protein is also expressed in a restricted fashion outside of the developing nervous system. The expression pattern of fasciclin II suggests that it might be involved in growth cone guidance. When grasshopper embryos are incubated with anti-fasciclin II antibodies before the MP1 growth cone contacts the MP1 fascicle, the MP1 growth cone stalls at the choice point, either takes longer than normal to join the appropriate axon fascicle or sometimes follows the wrong pathway, and often extends an aberrant second growth cone across the anterior commissure[3].

Fasciclin II cDNAs were initially isolated in grasshopper using oligonucleotides and cDNA expression cloning[2,3] and subsequently in *Drosophila* using PCR[4,7]. The cDNA ORF of grasshopper fasciclin II encodes a transmembrane glycoprotein of 897 amino acids; the extracellular domain contains 5 immunoglobulin C2-type domains and two **fibronectin** type III domains[3]; the *Drosophila* homologue has a similar extracellular domain structure[4,7] (see Figure page 135). The deduced grasshopper and *Drosophila* fasciclin II proteins have approximately 45% amino acid identity. Fasciclin II is expressed in a very similar fashion in the developing CNS of both insects[3,7]. In *Drosophila* at least two forms of the fasciclin II protein can be detected using anti-fasciclin II antibodies (Grenningloh et al., unpublished results). The long form has an apparent molecular weight of 110 kDa and is a transmembrane protein, whereas the short form has an apparent molecular weight of 98 kDa and is presumably linked to the membrane by a glycosyl-phosphatidylinositol-lipid moiety.

Two classes of mutations have been isolated in the *fasciclin II* gene: a set of viable, hypomorphic alleles, and a set of protein null, lethal alleles[7]. In homozygous null mutant embryos, the MP1 axon pathway does not form, although many other pathways develop normally. The protein null mutation alters the selective affinity of at least two identified growth cones: vMP2 and MP1[7]. Although this genetic analysis is at an early stage, it clearly shows that fasciclin II is involved in selective fasciculation and growth cone guidance.

Fasciclin II has a similar domain organization to vertebrate **NCAM**, and the grasshopper protein shares 28% amino acid identity over the entire extracellular domain to mouse or chicken NCAM[5,6]. Insect fasciclin II and vertebrate NCAMs probably evolved from a common ancestral molecule which predated the split in the two evolutionary lines leading to the arthropods and chordates.

PURIFICATION

Fasciclin II protein was isolated from grasshopper embryonic lysates by affinity chromatography on 8C6 MAb columns[1,2].

ACTIVITIES

Transfection of *Drosophila* S2 cells and *in vitro* cell aggregation experiments show that fasciclin II can mediate homophilic, Ca^{2+}-independent cell adhesion[4,7].

ANTIBODIES

The generation of monoclonal and polyclonal antibodies in mice and rats against grasshopper fasciclin II has been reported[1,2]. Rat antisera and mouse MAbs have also been generated using different *Drosophila* fasciclin II fusion proteins[7].

GENES AND MUTATIONS

The complete amino acid sequence of grasshopper fasciclin II has been published[3]; the *Drosophila* sequence is also published (GenBank M77165 and M77166)[7]. The *Drosophila* fasciclin II gene is located on the X chromosome at position 4B1-2[4,7]. Two classes of mutations have been isolated in the *Drosophila fasciclin II* gene: a set of viable, hypomorphic alleles, and a set of protein null, lethal alleles[7].

REFERENCES

1. Bastiani, M.J., Harrelson, A.L., Snow, P.M. and Godman, C.S. (1987) Cell 48, 745-755.
2. Snow, P.M., Zinn, K., Harrelson, A.L., McAllister, L., Schilling, J., Bastiani, M.J., Makk, G. and Goodman, C.S. (1988) Proc. Natl. Acad. Sci. (USA) 85, 5291-5295.
3. Harrelson, A.L. and Goodman, C.S. (1988) Science 242, 700-708.
4. Grenningloh, G., Bieber, A.J., Rehm, E.J., Snow, P.M., Traquina, Z.R., Hortsch, M., Patel, N.H. and Goodman, C.S. (1990) Cold Spring Habor Symp. Quant. Biol. 55, 327-340.
5. Cunningham, B.A., Hemperly, J.J., Murray, B.A., Prediger, E.A., Brackenbury, R. and Edelman, G.M. (1987) Science 236, 799-806.
6. Barthels, D., Santoni, M.J., Wille, W., Ruppert, C., Chaix, J.C., Hirsch, M.R., Fontecilla-Camps, J.C. and Goridis, C. (1987) EMBO J. 6, 907-914.
7. Grenningloh, G., Rehm, E.J. and Goodman, C.S. (1991) Cell 67, 45-57.

Michael Hortsch and Corey S. Goodman:*
Howard Hughes Medical Institute,
Department of Molecular and Cell Biology,
University of California, Berkeley, CA, USA
**Present address:*
Department of Anatomy and Cell Biology,
University of Michigan, Ann Arbor,
Michigan, USA

Fasciclin III

Fasciclin III[1,2] is a homophilic, Ca^{2+}-independent cell adhesion molecule[2] in Drosophila that is a member of the immunoglobulin (Ig) gene superfamily[3]; it has divergent Ig domains and is unrelated to any known vertebrate member of the Ig superfamily[2,3]. It is expressed on a subset of commissural axon fascicles and other axon pathways in the Drosophila embryo[1], as well as on a variety of other nonneuronal tissues[1,4-6]. The protein comes in at least two different forms generated by alternative splicing[1,2,6].

Drosophila fasciclin III is expressed on a subset of axon fascicles in the developing CNS (Figure), in a segmentally repeated pattern in the epidermis, and in the larval eye disc; because of this diverse pattern of expression, anti-fasciclin III monoclonal antibodies (mAbs) were isolated by several groups using a number of different antibody screens[1,4,5]. It has also been called the DENS-antigen[6]. During neurogenesis, fasciclin III is expressed in patches in the neurogenic region; during axon outgrowth, it is expressed on a subset of neuronal cell bodies and axons in the CNS as well as on a subset of nonneuronal (and presumably glial) cells[1]. Fasciclin III is expressed at high levels on five specific commissural axon fascicles and on the RP1 fascicle as it heads towards the intersegmental nerve root; it is absent from longitudinal axon pathways. Outside of the developing nervous system, fasciclin III is expressed in the visceral mesoderm and the luminal surface of the salivary gland epithelium[1,6]. In the larval eye disc it is expressed on the cone cells surrounding the clusters of eight photoreceptor cells which do not express fasciclin III[5].

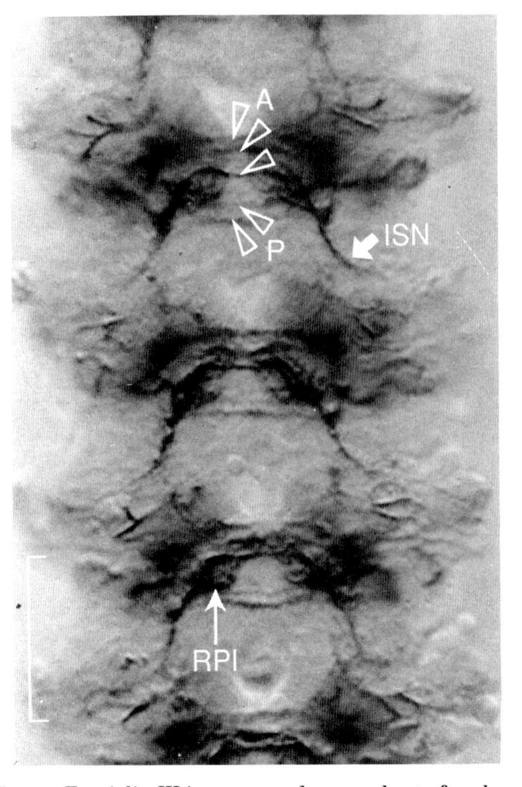

Figure. Fasciclin III is expressed on a subset of embryonic neurons and axon pathways. The figure shows four contiguous segmental neuromeres of the *Drosophila* CNS (anterior is up) at hour 12 of embryonic development. The embryo was dissected onto a glass slide and stained with the 7G10 mAb. At this stage fasciclin III protein is expressed on the surface of axons in five commissural fascicles (3 in the anterior or A and 2 in the posterior or P commissure) and on the continuation of one of the A commissure fascicles as it turns posteriorly and laterally (as the RP1 fascicle) towards one of the two peripheral nerves: the intersegmental nerve (ISN).

Anti-fasciclin III mAbs recognize at least two different forms of the protein with apparent molecular mass of 80 and 66 kDa[1,6]. The full length cDNA coding for the largest form of the fasciclin III protein (80 kDa) contains an open reading frame of 508 amino acids, 20 of which code for an N-terminal signal sequence[2]. A transmembrane segment of 24 hydrophobic amino acids divides the mature molecule into an extracellular domain of 326 amino acids and a cytoplasmic domain of 138 amino acids (See Figure on page 135). A full length cDNA that encodes the shorter form (66 kDa) of the protein reveals this form to be identical to the long form, except that it has a different cytoplasmic domain (Snow et al., unpublished).

Although initial analysis suggested that the fasciclin III protein appears unrelated to any other known protein, further analysis of its sequence reveals that the extracellular part of the molecule consists of three highly divergent Ig type domains[3]. The third most proximal Ig domain lacks the characteristic cysteine residues. The first domain appears to be of the V type, whereas assignment of the other two Ig domains is less certain. Both domains appear to be most related to the C2-type of Ig domains, although all three fasciclin III Ig domains are highly divergent from those found in other insect and vertebrate neural cell adhesion molecules.

■ PURIFICATION

Fasciclin III protein was isolated from *Drosophila* embryonic lysates by affinity chromatography on 2D5 mAb columns[1].

ACTIVITIES

Transfection of *Drosophila* S2 cells and *in vitro* cell aggregation experiments show that fasciclin III can mediate homophilic, Ca^{2+}-independent cell adhesion[2].

ANTIBODIES

The generation of monoclonal and polyclonal antibodies in mice and rats against *Drosophila* fasciclin III has been reported[1,4-6].

GENES AND MUTATIONS

A complete cDNA sequence encoding the 80 kDa form of *Drosophila* fasciclin III has been cloned and sequenced (GenBank M27813)[2]. The *Drosophila fasciclin III* gene maps to position 36E1 on the second chromosome[1]. A viable, protein null mutation in the *fasciclin III* gene has been isolated (Elkins et al., unpublished).

REFERENCES

1. Patel, N.H., Snow, P.M. and Goodman, C.S. (1987) Cell 48, 975-988.
2. Snow, P.M., Bieber, A.J. and Goodman, C.S. (1989) Cell 59, 313-323.
3. Grenningloh, G., Bieber, A.J., Rehm, E.J., Snow, P.M., Traquina, Z.R., Hortsch, M., Patel, N.H. and Goodman, C.S. (1990) Cold Spring Habor Symp. Quant. Biol. 55, 327-340.
4. Brower, D.L., Smith, R.J. and Wilcox, M. (1980) Nature 285, 403-405.
5. Zipursky, S.L., Venkatesh, T.R., Teplow, D.B. and Benzer, S. (1984) Cell 36, 15-26.
6. Gauger, A., Glicksman, M.A., Salatino, R., Condie, J.M., Schubiger, G. and Brower, D.L. (1987) Development 100, 237-244.

Michael Hortsch and Corey S. Goodman:*
Howard Hughes Medical Institute,
Department of Molecular and Cell Biology,
University of California, Berkeley, CA, USA
**Present address:*
Department of Anatomy and Cell Biology,
University of Michigan, Ann Arbor,
Michigan, USA

F11

F11 is a neural cell surface glycoprotein involved in neurite fasciculation[1]. The molecule comprises six domains related to the immunoglobulin domain type C2 and four resembling fibronectin repeat type III[2]. F11 interacts with the plasma membrane through covalently linked glycosyl-phosphatidylinositol (GPI)[3]. It is expressed in the central and peripheral nervous system.

When F11 is isolated from adult brain by immunoaffinity chromatography and analyzed in SDS-PAGE, it consists of a major 130 kDa polypeptide and a minor coisolating component at 170 kDa which is distinct from F11. The isoelectric point of the isolated F11 ranges from 7.0 to 6.1[1]. F11 contains two populations of N-linked carbohydrates, one containing mainly complex type carbohydrates and a second containing high-mannose/hybrid-type carbohydrates[4], but no O-glycosylation. F11 shares a common carbohydrate epitope with **NCAM** and **L1** (HNK-1/L2).

cDNA cloning reveals that F11 is composed of six immunoglobulin type C2 and 4 **fibronectin** Type III domains[2,5,6] (Figure 1). These domain types have also been found in other molecules implicated in axon-axon interactions, e.g. in mouse L1[7], in rat **TAG-1**[8], NCAM[9] and the invertebrate proteins **neuroglian**[10] and **fasciclin II**[11]. The two structural elements of F11 are separated by a short **collagen** like amino acid stretch that could serve as a hinge region to generate structural flexibility. The F11 primary structure shows a high degree of structural homology with TAG-1 with an overall amino acid identity of about 50% suggesting that F11 and TAG-1 form the first two members of a subgroup of the immunoglobulin superfamily. F11 exists in two forms, one interacts with the plasma membrane through covalently linked GPI[3] or a structurally similar lipid and the other is secreted into the extracellular matrix. Whether an additional transmembrane form of F11 exists is still controversial[2,6]. In the developing PNS and CNS, F11 is localized primarily in axon rich regions (Figure 2). It shows a restricted distribution and is transiently expressed which is consistent with its participation in neurite outgrowth and selective fasciculation. F11 can also be found in the literature under the names F3[5] (mouse) and contactin[6] (chick).

PURIFICATION

F11 can be purified by immunoaffinity chromatography using monoclonal anti-F11 antibodies from detergent, urea or from phosphatidylinositol specific phospholipase C (PI-PLC) extracts of plasma membrane preparations from brain tissue[1].

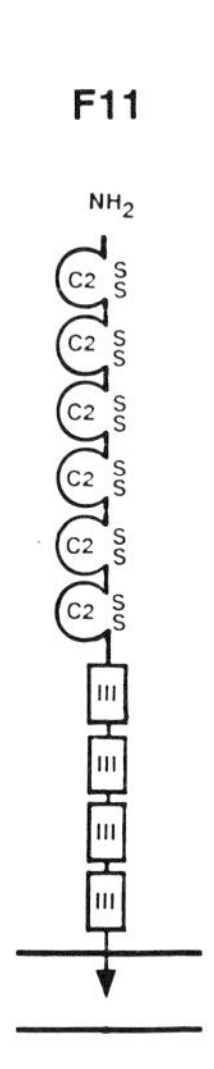

Figure 1. Schematic domain organization of F11 derived from its primary sequence. Immunoglobulin related domains are drawn as loops that are closed by putative disulphide bridges. Fibronectin type III-related domains are represented by boxes. Arrow indicates plasma membrane interaction by covalently attached lipid (modified from Ref. 2).

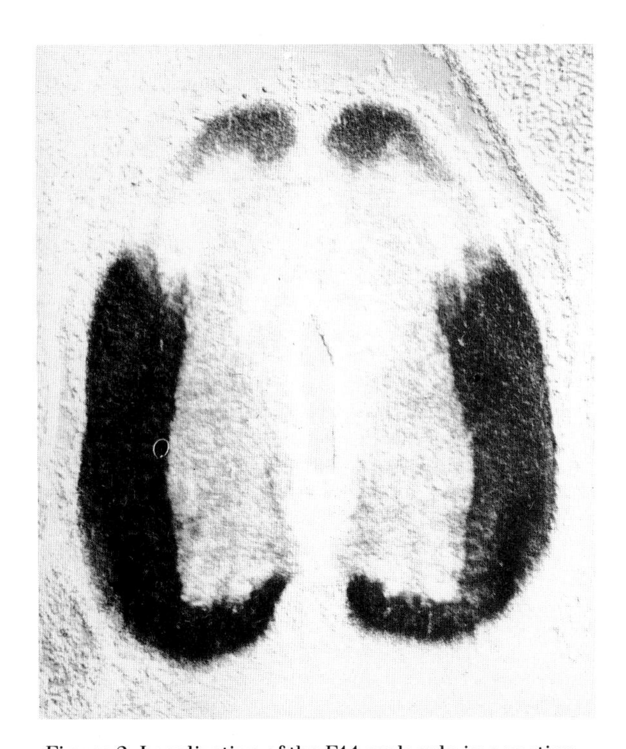

Figure 2. Localization of the F11 molecule in a section of the chick spinal cord from cervical regions at embryonic day eight. Cryostat section was stained indirectly with mAb F11. Abbreviations: NE, Neuroepithelial cells; DF, dorsal funiculus; DH, dorsal horn; VC, ventral commissure; VF ventral funiculus; VH, ventral horn.

ACTIVITIES

Based on antibody perturbation experiments using recently developed *in vitro* systems F11 appears to be involved in fasciculation of axons[1] and elongation[12] of axons on other axons. If F11 is expressed in CHO cells by transfection experiments it directly stimulates neurite outgrowth[13].

ANTIBODIES

Polyclonal antibodies and monoclonal antibodies to chick[1,4] and mouse F11[5] have been described. They generally do not crossreact with other species.

GENES

Complete cDNA sequences are available for chicken[1,6] (EMBL data library accession X14877) and mouse F11[5] (EMBL data accession library X14943F3). In the mouse the F11 gene maps to band F on chromosome 15[5].

REFERENCES

1. Rathjen, F.G., Wolff, J.M., Frank, R., Bonhoeffer, F. and Rutishauser, U. (1987) J. Cell Biol. 104, 343-353.
2. Brümmendorf, T., Wolff, J.M., Frank, R. and Rathjen, F.G. (1989) Neuron 2, 1351-1361.
3. Wolff, J.M., Brümmendorf, T. and Rathjen, F.G. (1989) Biochem. Biophys. Res. Comm. 161, 931-938.
4. Wolff, J.M., Rathjen, F.G., Frank, R. and Roth, S. (1987) Eur. J. Biochem. 168, 551-561.
5. Gennarini, G., Cibelli, G., Rougon, G., Mattei, M.-G. and Goridis, C. (1989) J. Cell Biol. 109, 775-788.
6. Ranscht, B. (1988) J. Cell Biol. 107, 1561-1573.
7. Moos, M., Tacke, R., Scherer, H., Teplow, D., Früh, K. and Schachner, M. (1988) Nature 334, 701-703.
8. Furley, A.J., Morton, S.B., Manalo, D., Karagogeos, D., Dodd, J. and Jessell, T.M. (1990) Neuron 61, 157-170.
9. Cunningham, B.A., Hemperly, J.J., Murray, B.A., Prediger, E.A., Backenbury, R. and Edelmann, G.M. (1987) Science 236, 799-806.
10. Bieber, A.J., Snow, P.M., Hortsch, M., Patel, N.H., Jacobs, J.R., Traquina, Z.R., Schilling, J. and Goodman, C.S. (1989) Cell 59, 447-460.
11. Harrelson, A.L. and Goodman, C.S. (1988) Science 242,700-708.
12. Chang, S., Rathjen, F. and Raper, J. (1987) J. Cell Biol. 104, 355-362.
13. Gennarini, G., Durbec, P., Boned, A., Rougon, G. and Goridis, C. (1991) Neuron 6, 595-606.

Fritz G. Rathjen:
Zentrum für Molekulare Neurobiologie,
Hamburg, Germany

GP Ib-IX Complex

The GP Ib-IX complex is a heterooligomeric platelet membrane glycoprotein essential for normal platelet adhesion and thrombus formation at sites of vascular injury. It functions as a receptor for the adhesive protein, von Willebrand factor (vWF), and for the platelet agonist, thrombin. Congenital deficiency of GP Ib-IX results in the hemorrhagic disorder known as Bernard-Soulier syndrome.

The GP Ib-IX complex[1] is composed of the equimolar association of three distinct gene products: the disulphide-linked α- and β-chains of GP Ib (also designated CD42b; gp135/25), and the noncovalently associated GP IX (CD42a; gp23). Electron microscopy demonstrates an elongated structure with a longitudinal axis of 59.5 nm and a globular N-terminal domain[2]. GP Ib has an apparent molecular mass of ~170 kDa (α-chain, ~140 kDa; β-chain, ~24 kDa); GP IX of ~18 kDa. All three proteins have transmembrane and cytoplasmic domains. Glycocalicin is a soluble fragment of GP Ibα corresponding essentially to its extracytoplasmic domain[3]. The GP Ib-IX complex is synthesized in megakaryocytes and expressed on the platelet surface in 16,000-30,000 copies per cell, second in density only to the GP IIb-IIIa complex. A molecule homologous, if not identical, to GP Ibα has been demonstrated in endothelial cells[4]. The sequence of biosynthetic steps involved in assembly and membrane insertion of the GP Ib-IX complex is not known in detail, but all three gene products need to be present for appropriate processing. Therefore, a lesion in each one of the three genes may cause absence of the complex from the platelet surface and cause Bernard-Soulier syndrome[5]. Variant forms of the disease caused by presence of a structurally abnormal GP Ibα chain have also been documented[6]. A schematic representation of the GP Ib-IX complex is shown in the Figure.

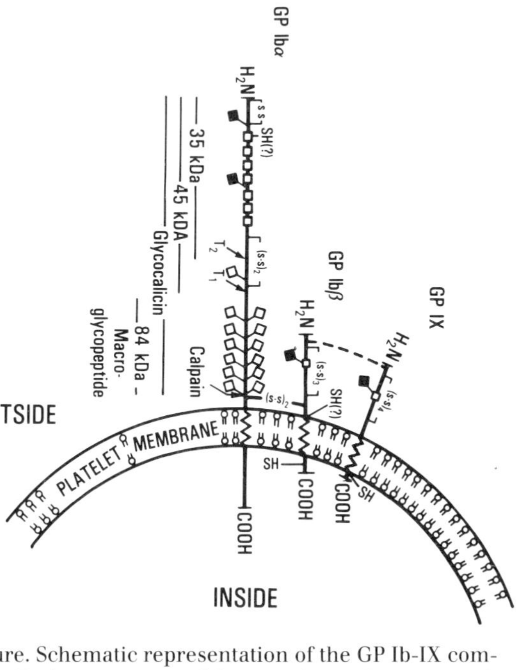

Figure. Schematic representation of the GP Ib-IX complex. Each component of the complex is drawn to scale proportional to the number of residues in the sequence. Domains of the extracytoplasmic portion of the α chain are identified. Glycosylation sites are indicated by closed diamonds (N-linked) or open diamonds (O-linked; exact number unknown); leucine-rich repeats by open squares; trypsin and calpain cleavage sites by arrows. The noncovalent bond between GP Ib and GP IX is indicated by a broken line; it is not known whether it involves the α-chain or β-chain, or both.

The GP Ibα cDNA codes for a 16 residue signal peptide and a mature 610 residue polypeptide[7]. GP Ibα can be divided in four distinct domains: the N-terminal (~45 kDa), residues 1-293, containing the binding sites for **vWF** and thrombin; the carbohydrate-rich macroglycopeptide (~84 kDa), residues 294-485; the transmembrane, residues 486-514 and the intracytoplasmic, residues 515-610. The N-terminal domain contains two N-linked and one O-linked carbohydrate chains[3], and seven copies of a 24-residue leucine-rich sequence, of still unknown functional significance, found in proteins of mammalian, insect, and yeast origin[8]. These proteins, with diverse functions, constitute the leucine-rich glycoprotein family that includes GP Ibβ and GP IX. The N-terminal domain of GP Ibα contains three disulphide loops and one unpaired cysteine[3]. The macroglycopeptide domain is very rich in O-linked carbohydrate (more than half of the molecular mass) and contains five repeats of a consensus sequence that may represent a conserved signal for O-linked glycosylation; it also contains two potential N-linked glycosylation sites[7]. Two Cys residues are located immediately N-terminal to the transmembrane domain and both may be involved in the interchain bond with GP Ibβ[1].

The GP Ibβ cDNA codes for a 25 residue signal peptide and a mature 181 residue polypeptide[9]. The extracytoplasmic domain of GP Ibβ is composed of 122 residues, with one N-linked glycosylation site and one leucine-rich repeat. It also contains nine cystein residues, of which three may form intrachain loops, two may be linked to cystein residues in the α chain, and one is unpaired or linked to membrane lipids[1]. The transmembrane domain of GP Ibβ is composed of 25 residues, and the intracytoplasmic domain, containing a free sulfhydryl, of 34 residues.

The GP IX cDNA codes for a mature 160 residue polypeptide[10]. The extracytoplasmic domain contains one N-linked glycosylation site and one copy of the leucine-rich motif; it also contains eight cystein residues, probably arranged in four intrachain loops. The 20 residue transmembrane domain contains one cystein residue, presumably involved in a thioester bond with a membrane lipid[11], and precedes a short, six residue intracytoplasmic tail. The physicochemical nature of the association between GP Ib and GP IX is not known.

PURIFICATION

The GP Ib-IX complex is purified from detergent extracts of platelet membranes by lectin (wheat germ agglutinin) affinity chromatography followed by immunoaffinity chromatography[12]. GP IX is separated from GP Ib during the last step of the procedure[13]. The α- and β-chains of GP Ib are separated by molecular sieve chromatography following disulphide reduction[12]. The soluble extracytoplasmic domain of GP Ibα (glycocalicin) ligand interaction sites of the molecule, is isolated following activation of an endogenous Ca^{2+}-dependent neutral protease (**calpain**)[3].

ACTIVITIES

The GP Ib-IX complex mediates the initiation and propagation of platelet responses to vascular injury by anchoring platelets to subendothelial surfaces or damaged endothelium. The specific interaction with surface bound von Willebrand factor[14], as well as the high affinity binding of thrombin[15] generated at sites of tissue injury, transmits intracellular signals causing propagation of the initial response through platelet activation. These functions require ligand recognition sites, located in the N-terminal extracytoplasmic domain of the α chain, and linkage to the membrane skeleton provided by one or more of the intracytoplasmic domains of the three chains[16]. The β chain of GP Ib also contains a phosphorylation site[17].

ANTIBODIES

Polyclonal sera against the GP Ib-IX complex are available. Monoclonal antibodies against the different components of the complex have been characterized in detail, particularly those against the N-terminus of the α-chain that can be used to study the structure-function relationships of the two distinct binding sites for vWF and thrombin[18].

GENES

cDNA sequences are known for GP Ibα (GenBank EMBL J02940), GP Ibβ (GenBank EMBL J03259) and GP IX (GenBank EMBL M25827). The GP Ibα gene is located on the distal short arm of chromosome 17, 17p12-ter[19]. The whole translated sequence is contained in a single exon of ~2.4 kB, with a single intron (233 Bp) present six Bp upstream of the translation initiation codon. The GP IX gene is located on chromosome 3[20]. The GP Ibβ gene appears to be located on chromosome 22 (Lopez, personal communication; unpublished).

REFERENCES

1. Ruggeri, Z.M. (1989) Progress in Hemostasis and Thrombosis 35-68.
2. Fox, J.E.B., Aggerbeck, L.P. and Berndt, M.C. (1988) J. Biol. Chem. 263, 4882-4890.
3. Titani, K., Takio, K., Handa, M. and Ruggeri, Z.M. (1987) Proc. Natl. Acad. Sci. (USA) 84, 5610-5614.
4. Konkle, B.A., Shapiro, S.S., Asch, A.S. and Nachman, R.L. (1990) J. Biol. Chem. 265, 19833-19838.
5. Ware, J., Russell, S.R., Vicente, V., Scharf, R.E., Tomer, A., McMillan, R. and Ruggeri, Z.M. (1990) Proc. Natl. Acad. Sci. (USA) 87, 2026-2030.
6. De Marco, L., Mazzucato, M., Fabris, F., De Roia, D., Coser, P., Girolami, A., Vicente, V. and Ruggeri, Z.M. (1990) J. Clin. Invest. 86, 25-31.
7. Lopez, J.A., Chung, D.W., Fujikawa, K., Hagen, F.S., Papayannopoulou, T. and Roth, G.J. (1987) Proc. Natl. Acad. Sci. (USA) 84, 5615-5619.
8. Roth, G.J. (1991) Blood 77, 5-19.
9. Lopez, J.A., Chung, D.W., Fujikawa, K., Hagen, F.S., Davie, E.W. and Roth, G.J. (1988) Proc. Natl. Acad. Sci. (USA) 85, 2135-2139.
10. Hickey, M.J., Williams, S.A. and Roth, G.J. (1989) Proc. Natl. Acad. Sci. (USA) 86, 6773-6777.
11. Muszbek, L. and Laposata, M. (1989) J. Biol. Chem. 264, 9716-9719.
12. Canfield, V.A., Ozols, J., Nugent, D. and Roth, G.J. (1987) Biochem. Biophys. Res. Commun. 147, 526-534.
13. Roth, G.J., Ozols, J., Nugent, D.J. and Williams, S.A. (1988) Biochem. Biophys. Res. Commun. 156, 931-939.
14. Vicente, V., Houghten, R.A. and Ruggeri, Z.M. (1990) J. Biol. Chem. 265, 274-280.
15. Harmon, J.T. and Jamieson, G.A. (1986) J. Biol. Chem. 261, 13224-13229.
16. Fox, J.E.B. (1985) J. Clin. Invest. 76, 1673-1683.
17. Wardell, M.R., Reynolds, C.C., Berndt, M.C., Wallace, R.W. and Fox, J.E.B. (1989) J. Biol. Chem. 264, 15656-15661.
18. Handa, M., Titani, K., Holland, L.Z., Roberts, J.R. and Ruggeri, Z.M. (1986) J. Biol. Chem. 261, 12579-12585.
19. Wenger, R.H., Wicki, A.N., Kieffer, N., Adolph, S, Hameister, H. and Clemetson, K.J. (1989) Gene 85, 517-524.
20. Hickey, M.J., Deaven, L.L. and Roth, G.J. (1990) FEBS Lett. 274, 189-192.

Zaverio M. Ruggeri and Jerry Ware:
Roon Research Center for Arteriosclerosis and Thrombosis,
Department of Molecular and Experimental Medicine,
Committee on Vascular Biology,
Research Institute of Scripps Clinic,
La Jolla, California, USA

Integrins

The integrins are a family of cell surface heterodimers which act as adhesion receptors for ligands on other cells or for extracellular matrix proteins. Inside the cell, the short cytoplasmic domains of integrins associate with various cytoskeletal proteins such as talin and α-actinin. Together, the transmembrane spanning binding functions of integrins help to regulate not only cell adhesion but also cell spreading, migration and shape changes. In addition, integrins may have a role in signal transduction and gene regulation.

Integrins are heterodimeric combinations of 14 different α-subunits and eight different β- subunits. Multiple α-subunits can associate with the same β-subunit, and to a lesser extent, different β-subunits can associate with the same α-subunit[1,2]. A tabulation of ligands recognized by different integrins is shown in Figure 1. Many of the ligands listed (e.g. **collagen, laminin, fibronectin**) are extracellular matrix (ECM) proteins, but some (e.g. **ICAM, VCAM**) are cell surface proteins. By having multiple integrins as receptors for common ECM proteins, cells have the flexibility to interact with different affinities at the same ligand site, and at different sites within the same ligand. Several integrins, particularly those containing the α_{IIb}, α_V, and α_5 subunits, bind to ligands at sites containing the amino acid sequence Arg-Gly-Asp (RGD)[3].

The ligand specificity for different integrins (Figure 1) can be altered depending on the type of divalent cation present[4], the surrounding lipid environment[5], and other unknown cell-type specific factors. Also, there are mechanisms probably involving receptor aggregation and/or conformational changes whereby the functions of an integrin can be rapidly upregulated without a change in the expression level[6,7]. These regulatory mechanisms give added flexibility to integrins with respect to ligand binding.

INTEGRIN SUBUNIT COMBINATIONS AND LIGANDS

	β1	β2	β3	β4	β5	β6	β7	β8
α1	Coll, Lm							
α2	Coll, Lm							
α3	Fn, Lm, Coll							
α4	Fn, VCAM-1						PP HEV	
αE							?	
α5	Fn							
α6	Lm			?				
α7	Lm							
α8	?							
αV	Fn,Vn?		Vn,Fb,Opn vWF,BSP,(Fn)		Vn,(Fn)	?		?
αIIb			Fb, Vn Fn, vWF					
αL		ICAM-1,2						
αM		C3bi, FX, Fb ICAM-1						
αX		(C3bi)						

Figure 1. Integrin α- and β-subunit combinations and ligands. The abbreviations used are: BSP, **bone sialoprotein**; Coll, collagen; Fb, **fibrinogen**; Fn, fibronectin; ICAM-1,2, intercellular adhesion molecule-1,2; Lm, laminin; Opn, **osteopontin**; PP HEV, Peyer's patch high endothelial venules; VCAM-1, vascular cell adhesion molecule-1; Vn, **vitronectin**; vWF, **von Willebrand's factor.** Empty boxes indicate αβ-subunit combinations which have not been observed.

One or more different integrin heterodimers are distributed on nearly all cell types, with the exception of red blood cells. Some integrins have a restricted distribution (e.g. α_{IIb} β_3 only appears on platelets, β_2 integrins only appear on leukocytes) whereas other integrins (e.g. those containing β_1, β_3, β_5) can appear on many different cell types. Cells of all animal species probably express some type of integrins, although invertebrate species such as *Drosophila* appear to have a more restricted repertoire of α- and β-subunits. Consistent with their widespread distribution, integrins have been found to play important roles in nearly all tissues and organ systems in the body. Studies to date have particularly stressed the roles of integrins in metastasis, inflammation, development, and thrombosis and hemostasis.

The importance of integrins is further emphasized by the biological defects which result when one or more subunits are genetically deficient. For example, the lack of a functional β_2-subunit in humans causes "Leukocyte Adhesion Deficiency" (LAD) disease, in which inflammatory responses are severely impaired and patients usually die at an early age[8]. In *Drosophila*, lack of a β_1-like subunit causes the "lethal myospheroid" mutation in which embryos do not progress past a late gastrulation stage[9]. Also, deficiencies in α_2, β_3, and α_{IIb} on platelets lead to bleeding disorders of varying degrees of severity.

Figure 2 shows a schematic diagram of a typical integrin αβ-subunit heterodimer. Each α-subunit has about 1000-1150 amino acids, and a mature size in the range of 140-210 kDa. Also each α-subunit contains three to four divalent cation sites which are essential for integrin function, and some α-subunits have a 200 amino acid insert called the "I-domain"[10,11]. Other α-subunits typically undergo a proteolytic cleavage, resulting in a disulphide linked C-terminal fragment of about 30 kDa. With the exception of β_4, each β-subunit is composed of about 740-780 amino acids and is 90-130 kDa in size. Most of the 56 cysteines in β-subunits are present within four repeating motifs that comprise the "cysteine-rich" region. The β_4-subunit is unusual because it has a very long cytoplasmic tail (over 1000 amino acids), whereas other β-subunits have cytoplasmic tails of only 40-50 amino acids.

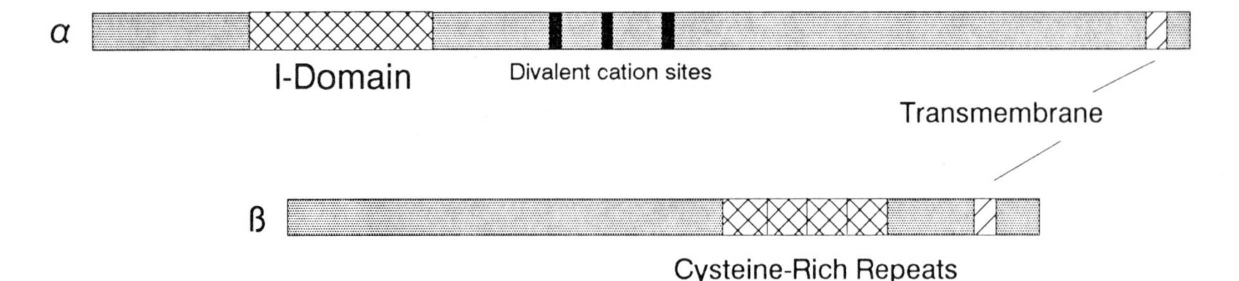

Figure 2. Schematic diagram of typical integrin α- and β-subunits.

Alternative splicing has been observed within the cytoplasmic domains of β_1[12], β_3[13], and β_4[14], whereas for α_{IIb} it occurs just external to the transmembrane region[15], and in the *Drosophila* α_{PS2} subunit, it occurs in the extracellular domain, near a probable ligand binding site[16]. In some cases there may be developmental or cell type specific regulation of alternative splicing, but the precise functional consequences of this are not yet known.

PURIFICATION

The purification of several integrins has been achieved by affinity chromatographic isolation using immobilized ligands. Typically, these columns have been eluted with EDTA, which inhibits nearly all integrin ligand binding functions, and elution sometimes has been achieved using RGD containing peptides. Ligand affinity columns have been used to isolate nearly all of the integrins which are composed of β_1-, β_3-, and β_5-subunits. Alternatively, integrins can be purified using immobilized monoclonal antibodies. While this approach is highly specific, it has the disadvantage of sometime requiring harsh elution conditions.

Table 1. References for cDNAs, Genbank#, and human chromosomal locations for different integrin α and β subunits.

Integrin Subunit	CD Name	cDNA reference; Species#	Genbank#	Human* Chromosome
α_1	CD49a	21,rat	--	5
α_2	CD49b	10	M28249;X17033	5
α_3	CD49c	55,56	M59911	17
α_3	CD49c	22,hamster	J05281	
α_4	CD49d	23	X16983;X15356	2
α_5	CD49e	24,25	X06256	12
α_6	CD49f	26	--	2
α_6	CD49f	57,chicken	--	
α_L	CD11a	11	Y00796	16
α_L	CD11a	58,mouse	--	
α_M	CD11b	27-29	J03925;J04145	16
	CD11b	30,mouse	X07640	
α_X	CD11c	31	--	16
α_V	CD51	32	J02826;M14648	2
	NA	33,chicken	--	
α_{IIb}	CD41	34,35	J02764	17
	CD41	36,rat	--	
α_{PS2}	NA	37,Drosophila	M19059	
β_1	CD29	24	X07979	10
	CD29	38,chicken	M14049	
	CD29	39,40,mouse	Y00769	
	NA	41,frog	J03736;M20140	
β_2	CD18	42,43	M15395;Y00057	21
β_3	CD61	44-46	J02703;M20311	17
β_4	--	14,47,48	X51841	17
β_5	--	49-51	X53002; M35011;J05633	--
β_6	--	52	J05522	--
β_7	--	53,59	M62880	--
β_8	--	60	M73780	--
β_8	--	60,rabbit	M73781	
β_{PS}	NA	9,Drosophila	J03251	

*Chromosome location only applies to human genes[2].
#Because the majority of cDNA's are human, only the non-human species are listed.
--; Information is not yet available.
NA; CD name does not apply to non-mammalian species.

ACTIVITIES

The adhesive activities of integrins have been determined (1) by using specific monoclonal antibodies to inhibit whole cell attachment to immobilized ligands, (2) by affinity chromatographic isolation of specific integrin heterodimers using ligand affinity columns, (3) by attachment of integrin containing liposomes to immobilized ligands, (4) by attachment of cells or ligands to immobilized integrin heterodimers, and (5) by using immunostaining to show the subcellular colocalization of individual integrins within regions that are in contact with ligands.

ANTIBODIES

A large number of monoclonal antibodies have been generated to most of the different α- and β-subunits[2,17]. In the absence of monoclonal or polyclonal antibodies to extracellular domains, polyclonal antibodies to C-terminal cytoplasmic domain peptides have been useful for both identification and immunostaining of various integrin subunits[18].

GENES

For all of the subunits listed in Figure 1, cDNA clones have been obtained, and references for these cDNAs as well as their GenBank numbers are listed in Table 1. Also, the chromosomal locations for several of the human subunits have been determined as indicated in Table 1. For α_X, α_{PS2}, β_3, and β2, the complete genomic organization has been determined[16,19,20,54].

REFERENCES

1. Hynes, R.O. (1987) Cell 48, 549-554.
2. Hemler, M.E. (1990) Ann. Rev. Immunol. 8, 365-400.
3. Ruoslahti, E. (1988) Ann. Rev. Biochem. 57, 375-413.
4. Elices, M.J., Urry, L.A. and Hemler, M.E. (1990) J. Cell Biol. 112, 169-181.
5. Conforti, G., Zanetti, A., Pasquali-Ronchetti, I., Quaglino, D. Jr., Neyroz, P. and Dejana, E. (1990) J. Biol. Chem. 265, 4011-4019.
6. Springer, T.A. (1990) Nature 346, 425-434.

7. Plow, E.F. and Ginsberg, M.H. (1989) Prog. Hemost. Thromb. 9, 117-156.
8. Anderson, D.C. and Springer, T.A. (1987) Ann. Rev. Med. 38, 175-194.
9. MacKrell, A.J., Blumberg, B., Haynes, S.R. and Fessler, J.H. (1988) Proc. Natl. Acad. Sci. (USA) 85, 2633-2637.
10. Takada, Y. and Hemler, M.E. (1989) J. Cell Biol. 109, 397-407.
11. Larson, R.S. Corbi, A.L., Berman, L. and Springer, T.A. (1989) J. Cell Biol. 108, 703-712.
12. Altruda, F., Cervella, P., Tarone, G., Botta, C., Balzac, F., Stefanuto, G. and Silengo, L. (1990) Gene 95, 261-266.
13. Van Kuppevelt, T.H., Languino, L.R., Gailit, J.O., Suzuki, S. and Ruoslahti, E. (1989) Proc. Natl. Acad. Sci. (USA) 86, 5415-5418.
14. Tamura, R.N., Rozzo, C., Starr, L., Chambers, J., Reichardt, L.F., Cooper, H.M. and Quaranta, V. (1990) J. Cell Biol. 111, 1593-1604.
15. Bray, P.F., Leung, C.S.-I. and Shuman, M.A. (1990) J. Biol. Chem. 265, 9587-9590.
16. Brown, N.H., King, D.L., Wilcox, M. and Kafatos, F.C. (1989) Cell 59, 185-195.
17. Knapp, W., Dorken, B., Gilks, W.R., Rieber, E.P., Schmidt, R.E., Stein, H. and von dem Borne, A.E.G.Kr. (eds.) (1989) In: Leucocyte Typing IV (Oxford University Press, New York).
18. Hynes, R.O. Marcantonio, E.E., Stepp, M.A., Urry, L.A. and Yee, G.H. (1989) J. Cell Biol. 109, 409-420.
19. Corbi, A.L., Garcia-Aguilar, J. and Springer, T.A. (1990) J. Biol. Chem. 265, 2782-2788.
20. Zimrin, A.B., Gidwitz, S., Lord, S., Schwartz, E., Bennett, J.S., White, G.C. and Poncz, M. (1990) J. Biol. Chem. 265, 8590-8595.
21. Ignatius, M.J., Large, T.H., Houde, M., Tawil, J.W., Barton, A., Esch, F., Carbonetto, S. and Reichardt, L.F. (1990) J. Cell Biol. 111, 709-720.
22. Tsuji, T., Yamamoto, F., Miura, Y., Takio, K., Titani, K., Pawar, S., Osawa, T. and Hakomori, S. (1990) J. Biol. Chem. 265, 7016-7021.
23. Takada, Y., Elices, M.J., Crouse, C. and Hemler, M.E. (1989) EMBO J. 8, 1361-1368.
24. Argraves, W.S., Suzuki, S., Arai, H., Thompson, K., Pierschbacher, M.D. and Ruoslahti, E. (1987) J. Cell Biol. 105, 1183-1190.
25. Fitzgerald, L.A., Poncz, M., Steiner, B., Rall, S.C., Jr., Bennetti, J.S. and Phillips, D.R. (1987) Biochem. 26, 8158-8165.
26. Tamura, R.N., Rozzo, C., Starr, L., Chambers, J., Reichardt, L.F., Cooper, H.M. and Quaranta, V. (1990) J. Cell Biol. 111, 1593-1604.
27. Corbi, A.L., Kishimoto, T.K., Miller, L.J. and Springer, T.A. (1988) J. Biol. Chem. 263, 12403-12411.
28. Arnaout, M.A., Gupta, S.K., Pierce, M.W. and Tenen, D.G. (1988) J. Cell Biol. 106, 2153-2158.
29. Hickstein, D.D., Hickey, M.J., Ozols, J., Baker, D.M., Back, A.L. and Roth, G.J. (1989) Proc. Natl. Acad. Sci. (USA) 85, 257-261.
30. Pytela, R. (1988) EMBO J. 7, 1371-1378.
31. Corbi, A.L., Miller, L.J., O'Connor, K., Larson, R.S. and Springer, T.A. (1987)EMBO J. 6, 4023-4028.
32. Suzuki, S., Argraves, W.S., Arai, H., Languino, L.R., Pierschbacher, M. and Ruoslahti, E. (1987) J. Biol. Chem. 262, 14080-14085.
33. Bossy, B. and Reichardt, L.F. (1990) Biochem. 29, 10191-10198.
34. Poncz, M., Eisman, R., Heidenreich, R., Silver, S.M., Vilaire, G., Surrey, S., Schwartz, E. and Bennett, J.S. (1987) J. Biol. Chem. 262, 8476-8482.
35. Uzan, G., Frachet, P., Lajmanovich, A., Frandini, M., Denarier, E., Duperray, A., Loftus, J., Ginsberg, M., Plow, E. and Marguerie, G. (1988) Eur. J. Biochem. 171, 87-93.
36. Poncz, M. and Newman, P.J. (1990) Blood 75, 1282-1289.
37. Bogaert, T., Brown, N. and Wilcox, M. (1988) Cell 51, 929-940.
38. Tamkun, J.W., DeSimone, D.W., Fonda, D., Patel, R.S, Buck, C., Horwitz, A.F. and Hynes, R.O. (1986) Cell 46, 271-282.
39. Tominaga, S. (1988) FEBS Lett. 238, 315-319.
40. Holers, V.M., Ruff, T.G., Parks, D.L., McDonald, J.A., Ballard, L.L. and Brown, E.J. (1989) J. Exp. Med. 169, 1589-1605.
41. DeSimone, D.W. and Hynes, R.O. (1988) J. Biol. Chem. 263, 5333-5340.
42. Kishimoto, T.K., O'Connor, K., Lee, A., Roberts, T.M. and Springer, T.A. (1987) Cell 48, 681-690.
43. Law, S.K.A., Gagnon, J., Hildreth, J.E.K., Wells, C.E., Willis, A.C. and Wong, A.J. (1987) EMBO J. 6, 915-919.
44. Fitzgerald, L.A., Steiner, B., Rall, S.C., Jr., Lo, S. and Phillips, D.R. (1987) J. Biol. Chem. 262, 3936-3939.
45. Rosa, J.P., Bray, P.F., Gayet, O., Johnston, G.I., Cook, R.G., Jackson, K.W., Shuman, M.A. and McEver, R.P. (1988) Blood 72, 593-600.
46. Zimrin, A.B., Eisman, R., Vilaire, G., Schwartz, E. and Bennett, J.S. (1988) J. Clin. Invest. 81, 1470-1475.
47. Hogervorst, F., Kuikman, I., von dem Borne, A.E.G.Kr. and Sonnenberg, A. (1990) EMBO J. 9, 765-770.
48. Suzuki, S. and Naitoh, Y. (1990) EMBO J. 9, 757-763.
49. Ramaswamy, H. and Hemler, M.E. (1990) EMBO J. 9, 1561-1568.
50. Suzuki, S., Huang, Z.-S. and Tanihara, H. (1990) Proc. Natl. Acad. Sci. (USA) 87, 5354-5358.
51. McLean, J.W., Vestal, D.J., Cheresh, D.A. and Bodary, S.C. (1990) J. Biol. Chem. 265, 17126-17131.
52. Sheppard, D., Rozzo, C., Starr, L., Quaranta, V., Erle, D.J. and Pytela, R. (1990) J. Biol. Chem. 265, 111502-11507.
53. Yuan, Q., Jiang, W.-M., Krissansen, G.W. and Watson, J.D. (1990) Inter. Immunol. 2, 1097-1108.
54. Weitzman, J.B., Wells, C.E., Wright, A.H., Clark, P.A. and Law, S.K.A. (1991) FEBS Lett. 294, 97-103.
55. Takada, Y., Murphy, E., Pil, P., Chen, C., Ginsberg, M.H. and Hemler, M.E. (1991) J. Cell Biol. 115, 257-266.
56. Tsuji, T., Hakomori, S. and Osawa, T. (1991) J. Biochem. 109, 659-665.
57. de Curtis, I., Quaranta, V., Tamura, R.N. and Reichardt, L.F. (1991) J. Cell Biol. 113, 405-416.
58. Kaufmann, Y., Tseng, E. and Springer, T.A. (1992) J. Immunol. 147, 369-374.
59. Erle, D.J., Ruegg, C., Sheppard, D. and Pytela, R. (1991) J. Biol. Chem. 266, 11009-11016.
60. Moyle, M., Napier, M.A. and McLean, J.W. (1991) J. Biol. Chem. 266, 19650-19658.

Martin E. Hemler:
Dana Farber Cancer Institute,
Boston, MA, USA

Intercellular Adhesion Molecules (ICAMs)

ICAM-1 (CD54) and ICAM-2 are adhesion molecules of the immune system that are members of the immunoglobulin (Ig) superfamily and bind to LFA-1 (CD11a/CD18) and Mac-1 (CD11b/CD18)[1]. Adhesion of leukocyte integrins to ICAMs is important in antigen specific responses of lymphocytes and in interaction of leukocytes with endothelium and emigration at inflammatory sites.

ICAM-1[2,3] and ICAM-2[4] are class I membrane proteins that contain five and two Ig-like domains, respectively. The two N-terminal domains of ICAM-1 are 35% identical to the two domains of ICAM-2, a much higher homology than found with other Ig family members. The extracellular domain of ICAM-1 is a bent rod 18 nm long, showing the Ig domains are arranged end to end rather than paired[5]. ICAM-1 is subverted as a receptor on nasal epithelium by the major group of rhinoviruses[6,7], and as a peripheral endothelium sequestration receptor for erythrocytes parasitized by *Plasmodium falciparum*[8]. The binding sites for the **integrin**, **LFA-1**, rhinovirus, and parasitized erythrocytes are located in distinct regions of the N-terminal, most membrane distal, Ig-like domain[5,9] (Figure). In contrast, the integrin Mac-1 binds to a site in the third Ig-like domain that is partially shielded by N-linked glycosylation[10].

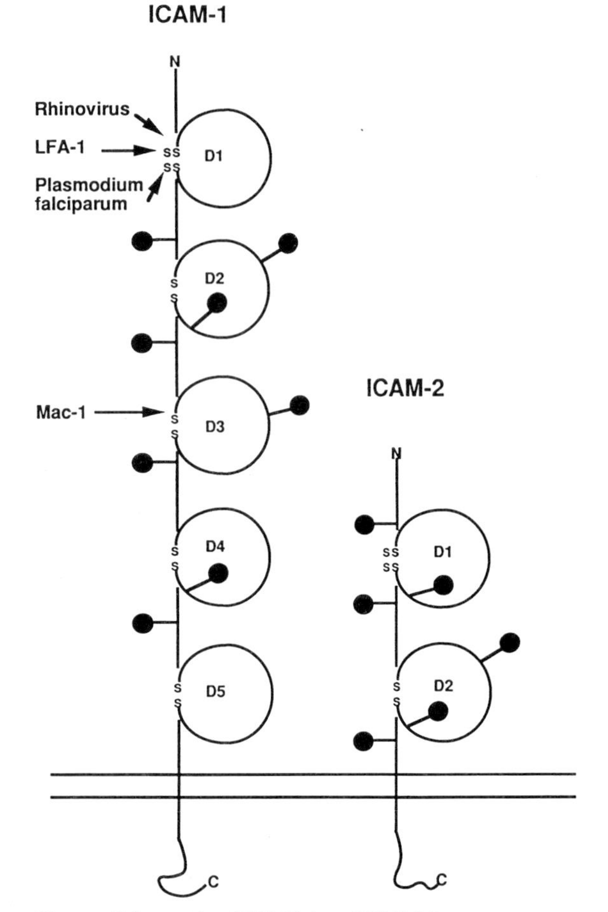

Figure. Schematic of ICAM-1 and ICAM-2.

ICAM-1 (85,000 - 110,000 M_r in SDS-PAGE and 55,319 M_r polypeptide chain)[11] and ICAM-2 (60,000 M_r in SDS-PAGE and 28,393 M_r polypeptide chain)[12] are each heavily glycosylated, all at N-linked sites (Figure). Recently, a third counter receptor for LFA-1, designated ICAM-3, has been defined with mAb that recognizes a heavily glycosylated protein of 124,000 M_r in SDS-PAGE[13].

ICAM-1 is basally expressed on endothelium, some epithelia, monocytes, and lymphocyte subpopulations[11], but inflammatory and immune stimuli induce greatly increased expression on these cell types as well as many other cell types such as connective tissue and epithelial cells that under basal conditions lack expression[11,14,15]. This facilitates leukocyte emigration in inflammatory responses and antigen specific cell interactions. In contrast, ICAM-2 is more strongly expressed basally on endothelium, lymphocytes, and monocytes, but is not inducible[12]. ICAM-3 is primarily found on leukocytes and is increased several fold upon stimulation of lymphocytes[13].

PURIFICATION

Purification of ICAM-1 from detergent lysates by mAb affinity chromatography and elution at high pH results in fully functional ICAM-1[16].

ACTIVITIES

Purified ICAM-1 can be incorporated into artificial phospholipid bilayers or absorbed onto plastic surfaces and assayed for binding to cells that express LFA-1 or Mac-1[16-19].

ANTIBODIES

ICAM-1 has been defined with mAb in the human (RR1/1[20], LB2[21], R6.5[22], CL203[23], 84H10[24]) and the mouse (YN1/1[25]). mAb to ICAM-1 are commercially available (R6.1; Bender Med. Systems, Dr. Boehringer Gasse 5-11, P.O.Box 73, A-1121, Vienna, Austria). Binding sites for mAb to ICAM-1 have been mapped by mutagenesis and correlated with inhibition of binding to ligands[5,10]. Both blocking and non-blocking mAb[12,26] are available to ICAM-2.

GENES

ICAM-1 cDNA have been cloned in the human[2,3] (accession No. J03132, X06990) and mouse[27-29] (accession No. M31585, X16624, X15372). Human ICAM-2 was first defined with cDNA[4] (accession No. X15606).

REFERENCES

1. Springer, T.A. (1990) Nature 346, 425-433.
2. Simmons, D., Makgoba, M.W. and Seed, B. (1988) Nature 331, 624-627.
3. Staunton , D.E., Marlin, S.D., Stratowa, C., Dustin, M.L. and Springer, T.A. (1988) Cell 52, 925-933.
4. Staunton, D.E., Dustin, M.L. and Springer, T.A. (1989) Nature 339, 61-64.
5. Staunton, D.E., Dustin, M.L., Erickson, H.P. and Springer, T.A. (1990) Cell 61, 243-254.
6. Staunton, D.E., Merluzzi, V.J., Rothlein, R., Barton, R., Marlin, S.D. and Springer, T.A. (1989) Cell 56, 849-853.
7. Greve, J.M., Davis, G., Meyer, A.M., Forte, C.P., Yost, S.C., Marlor, C.W., Kamarck, M.E. and McClelland, A. (1989) Cell 56, 839-847.
8. Berendt, A.R., Simmons, D.L., Tansey, J., Newbold, C.I. and Marsh, K. (1989) Nature 341, 57-59.
9. Ockenhouse, C.F., Betageri, R., Springer, T.A. and Staunton, D.E. (1992) Cell 68, 63-69.
10. Diamond, M.S., Staunton, D.E., Marlin, S.D. and Springer, T.A. (1991) Cell 65, 961-971.
11. Dustin, M.L., Rothlein, R., Bhan, A.K, Dinarello, C.A. and Springer, T.A. (1986) J. Immunol. 137, 245-254.
12. de Fougerolles, A.R., Stacker, S.A., Schwarting, R. and Springer, T.A. (1991) J. Exp. Med. 174, 253-267.
13. de Fougerolles, A.R. and Springer, T.A. (1991) J. Exp. Med. in press.
14. Kishimoto, T.K., Larson, R.S., Corbi, A.L., Dustin, M.L., Staunton, D.E. and Springer, T.A. (1989) Adv. Immunol. 46, 149-182.
15. Dustin, M.L., Staunton, D.E. and Springer, T.A. (1988) Immunol. Today 9, 213-215.
16. Marlin, S.D. and Springer, T.A. (1987) Cell 51, 813-819.
17. Dustin, M.L. and Springer, T.A. (1988) J. Cell Biol. 107, 321-331.
18. Dustin, M.L., García-Aguilar, J., Hibbs, M.L., Larson, R.S., Stacker, S.A., Staunton, D.E., Wardlaw, A.J. and Springer, T.A. (1989) Cold Spring Harbor Symp. Quant. Biol. 54, 753-765.
19. Diamond, M.S., Staunton, D.E., de Fougerolles, A.R., Stacker, S.A., García-Aguilar, J., Hibbs, M.L. and Springer, T.A. (1990) J. Cell Biol. 111, 3129-3139.
20. Rothlein, R., Dustin, M.L, Marlin, S.D. and Springer, T.A. (1986) J. Immunol. 137, 1270-1274.
21. Clark, E.A., Ledbetter, J.A., Holly, R.C., Dinndorf, P.A. and Shu, G. (1986) Hum. Immunol. 16, 100-113.
22. Rothlein, R., Czajkowski, M., O'Neil, M.M., Marlin, S.D., Mainolfi, E. and Merluzzi, V.J. (1988) J. Immunol. 141, 1665-1669.
23. Maio, M., Tessitori, G., Pinto, A., Temponi, M., Colombatti, A. and Ferrone, S. (1989) J. Immunol. 143, 181-188.
24. Makgoba, M.W., Sanders, M.E., Luce, G.E.G., Dustin, M.L., Springer, T.A., Clark, E.A., Mannoni, P. and Shaw, S. (1988) Nature 331, 86-88.
25. Takei, F. (1985) J. Immunol. 134, 1403-1407.
26. Nortamo, P., Salcedo, R., Timonen, T., Patarroyo, M. and Gahmberg, C.G. (1991) J. Immunol. 146, 2530-2535.
27. Siu, G., Hedrick, S.M. and Brian, A.A. (1989) J. Immunol. 143, 3813-3820.
28. Horley, K.J., Carpenito, C., Baker, B. and Takei, F. (1989) EMBO J. 8, 2889-2896.
29. Ballantyne, C.M., O'Brien, W.E. and Beaudet, A.L. (1989) Nucleic Acids Res. 17, 5853.

A. de Fougerolles and T.A. Springer:
Center for Blood Research,
Harvard Medical School,
Boston, MA, USA

L1

L1 is an integral membrane glycoprotein expressed by subpopulations of postmitotic neurons in the mammalian brain. It is an adhesion molecule of the immunoglobulin (Ig) superfamily and contains fibronectin type III domains. L1 mediates several types of neuron-neuron interactions, including fasciculation of neurites, cerebellar granule cell migration, adhesion of cell bodies and neurite extension.

Purified L1 from mouse brain consists of a 200 kDa glycoprotein which is readily degraded into two complementary molecules of 80 and 140 kDa[1]; the 140 kDa component comprises most of the N-terminal extracellular domain, while the 80 kDa component comprises some **fibronection** type III domains, the transmembrane part and the intracellular domain of the molecule[2,3]. Like the other L1 related molecules expressed by neurons (**TAG-1**, **F11**/F3 [contactin], **NCAM**), L1 consists of several Ig like domains at the N-terminus (6 for L1), as well as three more and two less conserved fibronectin type III domains situated between the transmembrane region and the Ig like domains[3]. A protein kinase activity associated with L1 specifically phosphorylates it[4]. Like NCAM and **J1 glycoproteins**, L1 is a member of the L2/HNK-1 carbohydrate family of adhesion molecules[5,6]. In the CNS, L1 is most prominently expressed by axons during axonal outgrowth, and less prominently by dendrites and cell bodies[7]. It is expressed by immature astrocytes in the CNS[8] and by Schwann cells in the PNS[9-14]. It is also present on some epithelial cells[15,16]. Expression of L1 by Schwann cells, immature astrocytes and neurons is enhanced by NGF, suggesting a role for L1 in promotion of neurite growth during development and regeneration[8, 17-21]. L1 is involved in fasciculation and outgrowth[22] of neurites *in vitro* and neuron-neuron[1] and neuron-Schwann cell[10] adhesion, probably in a homophilic binding manner[23]. L1 mediates expression of myelin specific proteins by Schwann cells and myelination[11,24,25] and promotes cerebellar granule cell migration[26]. Like NCAM, L1 is involved in signal transduction from the cell surface to the cytoplasm by a G-protein dependent influence on inositol phosphate turnover, intracellular Ca^{2+}-concentration and pH[27]. L1 interacts with NCAM in a so-called homophilic assisted binding mechanism which potentiates the adhesive properties of L1 in a carbohydrate dependent manner[23,28]. L1 also appears to act heterophilically[29,30].

L1 has been described in the literature under the names of Ng-CAM, (neuron-glial adhesion molecule) in the chicken[31] and NILE (NGF inducible large external glycoprotein) in the rat[32].

PURIFICATION

L1 has been purified from adult mouse brain by immunoaffinity chromatography using a monoclonal antibody column[1].

ACTIVITIES

The biological role of L1 is best assayed *in vitro* in neurite outgrowth[22], and neuron adhesion or aggregation tests[1,24].

ANTIBODIES

Polyclonal and monoclonal antibodies to mouse L1 have been described[1,24].

GENES

The full length cDNA sequence of mouse L1 has been published[3]. The mouse L1 gene (CamL1) has been mapped to the Rsvp region of the X chromosome. The L1 gene in the human has been mapped to the Xq 27-Xq 28 on the X chromosome[33,34]. EMBL Data Bank Accession number X12875.

REFERENCES

1. Rathjen, F.G. and Schachner, M. (1984) EMBO J. 3, 1-10.
2. Sadoul, K., Sadoul, R., Faissner, A. and Schachner, M. (1988) J. Neurochem. 50, 510-521.
3. Moos, M., Tacke, R., Scherer, H., Teplow, D., Früh, K. and Schachner, M. (1988) Nature 334, 701-703.
4. Sadoul, R., Kirchhoff, F. and Schachner, M. (1989) J. Neurochem. 53, 1471-1478.
5. Kruse, J., Mailhammer, R., Wernecke, H., Faissner, A., Sommer, I., Goridis, C. and Schachner, M. (1984) Nature 311, 153-155.
6. .Schachner, M., Antonicek, H., Fahrig, T., Faissner, A., Fischer, G., Künemund, V., Martini, R., Meyer, A., Persohn, E., Pollerberg, G.E., Probstmeier, R., Sadoul, K., Sadoul, R., Seilheimer, B. and Thor, G. (1990) (G.M. Edelman, B. Cunningham and J.-P. Thiery, eds.) John Wiley and Sons, Ltd., New York, 443-468.
7. Persohn, E. and Schachner, M. (1987) J. Cell Biol. 105, 569-576.
8. Saad, B., Constam, D.B., Ortmann, R., Moos, M., Fontana, A. and Schachner, M. (1991) J. Cell Biol. 115, 473-484.
9. Nieke, J. and Schachner, M. (1985) Differentiation 30, 141-151.
10. Seilheimer, B. and Schachner, M. (1988) J. Cell Biol. 107, 341-351.
11. Seilheimer, B., Persohn, E. and Schachner, M. (1989) J. Cell Biol. 109, 3095-3103.
12. Seilheimer, B., Persohn, E. and Schachner, M. (1989) J. Cell Biol. 108, 1909-1915.
13. Mirsky, R., Jessen, K.R., Schachner, M. and Goridis, C. (1986) J. Neurocytol. 15, 799-815.
14. Martini, R. and Schachner, M. (1986) J. Cell Biol. 103, 2439-2448.
15. Thor, G., Probstmeier, R. and Schachner, M. (1987) EMBO J. 6, 2581-2586.
16. Probstmeier, R., Tacke, R., Martini, R. and Schachner, M. (1990) Differentiation 44, 42-55.
17. Seilheimer, B. and Schachner, M. (1987) EMBO J. 6, 1611-1616.
18. Miragall, F., Kadmon, G., Husmann, M. and Schachner, M. (1988) Devel. Biol. 129, 516-531.
19. Miragall, F., Kadmon, G. and Schachner, M. (1989) Devel. Biol. 135, 272-286.
20. Martini, R. and Schachner, M. (1988) J. Cell Biol. 106, 1735-1746.
21. Kleitman, N., Simon, D.K., Schachner, M. and Bunge, R.P. (1989) Exp. Neurol. 102, 298-306.
22. Fischer, G., Künemund, V. and Schachner, M. (1986) J. Neuroscience 6, 605-612.
23. Kadmon, G., Kowitz, A., Altevogt, P. and Schachner, M. (1990) J. Cell Biol. 110, 193-208.
24. Wood, P.M., Moya, F., Eldridge, C., Owes, G., Ranscht, B., Schachner, M., Bunge, M. and Bunge, R.P. (1990) Ann. N.Y. Acad. Sci. 605, 1-14.
25. Wood, P.M., Schachner, M. and Bunge, R.P. (1990) J. Neurosci. 10, 3635-3645.
26. Lindner, J., Rathjen, F.G. and Schachner, M. (1983) Nature 305, 427-430.
27. Schuch, U., Lohse, M.J. and Schachner, M. (1989) Neuron 3, 13-20.
28. Kadmon, G., Kowitz, A., Altevogt, P. and Schachner, M. (1990) J. Cell Biol. 110, 209-218.
29. Werz, W. and Schachner, M. (1988) Devel. Brain Res. 43, 225-234.
30. Bartsch, U., Kirchhoff, F. and Schachner, M. (1989) J. Comp. Neurol. 284, 451-462.
31. Grumet, M. and Edelman, G.M. (1988) J. Cell Biol. 106, 487-503.
32. Bock, E., Richter-Landsberg, C., Faissner, A. and Schachner, M. (1985) EMBO J. 4, 2765-2768.
33. Djabali, M, Mattei, M.-G., Nguyen, C., Roux, D., Demengeot, J., Denizot, F., Moos, M., Schachner, M., Goridis, C. and Jordan, B.R. (1990) Genomics 7, 587-593.
34. Chapman, V.M., Keitz, B.T., Stephenson, D.A., Mullins, L.J., Moos, M. and Schachner, M. (1990) Genomics 8, 113-118.

M. Schachner:
Department Neurobiology,
Swiss Federal Institute of Technology,
Hönggeberg, 8093 Zurich,
Switzerland

Leukocyte Common Antigen Protein Tyrosine Phosphatase(LCA, CD45)

Leukocyte common antigen (LCA, CD45) is a transmembrane protein tyrosine phosphatase uniquely expressed by all nucleated cells of hematopoietic origin[1].Various isoforms of the glycoprotein exist. The molecular basis of these isoforms is due to the alternative splicing of three consecutive exons that encode a region of O-linked glycosylation near the N-terminus of the protein. The biological role of LCA is as yet not fully understood but it is required for antigen induced T lymphocyte activation.

LCA is a large, single chain, cell surface glycoprotein of which there are multiple isoforms (180-220 kDa). It consists of an N-terminal exterior domain of between 390-550 amino acids (depending upon the isoform), a short 22 amino acid membrane spanning domain and a 700 amino acid cytoplasmic region. The exterior domain can be loosely divided into three regions: an immediate N-terminal region of between 40 and 200 amino acids that contains O-linked carbohydrates and two regions of cysteine clusters; one of approximately 100 amino acids and the next of approximately 220 amino acids. Each of the cysteine cluster regions contains eight cysteines and numerous N-linked carbohydrate sites; 25% by weight of the glycoprotein is carbohydrate[2]. The region with O-linked carbohydrates is encoded by six exons (3-8). The middle three exons (4-6) are variably spliced at the mRNA level to yield eight different protein isoforms[3-7]. The expression of these isoforms is precisely controlled in both hematopoietic differentiation and cellular activation[7-9]. Thus, the size of the O-link glycosylation region expands (or contracts) in a very precise fashion during differentiation and activation, particularly within the lymphoid lineage. LCA glycosylation patterns also change during differentiation and activation, thus increasing the complexity of the glycoprotein isoforms[10]. However, as of yet, there is no direct correlation between changes in exon usage and glycosylation patterns. The interactions of the exterior domain have not been defined.

The 80 kDa cytoplasmic domain of LCA consists of two subdomains of approximately 260 amino acids each that share 33% identical residues[11]. Sequence analysis of a protein tyrosine phosphatase from human placenta revealed similar percentage of identical residues to each of the previously determined LCA cytoplasmic subdomain sequences[12], and LCA has protein tyrosine phosphatase activity[13]. However, only the first of the two protein tyrosine phosphatase domains have activity and the function of the second domain is unknown[14]. The physiologically relevant substrates are unknown. However, examination of LCA expressing and nonexpressing lymphoma cell lines, as well as antibody crosslinking studies between anti-**CD4** or anti-**CD8** and anti-CD45 antibodies have indicated that a possible substrate may be p56*lck*[15-17] (p56*lck* is a member of the ***src* protein tyrosine kinase** gene family which is associated with CD4 and CD8 and whose activity is affected by tyrosine phosphorylation). LCA deficient T cell clones have been generated[18]. Importantly, T cell clones are not transformed and require antigen, antigen presenting cells, and the growth factor IL-2 for *in vitro* maintenance. These LCA deficient cells are greatly diminished in their ability to respond to antigen, however, they can still respond to IL-2. Revertants regain the ability to respond to antigen. The data indicates that LCA is required to induce a response to antigen and suggests that the LCA protein tyrosine phosphatase is a positive regulator, rather than a negative regulator, of cell growth (in contrast to what one might have thought as a possible antagonist to the tyrosine kinases). In addition, LCA functions, at least in the T lymphocytes, in a specific signalling mechanism (through the T cell antigen receptor). The second point is difficult to reconcile, since LCA is expressed by all leukocytes, but may indicate that for some leukocytes at least two signals are required for response: a transitory environmental signal such as antigen, which is coupled with a transmembrane tyrosine phosphatase signal.

PURIFICATION

LCA can be purified from either spleen or thymus by detergent extraction of a membrane fraction and monoclonal antibody affinity chromatography[19]. Frequently, a second step of gel filtration chromatography is needed. For studies that require functionally active molecules it is best to start with fresh tissue or frozen membranes stored under reducing conditions since aggregation can occur and a free sulfhydryl group is required for enzymatic activity[14,20].

ACTIVITIES

Protein tyrosine phosphatase activity can be measured against a variety of tyrosine phosphorylated substrates including reduced, carboxyamidomethylated, maleylated lysozyme, myelin basic protein, p56*lck*, EGF receptor and various peptides[14,20]. The physiological substrate(s) are not known. Mutational analyses indicate that for nonphysiological substrates only the first of the two protein tyrosine phosphatase homology domains has activity and the highly conserved GPIVVHCSAGVGRTG sequence is required for optimum activity[14]. Assays are usually performed under neutral pH in a buffer containing EDTA and either dithiothreotol or β-mercaptoethanol.

ANTIBODIES

Antibodies against LCA to either human, mouse or rat are available commercially (Boehringer, Mannheim; Bioproducts for Science (Indianapolis, IN, USA)) or through individual investigators. Some antibodies are also available through the ATCC. The Leucocyte Typing Workshop uses the cluster of differentiation (CD) nomenclature to group antibodies that map to the same human molecules. Under this nomenclature LCA has been designated CD45. The usefulness of this nomenclature diminishes as more randomly chosen molecules are assigned CD numbers. Since there are multiple isoforms of LCA, antibodies can be made that recognize protein or carbohydrate differences between isoforms[21]. The antibodies are designated CD45R (for restricted, based on the restricted distribution of particular isoforms). Thus, those antibodies that recognize an epitope encoded by the first variable exon, exon 4, are designated CD45RA, those that recognized an exon 5 encoded epitope are designated CD45RB, etc. An antibody that recognizes just the low molecular weight form of LCA (encoded by an mRNA that joins exons 3 and 7, removing the three variable exons) are designated CD45RO.

GENES

cDNAs have been isolated and sequenced for human mouse and rat and the genomic organization elucidated for human and mouse[3,4,6-8,22-25] (GenBank Y00062, J04640, J04691, M10072, K03039). The genes have been mapped to chromosome 1 for both human and mouse (1q3.2 for human) to a region that is synthetic between these two species. Clones are available to interested investigators.

REFERENCES

1. Thomas, M.L. (1989) Ann. Rev. Immunol. 7, 339-369.
2. Brown, W.R.A., Barclay, A.N., Sunderland, C.A. and Williams, A.F. (1981) Nature 289, 456.
3. Ralph, S.J., Thomas, M.L., Morton, C.C. and Trowbridge, I.S. (1987) EMBO J. 6, 1251.
4. Barclay, A.N., Jackson, D.I., Willis, A.C. and Williams, A.F. (1987) EMBO J. 6, 1259.
5. Thomas, M.L., Reynolds, P.J., Chain, A., Ben-Neriah, Y. and Trowbridge, I.S. (1987) Proc. Natl. Acad. Sci. (USA) 84, 5360.
6. Streuli, M., Hall, L.R., Saga, Y., Schlossman, S.F. and Saito, H. (1987) J. Exp. Med. 166,1548.
7. Saga, Y., Furukawa, K., Rogers, P., Tung, J.-S., Parker, D. and Boyse, E.A. (1990) Immunogenetics 31, 296-306.
8. Sanders, M.E., Makgoba, M.W. and Shaw, S. (1988) Immunol. Today 9, 195-199.
9. Mason, D.W. (1988) Immunol. Lett. 14, 169.
10. Brown, W.R.A. and Williams A.F. (1982) Immunology 46, 713.
11. Thomas, M.L., Barclay, A.N., Gagnon, J. and Wiliams, A.F. (1985) Cell 41, 83.
12. Charbonneau, H., Tonks, N.K., Walsh, K.A. and Fisher, E.H. (1988) Proc. Natl. Acad. Sci. (USA) 85, 7182-7186.
13. Tonks, N.K., Charbonneau, H., Diltz, C.D., Fischer, E.H. and Walsh, K.A. (1988) Biochemistry 27, 8694-8701.
14. Streuli, M., Krueger, N.X., Thai, T., Tang, M. and Saito, H. (1990) EMBO J. 9, 2399-2407.
15. Mustelin, T., Coggeshall, K.M. and Altman, A. (1989) Proc. Natl. Acad. Sci. (USA) 86, 6302-6306.
16. Ostergaard, H.L., Schackelford, D.A., Hurley, T.R., Johnson, P., Hyman, R., Sefton, B.M. and Trowbridge, I.S. (1989) Proc. Natl. Acad. Sci. (USA) 86, 8959-8963.
17. Ostergaard, H.L. and Trowbridge, I.S. (1990) J. Exp. Med. 172, 347-350.
18. Pingel, J.T. and Thomas, M.L. (1989) Cell, 58, 1055-1065.
19. Sunderland, C.A., McMaster, W.R. and Williams, A.F. (1979) Eur. J. Immunol. 9, 155.
20. Tonks, N.K., Diltz, C.D. and Fischer, E.H. (1990) J. Biol. Chem. 265, 10674-10680.
21. Streuli, M., Morimoto, C., Schreiber, M., Schlossman, S.F. and Saito, H. (1988) J. Immunol. 141, 3910-3914.
22. Saga, Y., Tung, J.-S., Shen, F.-W. and Boyse, E.A. (1987) Proc. Natl. Acad. Sci. (USA) 84, 5364.
23. Hall, L.R., Streuli, M., Schlossman, S.F. and Saito, H. (1988) J. Immunol. 141, 2781-2787.
24. Saga, Y., Tung, J.-S., Shen, F.-W., Pancoast, T.C. and Boyse, E.A. (1988) Mol. Cell. Biol. 8, 4889-4895.
25. Johnson, N.A., Meyer, C.M., Pingel, J.T. and Thomas, M.L. (1989) J. Biol. Chem. 264, 6220-6229.

Matthew L. Thomas:
Department of Pathology,
Washington University, School of Medicine,
St. Louis, MO 63110, USA

LFA-1 (CD11a/CD18)

LFA-1 (lymphocyte function associated antigen 1) is an adhesion receptor on leukocytes that is a member of the integrin family[1]. Originally identified with monoclonal antibodies (mAb) that inhibit lymphocyte mediated antigen specific killing[2-5], LFA-1 has been shown to participate in a wide variety of cell adhesion interactions, including neutrophil, monocyte and lymphocyte adhesion to endothelium[6].

LFA-1 binds to intercellular adhesion molecules (**ICAM**s) that are immunoglobulin (Ig) superfamily members[6]. Adhesiveness of LFA-1 for ICAMs is regulated by cellular stimulation[7]. LFA-1 contains an αL, or CD11a subunit of 180 kDa and a β, or CD18 subunit of 95 kDa that are non-covalently associated in an $\alpha_1\beta_1$ complex[8,9] (Figure). Both subunits are transmembrane proteins and contain high mannose and complex-type N-linked carbohydrate[10,11]. The β subunit is shared with two other leukocyte **integrins**, Mac-1 and p150,95. Surface expression of LFA-1, Mac-1 and p150,95 is lacking in leukocyte adhesion deficiency, an inherited disease caused by β subunit mutations and characterized by recurrent life threatening infections and lack of neutrophil emigration at inflammatory sites[6]. LFA-1 is expressed on all leukocytes except resident macrophages and is not found on other cell types[12,13].

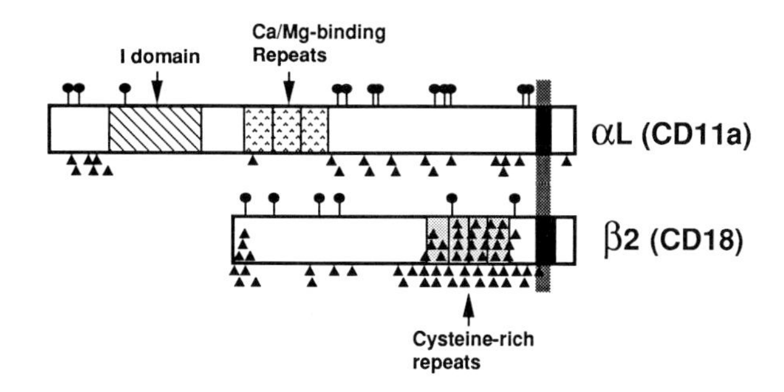

Figure. Schematic of LFA-1, based on α[21] and β subunit[1] cDNA sequences.

■ PURIFICATION

LFA-1 can be purified from detergent lysates by mAb affinity chromatography. With Mg^{2+} and the right choice of mAb, high pH elution conditions can be found that retain adhesive activity and subunit association[7].

■ ACTIVITIES

Adhesive activity of LFA-1 can be measured in four ways. 1. Certain mAb to LFA-1 α and common β subunit block lymphocyte, neutrophil and monocyte adhesion and adhesion-dependent functions[2-6,14,15]. 2. Adhesive activity of LFA-1 on the surface of cells can be measured by binding of these cells to purified ICAM-1 absorbed on plastic surfaces or in artificial membrane bilayers[7,16]. 3. Adhesive activity of purified LFA-1 in protein micelles can be measured by binding to purified ICAM-1 adsorbed to a substrate[7]. 4. Adhesiveness of purified LFA-1 bound to a substrate can be measured by binding of cells bearing ICAMs[7].

■ ANTIBODIES

MAb to the LFA-1 α subunit are specific for LFA-1, whereas mAb to the β subunit crossreact with Mac-1 and p150,95[9,17-19]. The M17/4 and M18/2.a mAb to the murine α and β subunits, respectively[20], and the TS1/22 and TS1/18 mAb to the human α and β subunits, respectively[4,9], have been deposited in the American Type Culture Collection and Developmental Biology Hybridoma Bank, and many of these mAb are commercially available (Boehringer-Mannheim for mAb to mouse and Boehringer Ingelheim for mAb to human).

■ GENES

The cDNAs have been cloned for human LFA-1 α[21] (Y00796), human β[1] (M15395), murine α[22] (M60778) and murine β[23] (X14951) and expressed[22,24].

■ REFERENCES

1. Kishimoto, T.K., O'Connor, K., Lee, A., Roberts, T.M. and Springer, T.A. (1987) Cell 48, 681-690.
2. Davignon, D., Martz, E., Reynolds, T., Kurzinger, K. and Springer, T.A. (1981) Proc. Natl. Acad. Sci. (USA) 78, 4535-4539.
3. Pierres, M., Goridis, C. and Goldstein, P. (1982) Eur. J. Immunol. 12, 60-69.
4. Sanchez-Madrid, F., Krensky, A.M., Ware, C.F., Robbins, E., Strominger, J.L., Burakoff, S.J. and Springer, T.A. (1982) Proc. Natl. Acad. Sci. (USA) 79, 7489-7493.
5. Hildreth, J.E.K., Gotch, F.M., Hildreth, P.D.K. and McMichael, A.J. (1983) Eur. J. Immunol. 13, 202-208.
6. Larson, R.S. and Springer, T.A. (1990) Immunol. Rev. 114, 181-217.
7. Dustin, M.L. and Springer, T.A. (1989) Nature, 341, 619-624.
8. Kürzinger, K. and Springer, T.A. (1982) J. Biol. Chem. 257, 12412-12418.
9. Sanchez-Madrid, F., Nagy, J., Robbins, E., Simon, P. and Springer, T.A. (1983) J. Exp. Med. 158, 1785-1803.
10. Miller, L.J. and Springer, T.A. (1987) J. Immunol. 139, 842-847.
11. Dahms, N.M. and Hart, G.W. (1986) J. Biol. Chem. 261, 13186-13196.
12. Kürzinger, K., Reynolds, T., Germain, R.N., Davignon, D., Martz, E. and Springer, T.A. (1981) J. Immunol. 127, 596-602.
13. Krensky, A.M., Sanchez-Madrid, F., Robbins, E., Nagy, J., Springer, T.A. and Burakoff, S.J. (1983) J. Immunol. 131, 611-616.

14.Springer, T.A., Dustin, M.L., Kishimoto, T.K. and Marlin, S.D. (1987) Annu. Rev. Immunol. 5, 223-252.
15.Diamond, M.S., Johnston, S.C., Dustin, M.L., McCaffery, P. and Springer, T.A. (1989) In: Leukocyte Typing IV (eds. Knapp, W., Dorken, B., Gilks, W.R., Rieber,E.P., Schmidt, R.E., Stein, H. and von dem Borne, A.E.G.Kr.) Vol. 4th, pp 570-574 (Oxford University, London)
16.Marlin, S.D. and Springer, T.A. (1987) Cell 51, 813-819.
17.Sanchez-Madrid, F., Simon, P., Thompson, S. and Springer, T.A. (1983) J. Exp. Med. 158, 586-602.
18.Hildreth, J.E.K. and August, J.T. (1985) J. Immunol. 134, 3272-3280.
19.Larson, R.S., Hibbs, M.L., Corbi, A.L., Luther, E., García-Aguilar, J. and Springer, T.A. (1989) In Leukocyte Typing IV (eds. Knapp, W., Dorken, B., Gilks, W.R., Rieber, E.P., Schmidt, R.E., Stein, H. and von dem Borne, A.E.G.Kr.) Vol. 4th, pp 566-570 (Oxford University, London)
20.Sanchez-Madrid, F., Davignon, D., Martz, E. and Springer, T.A. (1982) Cell. Immunol. 73, 1-11.
21.Larson, R.S., Corbi, A.L., Berman, L. and Springer, T.A. (1989) J. Cell Biol. 108, 703-712.
22.Kaufman, Y., Tseng, E. and Springer, T.A. (1991) J. Immunol. 147, 369-374.
23.Wilson, R., O'Brien, W. and Beaudet, A. (1989) Nucl. Acids Res. 17, 5397.
24.Larson, R.S, Hibbs, M.L. and Springer, T.A. (1990) Cell Reg. 1, 359-367.

T.A. Springer:
Center for Blood Research,
Harvard Medical School,
Boston, MA, USA

LFA-3 (CD58)

LFA-3 (lymphocyte function associated antigen 3) is an adhesion molecule that is a member of the immunoglobulin (Ig) superfamily and that is important in immune cell interactions[1]*. LFA-3 was originally defined with monoclonal antibodies (mAb) that inhibit T lymphocyte mediated killing*[2] *and subsequently was shown to bind to the CD2 (LFA-2, T11, or erythrocyte rosette receptor) molecule on T lymphocytes*[3,4]*.*

The LFA-3 molecule is expressed on monocytes, granulocytes, B lymphocytes, the memory subpopulation of T lymphocytes, platelets, endothelial cells, epithelial cells, connective tissue cells and in lower amount ($4x10^3$ molecules/cell) on erythrocytes, but not on thymocytes[1,5-7]. A similar molecule termed T11TS has been described in sheep[8]. Adhesion of the **CD2** molecule on T lymphocytes to the LFA-3 molecule on antigen presenting cells and target cells facilitates antigen specific responses[1,2,5,8,9]. Human T lymphocytes can rosette with erythrocytes; rosetting with human and sheep erythrocytes is mediated by interaction of CD2 with human LFA-3 and its sheep homologue T11TS, respectively[6-8].

or

LFA-3
CD58

Figure. Schematic of the two isoforms of LFA-3

LFA-3 is a member of the Ig superfamily with two Ig-like domains[10,11]. It exists with both phosphatidylinositol glycan and polypeptide membrane anchors on all examined cell types except erythrocytes, which express only the lipid anchored isoform[12]. The two isoforms (Figure) arise from alternative mRNA splicing[10,11] and their significance is unknown[13]. LFA-3 migrates as a diffuse band in SDS-PAGE; the average size varies from 43,000 to 76,000 M_r depending on the cell type on which it is expressed[12]. This heterogeneity is from N-linked glycosylation at six sites; the polypeptide chain backbones are 29 and 25.5 kDa for the polypeptide and lipid anchored isoforms, respectively.

■ PURIFICATION

LFA-3 has been purified by mAb affinity chromatography from detergent lysates of erythrocytes and nucleated cells, or after release by phosphatidylinositol specific phospholipase C[14,15]. The polypeptide anchored isoform may be selectively purified from mutant cells that lack the lipid anchored isoform[13,15].

■ ACTIVITIES

Adhesiveness due to LFA-3 may be blocked with the mAb TS2/9[2]. Purified detergent-released lipid anchored LFA-3 forms octameric protein micelles that bind to CD2 on the surface of cells with K_d=1 nM; phospholipase-released LFA-3 is monomeric and has a K_i=1 μM[15]. Purified LFA-3 may be bound to plastic surfaces or incorporated in artificial membrane bilayers and assayed by its ability to bind CD2+ cells[15]. Purified lipid-anchored LFA-3 mediates agglutination of CD2+ cells[6] and both multimeric and monomeric LFA-3 preparations inhibit rosetting of T lymphocytes with erythrocytes[15].

ANTIBODIES

The TS2/9 hybridoma (IgG1) to LFA-3 is available from the American Type Culture Collection and the Developmental Biology Hybridoma Bank.

GENES

The cDNAs for the polypeptide chain anchored isoform[10] (accession No.Y00636) and the lipid-anchored isoform[11] (accession No. X06296) of LFA-3 are published.

REFERENCES

1. Springer, T.A. (1990) Nature 346, 425-433.
2. Sanchez-Madrid, F., Krensky, A.M., Ware, C.F., Robbins, E., Strominger, J.L., Burakoff, S.J. and Springer, T.A. (1982) Proc. Natl. Acad. Sci. (USA) 79, 7489-7493.
3. Selvaraj, P., Plunkett, M.L., Dustin, M., Sanders, M.E., Shaw, S. and Springer, T.A. (1987) Nature 326, 400-403.
4. Shaw, S., Luce, G.E.G., Quinones, R., Gress, R.E., Springer, T.A. and Sanders, M.E. (1986) Nature 323, 262-264.
5. Krensky, A.M., Sanchez-Madrid, F., Robbins, E., Nagy, J., Springer, T.A. and Burakoff, S.J. (1983) J. Immunol. 131, 611-616.
6. Plunkett, M.L., Sanders, M.E., Selvaraj, P., Dustin, M.L. and Springer, T.A. (1987) J. Exp. Med. 165, 664-676.
7. Selvaraj, P., Dustin, M.L., Mitnacht, R., Hünig, T., Springer, T.A. and Plunkett, M.L. (1987) J. Immunol. 139, 2690-2695.
8. Hünig, T. (1985) J. Exp. Med. 162, 890-901.
9. Palacios, R. and Martinez-Maza, O. (1982) J. Immunol. 129, 2479-2485.
10. Wallner, B.P., Frey, A.Z., Tizard, R.J., Hession, C., Sanders, M.E., Dustin, M.L. and Springer, T.A. (1987) J. Exp. Med. 166, 923-934.
11. Seed, B. (1987) Nature 329, 840-842.
12. Dustin, M.L., Selvaraj, P., Mattaliano, R.J. and Springer, T.A. (1987) Nature 329, 846-848.
13. Hollander, N., Selvaraj, P. and Springer, T.A. (1988) J. Immunol. 141, 4283-4290.
14. Dustin, M.L., Sanders, M.E., Shaw, S. and Springer, T.A. (1987) J. Exp. Med. 165, 677-692.
15. Dustin, M.L., Olive, D. and Springer, T.A. (1989) J. Exp. Med. 169, 503-517.

T.A. Springer:
Center for Blood Research,
Harvard Medical School,
Boston, MA, USA

Mannose Binding Proteins (MBP)

Mannose binding proteins (MBP) are hepatocyte derived serum proteins that have been isolated from the liver and serum of a number of mammalian species[1]. Oligomerization of a 32 kDa subunit appears to be mediated via a cysteine-rich N-terminus that stabilized the helical formation of a collagen like domain which is followed by the carbohydrate recognition domain (CRD). MBP is a member of an ever growing family of animal lectin like proteins[1] and MBP's predominant biological role appears to be as a first line host defense against a variety of pathogens[2].

Mannose binding proteins have been isolated from the liver and serum of rabbits, chickens, rats, and humen where they exist as multimers of a 32 kDa subunit (references within[3]). Homologous molecules from rat and man are the best characterized[1]. Two homologous rat MBPs designated MBP-A and MBP-C have been structurally characterized and their corresponding cDNAs isolated[4]. MBP-A appears to be the predominant serum form and assembles into oligomers of ~650 kDa whilst MBP-C forms multimers of ~200 kDa[5]. In addition, cDNAs from mouse that correspond to rat MBP-A and C respectively have been isolated[6]. However, no information about the native mouse proteins exists. Although native human MBP isolated by mannan affinity chromatography appeared to resolve into two fractions of ~650 kDa and ~200 kDa[7], another report indicates that the lower molecular weight species may indeed be MBP-C which binds mannan sepharose[8]. Amino acid sequences of the N-terminus of the predominant serum form of human MBP is represented in a human MBP cDNA[9]. This sequence appeared to be most homologous to rat MBP-C, however, repeat sequencing of these original clones as well as four other full length human cDNAs have revealed several errors in the reported sequence[9]. The corrected sequence (EMBL Accession 15422), also confirmed by analysis of the human MBP gene[10,11], reveals that human MBP shows 62% and 61% identity with rat MBP-C and A respectively. Despite extensive search, no other human MBP cDNAs have been identified and there appears to be only one human MBP gene. These findings do not exclude the presence of another MBP homologue in humen but are suggestive that only one human MBP exists. Truncated recombinant forms of MBP-C and MBP-A monomers appear to have distinct and overlapping binding specificities[12], but it is unclear at present how these findings relate to human MBP. The rat MBP-C gene and the human MBP gene have similar struc-

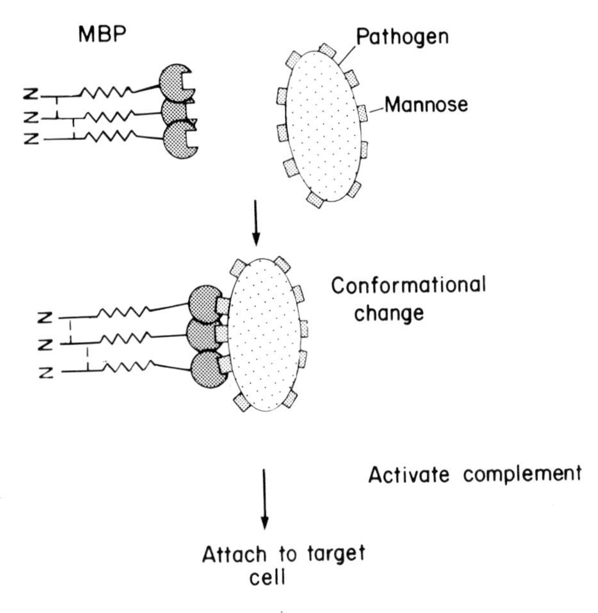

Figure 1. Proposed model describing the role of serum mannose-binding proteins as a direct opsonin or as an opsonin in the presence of complement.

tures and have an organization similar to human surfactant Sp-A gene[10,13]. The functional domains of MBPs and Sp-A appear to have evolved by juxtaposition of exons encoding genes for nonfibrillar **collagen** with those encoding carbohydrate recognition domains[10,13]. The gene for human MBP is located on chromosome 10 at 10q11.2-q21[10] and is an informative marker in some families with multiple endocrine neoplasma type 2A[14].

The CRD of rat MBP-A has been expressed in *E. coli* and the 3-dimensional structure was determined by multi-wavelength anomalous dispersion phasing using crystals in which the trivalent lanithanide ion, holmium, was substituted for calcium[15]. The structure demonstrates that the CRD is a compact unit which is divided in half by a β strand. The upper half contains two holmium sites that represent calcium binding regions. The location of the binding site is not obvious and the solution of the MBP-CRD ligand complex is yet to be determined.

PURIFICATION

MBPs are routinely purified by conventional mannan affinity chromatography[16] or antibody affinity purification (unpublished). Human MBP has been expressed in mammalian cells in an active form[17] and functional truncated rat MBP-C has been expressed in *E. coli*[18].

ACTIVITIES

The predominant biological role appears to be as a first line host defense against a wide range of pathogens that have mannose rich cell walls. MBPs bind mannans[19], can serve as opsonins, by enhancing clearance of certain gram negative organisms by phagocytes (Figure 1)[17], prevent HIV infection of H9 lymphoblasts[20] by binding to the high mannose glycans in the envelope glycoprotein of HIV[20,21], and may indicate natural resistance against certain strains of influenza virus in mice[22]. In addition, MBP ligand complexes are able to activate both the classical[23-25] and alternative complement pathways[26]. The activation of the classical pathway is most likely based on structural similarity with C1q as shown by electron microscopy rotary shadowing (Figure 2). In addition, inadequate serum levels in human serum appear to predispose children to recurrent infections[27].

ANTIBODIES

Polyclonal sera against rat[17,28] and human MBP is available[27,29]. In addition, we have recently prepared a panel of monoclonal antibodies to human MBP (unpublished).

GENES

Two rodent forms of MBP have been cloned and sequenced[17]. Human MBP cDNAs have been characterized (EMBL Accession 15422) as have the rat MBP-C gene[10] and the human MBP gene[8].

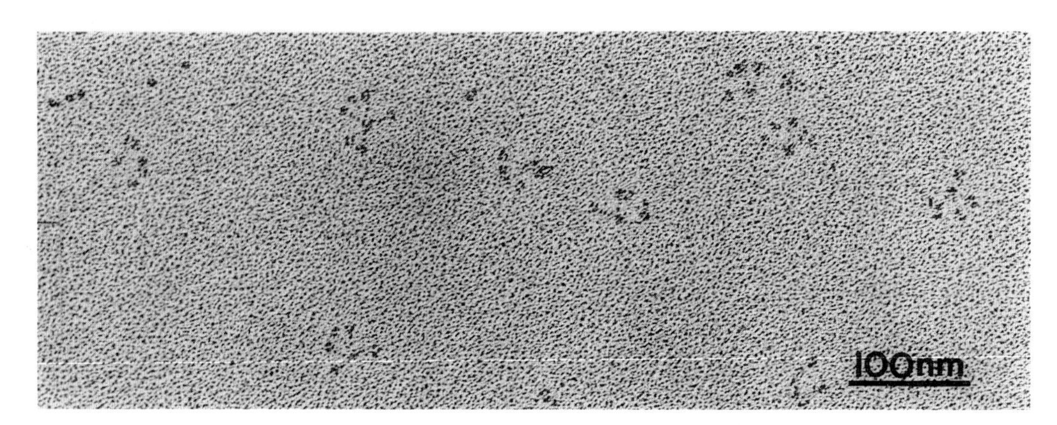

Figure 2. Electron microscopy of human MBP. (Picture courtesy of Dr. K. Reid.)

REFERENCES

1.Drickamer, K. (1988) J. Biol. Chem. 263, 9557-9560.
2.Ezekowitz, R.A.B. (1991) Current Biol. 1, 60-62.
3.Ezekowitz, R.A.B. and Stahl, P.D. (1988) J. Cell Science Supple. 9, 121-133.
4.Drickamer, K., Dordal, M.S. and Reynolds, L. (1986) J. Biol. Chem. 261, 6878-6887.
§5.Wild, J., Robinson, D. and Winchester, B. (1983) Biochem J. 210,167-174.
6.Sastry, K., Zahedi, K., Lelias, J.M., Whitehead, A. and Ezekowitz, R.A.B. (1991) J. Immun. 147, 692-697.
7.Taylor, M.E. and Summerfield, J.A. (1987) Biochem. Biophys. Acta 915, 60-67.
8.Colley, K.M., Beranek, C. and Baenziger, J.U. (1988) Biochem. J. 256, 61-68.
9.Ezekowitz, R.A.B., Day, L. and Herman, G. (1988) J. Exp. Med. 167, 1034-1046.
10.Sastry, K., Herman, G., Day, L., Deignan, E., Burns, G., Morton, C. and Ezekowitz, R.A.B. (1989) J. Exp. Med. 170, 1175-1189.
11.Taylor, M.E., Brickell, P.M., Craig, R.K. and Summerfield, J.A. (1989) Biochem. J. 262, 763-771.
12.Childs, R.A., Feizi, T., Yuen, C.T., Drickamer, K. and Queenberry, M.S. (1990) J. Biol. Chem. 265, 20770-20777.
13.Drickamer, K. and McCreary, V. (1987) J. Biol. Chem. 262, 2582-2589.
14.Schuffenecker, I., Narod, S.A., Ezekowitz, R.A.B., Sobol, H., Feunteum, J. and Lenoir, G.M. (1991) Cytogenetics and Cell Genetics, in press.
15.Weis, I.W., Kahn, R., Foume, R., Drickamer, K. and Hendrickson, W.A. (1991) Science 254, 1608-1615.
16.Maynard, Y. and Baenziger, J.U. (1982) J. Biol. Chem. 257, 3788-3794.
17.Kuhlman, M., Joiner, K. and Ezekowitz, R.A.B. (1989) J. Exp. Med. 169, 1733-1745.
18.Childs, R.A., Drickamer, K., Kawasaki, T., Thiel, S., Mizuochi, T. and Feizi, T. (1989) Biochem. J. 262, 131-138.
19.Kawasaki, T., Etoh, R. and Yamashura, I. (1978) Biochem. Biophys. Res. Comm. 81, 1018-1024.
20.Ezekowitz, R.A.B., Kuhlman, M., Groopman, J. and Byrn, R. (1989) J. Exp. Med. 169, 185-196.
21.Feizi, T. and Larkin, M. (1990) Glycobiology 1, 17-33.
22.Anders, E.M., Hartley, C.A. and Jackson, D.C. (1990) Proc. Natl. Acad. Sci. (USA) 87, 4485-4489.
23.Ikeda, K., Sannoh, T., Kawasaki, N., Kawasaki, T. and Yamashina, I. (1987) J. Biol. Chem. 262, 7451-7451.
24.Lu, J.H., Thiel, S., Wiedmann, H., Timpl, R. and Reid, K.B. (1990) J. Immunol. 144, 2287-2294.
25.Super, M., Levinsky, R.J. and Turner, M.W. (1990) Clin. Exp. Immunol. 79, 144- 150.
26.Schweinle, J., Ezekowitz, R.A.B., Tenner, A. and Joiner, K. (1989) J. Clin. Invest. 84, 1821-1829.
27.Super, M., Thiel, S., Lu, J. and Turner, M.W. (1989) Lancet 2 (8674), 1236-1238.
28.Colley, J.J. and Baenziger, J. (1987) J. Biol. Chem. 262, 3415.
29.Colley, K.J. and Baenziger, J.U. (1987) J. Biol. Chem. 262, 3415-3421.

R. Alan B. Ezekowitz:
Division of Hematology/Oncology,
Children's Hospital and Dana Farber Cancer Institute,
Dept. of Pediatrics, Harvard Medical School,
Boston, MA, USA

MUC18

MUC18[1,2] is a cell surface glycoprotein of human melanoma cells. Within the melanocyte lineage, MUC18 is expressed on advanced primary tumours and metastases, indicating that it is a marker of tumour progression in melanomas. MUC18 belongs to the immunoglobulin (Ig) supergene family and shows the highest sequence similarity to molecules which have been shown to mediate intercellular adhesion.

MUC18 is a single chain cell surface molecule which migrates on SDS-PAGE with an apparent molecular mass of 113 kDa under reducing conditions. Under nonreducing conditions it migrates with an apparent molecular mass of 98 kDa, indicating the presence of intrachain disulphide bonds. MUC18 expresses the sulphated glucuronic acid epitope defined as CD57, a characteristic shared with the neural cell adhesion molecules **NCAM**, **L1** and **myelin associated glycoprotein** (MAG)[3]. Deglycosylated MUC18 has an apparent molecular mass of 65 kDa under reducing conditions.

A single mRNA species of 3.3 kB is detected in melanoma cell lines. Sequence analysis of the MUC18 cDNA indicates that it is a member of the Ig supergene family[4]. MUC18 has five Ig domains; domains I and II belong to the V-set sequences while domains III-V belong to the C2-set sequences. MUC18 is most closely related to DCC (22.5% identity over 377 amino acids), a putative tumour suppressor gene which is deleted or mutated in colorectal carcinomas[5], carcinoembryonic antigen (20% identity over 297 amino acids) and a series of neural cell adhesion molecules including NCAM (22% identity over 196 amino acids) and L1 (22.6% identity over 288 amino acids). As each of these molecules has been shown to mediate heterotypic and/or homotypic intercellular adhesion, it is likely that MUC18 has a similar function.

Investigation of a wide range of normal adult tissues with immunohistochemical methods on frozen tissue sections revealed that the expression of MUC18 is limited to some smooth muscle, particularly that in small blood vessels[1]. The intriguing aspect of MUC18 is its expression on cutaneous malignant melanomas (Figure)[6]. The molecule cannot be detected on normal epidermal melanocytes and is very rare on the benign melanocytic tumours

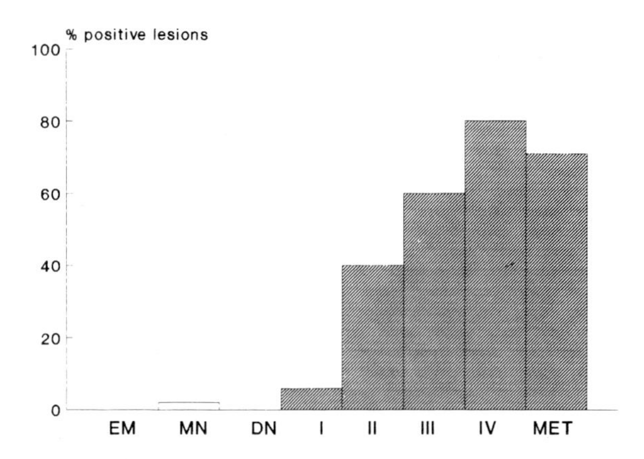

Figure. The expression of MUC18 on human melanocytes *in situ*. EM, epidermal melanocytes; MN, melanocytic nevi; DN, dysplastic nevi; I, primary melanomas < 0.75 mm thick, 5 year mortality 2%; II primary melanomas 0.76-1.5 mm thick, 5 year mortality 10%; III, primary melanomas 1.51-3.0 mm thick, 5 year mortality 23%; IV, primary melanomas > 3.0 mm thick, 5 year mortality 54%; MET, metastases.

known as nevi or moles. Its expression is however characteristic of the majority of malignant melanoma metastases. In cutaneous melanoma the probability of eventual development of metastatic disease can be predicted from the vertical thickness of the primary tumour[7]. With increasing tumour thickness both the frequency and intensity of MUC18 expression by melanoma cells increases indicating that the expression of this molecule correlates with tumour progression and the development of metastatic potential.

PURIFICATION

MUC18 can be purified from nonionic detergent lysates of melanomas and melanoma cell lines by affinity chromatography using monoclonal antibodies[2].

ACTIVITIES

The biological function of MUC18 remains unknown. However, from the sequence similarities to other cell adhesion molecules, MUC18 is predicted to mediate intercellular adhesion.

ANTIBODIES

The original MUC18 monoclonal antibody is suitable for immunohistochemistry and immunoprecipitation[1]. Five additional monoclonal antibodies were produced against the denatured, isolated antigen[2]. These antibodies are suitable for immunoprecipitation and for detection of the antigen by immunoblotting. MUC BA 18.4, like MUC18, reacts with the cell surface of unfixed cells while the other antibodies react only with fixed cells. All of the second generation antibodies show a strong reactivity with smooth muscle[2].

GENES

The complete cDNA sequence of the human MUC18 has been published (GenBank M28882)[2].

REFERENCES

1. Lehmann, J.M., Holzmann, B., Breitbart, E.W., Schmiegelow, P., Riethmüller, G. and Johnson, J.P. (1987) Cancer Res. 47, 841-845.
2. Lehmann, J.M., Riethmüller, G. and Johnson, J.P. (1989) Proc. Natl. Acad. Sci. (USA) 86, 9891-9895.
3. Kruse, J., Mailhammer, R., Wernecke, H., Faissner, A., Sommer, I., Goridis, C. and Schachner, M. (1984) Nature 311, 153-155.
4. Williams, A.F. and Barclay, A.N. (1988) Ann. Rev. Immunol. 6, 381-405.
5. Fearon, E.R., Cho, K.R., Nigro, J.M., Kern, S.E., Simons, J.W., Ruppert, J.M., Hamilton, S.R., Preisinger, A.C., Thomas, G., Kinzler, K.W. and Vogelstein, B. (1990) Science 247, 49-56.
6. Holzmann, B., Bröcker, E.B., Lehmann, J.M., Ruiter, D.J., Sorg, C., Riethmüller, G. and Johnson, J.P. (1987) Int. J. Cancer 39, 466-471.
7. Breslow, A. (1970) Ann. Surg. 172, 902-908.

Judith P. Johnson:
Institute for Immunology,
Goethestraße 31, D-8000 Munich 2,
Germany

Myelin Associated Glycoprotein (MAG)

MAG is a transmembrane glycoprotein of the immunoglobulin (Ig) gene superfamily expressed by glial cells of the CNS and PNS. It is postulated to have five Ig domains that mediate binding of Schwann cells and oligodendrocytes to neurons during the development of the myelin sheath.

MAG[1,2] runs as a broad band of ~100 kDa on SDS-PAGE. 30% of its weight is contributed by N-linked complex carbohydrates; the protein backbone is either 67 or 72 kDa (p67, p72) due to alternative splicing of the cytoplasmic domain[3,4]. p72 is predominant in oligodendrocytes during the period of myelination, switching to p67 in the adult. In the PNS, p67 is the major form throughout development although p72 mRNA is present at low levels[5,6]. The high percentage of acidic amino acids and the content of sialic acid and sulphate residues contribute to a pI in the range of 3-4.5. MAG is also palmitylated[7] and phosphorylated on tyrosine, serine and threonine both *in vivo* and *in vitro*[8,9]. The extracellular domain contains no free cysteines: the position of seven disulphide bridges have been tentatively mapped[7].

By immunocytochemistry, MAG staining is seen throughout the wraps of loose myelin; after compaction, MAG is restricted to the first wrap of myelin i.e. the periaxonal space, the Schmidt-Lanterman incisures, the paranodal loops, and the inner and outer mesaxons[10,11]. Using an *in vitro* myelination system, MAG expression is seen to correlate with contact of the Schwann cell with the axon, and subsequent elongation of the glial cell process along the axon and investment[12]. Schwann cells expressing antisense MAG mRNA ensheath axon fascicles but cannot proceed to segregation of the large caliber axons and formation of an individual sheath[13].

IgM antibodies, binding to a carbohydrate epitope shared by MAG, other adhesion molecules and glycolipids, have been found in cases of plasma cell dyscrasias with associated peripheral neuropathy[14]. It also appears that MAG is among the first proteins broken down in multiple sclerosis (MS) plaques and some studies reveal the presence of anti-MAG antibodies in patients with MS[2].

PURIFICATION

MAG was first identified as the major glycoprotein of CNS myelin by ^{3}H-fucose labelling[15]. It can be purified from myelin by extraction with lithium diiodosalicyte and phenol[16] or by immunoaffinity chromatography using a monoclonal antibody, 513[17].

ACTIVITIES

Direct binding of MAG to neuronal surfaces was first shown by Poltorak et al.[17] using affinity purified MAG incorporated into liposomes. Specific binding was shown to dorsal root ganglion and spinal cord cells[18]. Another assay of MAG binding activity has been a neurite extension assay on MAG transfected fibroblasts; MAG transfectants stimulated a 2-fold increase in neurite length compared to nontransfected fibroblasts[19]. Binding of a naturally cleaved form of MAG (dMAG) to extracellular matrix constituents has been reported[20]. MAG has also been found to affect the polymerization of collagen type I fibrils, leading to filaments 3x the normal diameter[21].

ANTIBODIES

The L2/HNK-1 antibodies recognizes a carbohydrate epitope of MAG; this sulphated-3-glucuronyl moiety[22] is also found on several other cell adhesion molecules and glycolipids[23]. The monoclonal antibody 513 is a function-blocking antibody[17], and its binding is conformation dependent. Other monoclonals have been raised that recognize the peptide backbone of MAG[24,25], and their epitopes have been mapped[26]. Polyclonal antibodies recognizing synthetic peptides from the deduced MAG sequence have also been produced[27].

GENES

Sequences have been published for rat[3,4,28] (M14871, X05301, M16800), mouse[29] (X07849) and human[30,31] (M29273). All are highly similar, with >90% homology. The rat MAG gene consists of 13 exons over ~16 kB; inclusion of exon 12, which has a stop codon, leads to the p67 transcript; exclusion leads to the production of p72[3]. The MAG locus maps to mouse chromosome 7 and human chromosome 19[32].

REFERENCES

1. Quarles, R.H. (1983/1984) Dev. Neurosci. 6, 285-303.
2. Quarles, R.H. (1989) Crit. Rev. Neurobiol. 5, 1-28.
3. Lai, C., Brow, M.A., Nave, K-A., Noronha, A.B., Quarles, R.H., Bloom, F.E., Milner, R.J. and Sutcliffe, J.G. (1987) Proc. Natl. Acad. Sci. (USA) 84, 4337-4341.
4. Salzer, J.L., Holmes, W.P. and Colman, D.R. (1987) J. Cell Biol. 104, 957-965.
5. Tropak, M.B., Johnson, P.W., Dunn, R.J. and Roder, J.C. (1988) Mol. Brain Res. 4, 143-155.
6. Pedraza, L., Frey, A.B., Hempstead, D.R., Colman, D.R. and Salzer, J.L. (1991) J. Neurosci. Res. 29, 141-148.
7. Pedraza, L., Owens, G.C., Green, L.A.D. and Salzer, J.L. (1990) J. Cell Biol. 111, 2651-2661.
8. Afar, D.E.H., Salzer, J.L., Roder, J., Braun, P.E. and Bell, J.C. (1990) J. Neurochem. 55, 1418-1426.
9. Edwards, A.M., Arquint, M., Braun, P.E., Roder, J.C., Dunn, R.J., Pawson, T. and Bell, J.C. (1988) Mol. Cell. Biol. 8, 2655-2658.

10.Sternberger, N.H., Quarles, R.H., Itoyama, Y. and Webster, H.D. (1979) PNAS 76, 1510-1514.
11.Martini, R. and Schachner, M. (1986) J. Cell Biol. 103, 2439-2448.
12.Owens, G.C. and Bunge, R.P. (1989) Glia 2, 119-128.
13.Owens, G.C. and Bunge, R.P. (1991) Neuron 7, 565-575.
14.Steck, A.J., Murray, N., Dellagi, K., Brouet, J.-C. and Seligmann, M. (1987) Neurology 22, 764-767.
15.Quarles, R.H., Everly, J.L. and Brady, R.O. (1973) J. Neurochem. 21, 1177-1191.
16.Quarles, R.H., Barbarash, G.R., Figlewicz, D.A. and McIntyre, L.J. (1983) Biochim. Biophys. Acta 757, 140-143.
17.Poltorak, M., Sadoul, R., Keilhauer, G., Landa, C., Fahrig, T. and Schachner, M. (1987) J. Cell Biol. 105, 1893-1899.
18.Sadoul, R., Fahrig, T., Bartsch, U. and Schachner, M. (1990) J. Neurosci. Res. 25, 1-13.
19.Johnson, P.W., Abramow-Newerly, W., Seilheimer, B., Sadoul, R., Tropak, M.B., Arquint, M., Dunn, R.J., Schachner, M. and Roder, J.C. (1989) Neuron 3, 377-385.
20.Fahrig, T., Landa, C., Pesheva, P., Kuhn, K. and Schachner, M. (1987) EMBO J. 6, 2875-2883.
21.Probstmeier, R., Fahrig, T., Spiess, E. and Schachner, M. (1992) J. Cell Biol. 116, 1063-1070.
22.Chou, D.H.K., Ilyas, A.A., Evans, J.E., Costello, O., Quarles, R.H. and Jungalwala, F.B. (1986) J. Biol. Chem. 261, 11717-11723.
23.Kruse, J., Mailhammer, R., Wernecke, H., Faissner, A., Sommer, I., Goridis, C., Schachner, M. (1984) Nature 311, 153-155.
24.Nobile-Orazio, E, Hays, A.P., Latov, N., Perman, G., Golier, J., Shy, M.E. and Freddo, L. (1984) Neurology 34, 1336-1342.
25.Dobersen, M.J., Hammer, J.A., Noronha, A.B., MacIntosh, T.D., Trapp, B.D., Brady, R.O. and Quarles, R.H. (1985) Neurochem. Res. 10, 499-513.
26.Noronha, A.B., Hammer, J.A., Lai, C., Kiel, M., Milner, R.J., Sutcliffe, J.G. and Quarles, R.H. (1989) J. Mol. Neurosci. 1, 159-170.
27.Sutcliffe, J.G., Milner, R.J., Shinnick, T.M. and Bloom, F.E. (1983) Cell 33, 671-682.
28.Arquint, M., Roder, J.C., Chia, L.-S., Down, J., Wilkinson, D., Bayley, H., Braun, P. and Dunn, R.J. (1987) Proc. Natl. Acad. Sci. (USA) 84, 600-604.
29.Nakano, R., Fujita, N., Sato, S., Inuzuka, T., Sakimura, K., Ishiguro, H., Mishina, M. and Miyatake, T. (1991) Biochem. Biophys. Res. Comm. 178, 282-290.
30.Spagnol, G., Williams, M., Srinivasan, J., Golier, J., Bauer, D., Lebo, R.V. and Latov, N. (1989) J. Neurosci. Res. 24, 137-142.
31.Sato, S., Fujita, N., Kurihara, T., Kuwano, R., Sakimura, K., Takahashi, Y. and Miyatake, T. (1989) Biochem. Biophys. Res. Comm. 163, 1473-1480.
32.Barton, D.E., Arquint, M., Roder, J., Dunn, R. and Francke, U. (1987) Genomics 1, 107-112.

■ *John Attia*
and Robert Dunn
Center for Research in Neuroscience,
Montreal General Hospital Research Institute,
1650 Cedar Ave.,
Montreal, Quebec,
Canada H3G 1A4

■ *Mike Tropak and John Roder*
Samuel Lunenfeld Research Institute,
Mount Sinai Hospital,
600 University Ave.,
Toronto, Ontario,
Canada M5G 1X5

Neural Cell Adhesion Molecule (NCAM)

NCAM[1,2] is a large cell surface membrane glycoprotein that is abundant and broadly expressed in developing and adult tissues. It is a single polypeptide with multiple domains including five C2 Ig-like repeats and three fibronectin type III repeats (Figure). The molecule is synthesized as several polypeptide and carbohydrate variants. NCAM is proposed to participate in adhesion between a wide variety of neural and nonneural cell types.

Polypeptide variants of NCAM are produced by alternative splicing and include two major forms of 140 and 180 kDa (NCAM-140 and NCAM-180) having a transmembrane region and two sizes of cytoplasmic domains; a third form of 120 kDa (NCAM-120) has no intracellular region and is linked to the membrane via a glycolipid anchor[3,4]. The carbohydrate heterogeneity largely reflects differential polysialylation of one or more asparaginyl residues in the fifth Ig domain, and the presence of this large linear carbohydrate homopolymer appears to decrease cell-cell interactions[5]. The role of the cytoplasmic domains is as yet poorly understood, but they may influence intracellular chemistry[6] or associations with cytoskeletal elements[7].

NCAM has been implicated in a wide variety of contact-dependent biological events. In the nervous system, these include retinal histogenesis, axon guidance and bundling, and nerve-target interactions[1]. NCAM may also regulate overall membrane-membrane apposition, and to thereby influence the establishment of cell-cell interactions that do not themselves directly involve NCAM, such as gap junctions[8]. In this respect, the polysialic acid moiety may serve as a negative regulator of contact, and studies *in vivo* suggest that changes in NCAM sialylation play a key role in producing different patterns of axon branching and sprouting[9].

■ PURIFICATION

NCAM has been purified in milligram quantities from non-ionic detergent extracts of embryonic brain membrane vesicles by immunoaffinity chromatography[10] using a variety of different monoclonal antibodies according to their species specificity.

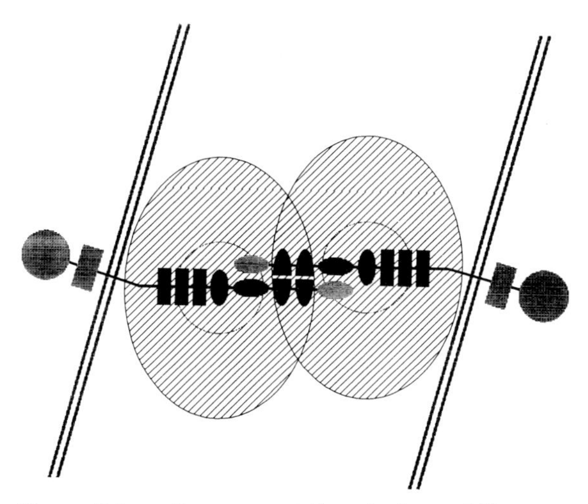

Figure. Schematic representation of a homophilic NCAM-NCAM bond. Small elipses represent Ig-type domains; rectangles are fibronectin type III repeats; the large hatched elipses represent the volume occupied by polysialic acid chains.

■ ACTIVITIES

A variety of evidence suggests that NCAM influences cell-cell adhesion through the formation of homophilic (NCAM-NCAM) bonds via the second and third Ig domains[1]. The precise mechanism of adhesion, remains undefined and could involve other components, such as heparan sulphate[11]. Much of the functional characterization of the molecule has involved the ability of polyclonal anti-NCAM antibody Fab-fragments to block cell-cell adhesion. Transfection studies indicate the ability of the molecule to confer adhesive properties to a nonadherent cell[12].

■ ANTIBODIES AND ENZYMES

Both polyclonal and monoclonal reagents are widely available for use in chick, mouse, frog and human studies. Some reagents are available from the Developmental Biology Hybridoma Bank (Johns Hopkins University). Species specificity exists and is variable, depending on the reagent. Polyclonal Fab-fragments are most effective for perturbation of function. Monoclonal antibodies are useful for identification of some polypeptide variants and detection of polysialic acid, both in immunoblots and immunohistology. A phage endoneuraminidase capable of selectively removing polysialic acid from NCAM has been described[13].

■ GENES

NCAM is encoded by a single gene on chromosome 9 in mouse and band q23 on chromosome 11 in human[1]; in both cases linked to or near Thy-1. The gene has been cloned in chicken (CHKNCAM 1-20)[2], mouse (Y00051)[14], rat (X06564)[15] and frog[16]. No clear counterpart, other than Ig superfamily members, has been identified in invertebrates.

■ REFERENCES

1. Rutishauser, U. and Jessel, T. (1988) Physiological Reviews 68(3), 819-857.
2. Cunningham, B.A., Hemperly, J.J., Murray, B.A., Prediger, E.A., Brackenbury, R. and Edelman, G.A. (1987) Science 236, 799-806.
3. Santoni, M.J., Barthels, D., Vopper, G., Boned, A., Goridis, C. and Wille, W. (1989) EMBO 8, 385-392.
4. Walsh, F.S. and Dickson, G. (1989) BioEssays 11(4) 83-88.
5. Rutishauser, U. (1989) Neurobiology of Glycoconjugates (Margolis and Margolis, eds.) Plenum Publishing, New York, p. 367.
6. Schuch, U., Lohse, M.J. and Schachner, M. (1989) Neuron 3, 13-20.
7. Pollerberg, G.E., Burridge, K., Krebs, K.E., Goodman, S.R. and Schachner, M. (1987) Cell Tissue Res. 250, 227-236.
8. Rutishauser, U., Acheson, A., Hall, A.K., Mann, D.M. and Sunshine, J. (1988) Science 240, 53-57.
9. Landmesser, L., Dahm, L., Tang, J. and Rutishauser, U. (1990) Neuron 4, 655-667.
10. Hoffman, S., Sorkin, B.C., White, P.C., Brackenbury, R., Mailhammer, R., Rutishauser, U., Cunningham, B.A. and Edelman, G.M. (1982) J. Biol. Chem. 257, 7720-7729.
11. Cole, G.J., Loewy, A., Cross, N.V, Akeson, R. and Glaser, L. (1986) J. Cell Biol. 103, 1739-1744.
12. Doherty, P., Fruns, M., Seaton, P., Dickson, G., Barton, C.H., Sears, T.A. and Walsh, F.S. (1990) Nature 343, 464-466.
13. Rutishauser, U., Watanabe, M., Silver, J., Troy, F.A. and Vimr, E.R. (1985) J. Cell Biol. 101, 1842-1849.
14. Barbas, J.A., Chaix, J.-C., Steinmetz, M. and Goridis, C. (1988) EMBO 7, 625-632.
15. Small, S.J., Shull, G., Santoni, M.-J. and Akeson, R. (1987) J. Cell Biol. 105, 2335-2345.
16. Kreig, P.A., Sakaguchi, D.S. and Kinter, C.R. (1989) Nucleic Acids Research 17, 10321-10335.

■ *Urs Rutishauser:*
Case Western Reserve University,
School of Medicine,
Cleveland, Ohio, USA

Neurofascin

Neurofascin is a transmembrane cell surface glycoprotein implicated in axonal fasciculation and elongation[1]. It is localized primarily in axon rich areas of the developing and adult nervous system, where it is transiently expressed with a restricted distribution pattern.

Neurofascin can be isolated from nervous tissue in different molecular weight forms[1]. In immunotransfers of chick brain homogenates prepared under conditions minimizing proteolytical degradation, two major forms at 160 and 185 kDa are detected. In contrast, when neurofascin is purified by immunoaffinity chromatography from detergent extracts of plasma membrane preparations, a complex pattern of major bands is obtained ranging from 110 to 185 kDa. Minor bands in these isolates are visible at 55, 64, 72 and 77.5 kDa. Furthermore, the major components show a considerable charge heterogeneity when analyzed by isoelectric focusing followed by SDS-PAGE with overlapping regions ranging from 4.5 to 6.0. Tryptic finger prints reveal a clear relationship of all neurofascin components. All peptides of the lower molecular weight forms are included in the fingerprint of the 185 kDa component, suggesting that they might be proteolytic breakdown products of the 185 kDa polypeptide.

Neurofascin is a glycoprotein with N- and O-linked carbohydrates. Like other vertebrate cell surface proteins involved in axonal growth including **L1**[2,3], **TAG-1**[4], **F11**[5-7], and axonin-1[8], neurofascin is primarily found in developing fiber tracts. In retina, spinal cord and cerebellum, neurofascin antigenic determinants are expressed transiently during neural development which is consistent with its proposed function.

■ PURIFICATION

Neurofascin can be purified by immunoaffinity chromatography from detergent extracts of plasma membrane preparations of embryonic or adult neural tissue[1].

■ ACTIVITIES

Neurofascin has been indirectly shown by antibody perturbation experiments combined with recently developed in vitro bioassays[9,10] to be implicated in the fasciculation of extending axons (Figure). In one assay polyclonal antibodies to neurofascin cause debundling of retinal axons growing on a tectal plasma membrane preparation. In another assay, which measures the extension of sympathetic growth cones on sympathetic axons, polyclonal antibodies to neurofascin decreases the average neurite length to 70% of control values[1]. As expected from its predominant localization, neurofascin does not participate in the Ca^{2+}-dependent and Ca^{2+}-independent reaggregation of dissociated neural cells.

■ ANTIBODIES

Polyclonal and monoclonal antibodies to chicken neurofascin have been characterized. They do not crossreact with other species[1].

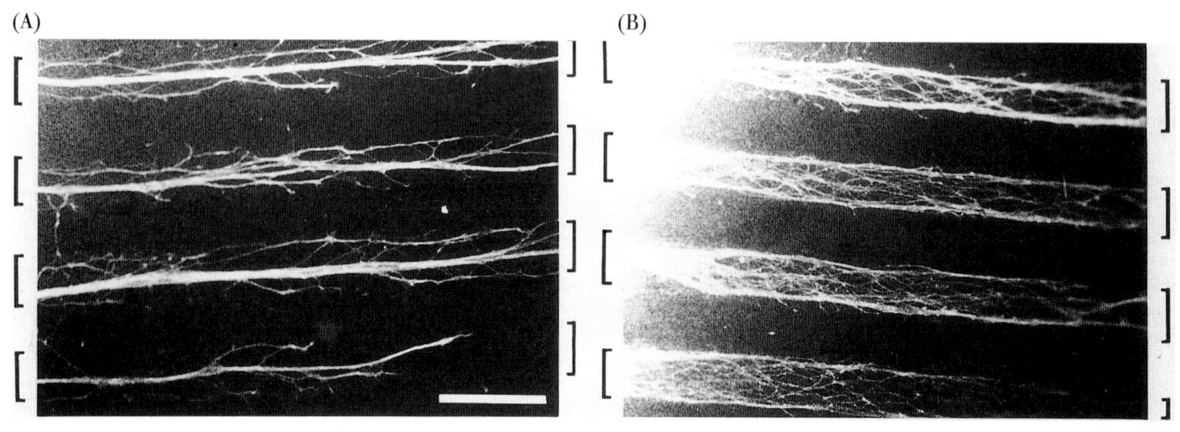

Figure. Inhibition of retinal axons by Fab-fragments of polyclonal antibodies to neurofascin. Retinae from six days old chicken embryos were labelled with rhodamine isothiocyanate and explanted on filters coated with stripes of anterior tectal plasma membrane preparations. Stripes, 90 μm in width, were formed (indicated by brackets) interspersed with noncoated stripes (not marked) of the same width. Retinal explants were cultured in the presence of Fab-fragments of nonimmune serum (A) or of polyclonal antibodies to neurofascin (B). Fluorescent photomicrographs of rhodamine labelled fascicles of temporal axons are shown. The retinal explant is at the left of each photomicrograph (not shown). (Taken from Ref. 1).

GENES

It contains six immunoglobulin (Ig) like motifs of the C2 subcategory at the N-terminus which are followed by four fibronectin type III (FNIII)-related repeats. Between the FNIII like repeats and the plasma membrane spanning region neurofascin contains a domain 75 amino acid residues in length rich in proline, alanine and threonine which might be the target of extensive O-linked glycosylation. A transmembrane segment is followed by a 113 amino acid residues long cytoplasmic domain[11]. Sequence comparisons indicate that neurofascin is most closely related to chick Nr-CAM[12] and forms with L1[3] (NgCAM[13]) and NrCAM a subgroup within the vertebrate Ig superfamily. Complete cDNA sequences are available for chicken neurofascin[11] (EMBL data library accession X65224).

REFERENCES

1. Rathjen, F.G., Wolff, J.M., Chang, S., Bonhoeffer, F. and Raper, J.A. (1987) Cell 51, 841-849.
2. Rathjen, F.G. and Schachner, M. (1984) EMBO J. 3, 1-10.
3. Moos, M., Tacke, R., Scherer, H., Teplow, D., Früh, K. and Schachner, M. (1988) Nature 334, 701-703.
4. Furley, A.J., Morton, S.B., Manalo, D., Karagogeos, D., Dodd, J. and Jessell, T.M. (1990) Cell 61, 157-170.
5. Rathjen, F.G., Wolff, J.M., Frank, R., Bonhoeffer, F. and Rutishauser, U. (1987) J. Cell Biol. 104, 343-353.
6. Brümmendorf, T., Wolff, J.M., Frank, R. and Rathjen, F.G. (1989) Neuron 2, 1351-1361.
7. Gennarini, G., Cibelli, G., Rougon, G., Mattei, M.-G. and Goridis, C. (1989) J. Cell Biol. 109, 775-788.
8. Ruegg, M.A., Stoeckli, E.T., Lanz, R.B., Streit, P. and Sonderegger, P. (1989) J. Cell Biol. 109, 2363-2378.
9. Walter, J., Kern-Veits, B., Huf, J., Stolze, B. and Bonhoeffer, F. (1987) Development 101, 685-696.
10. Chang, S., Rathjen, F. and Raper, J. (1987) J. Cell Biol. 104, 355-362.
11. Volkmer, H., Hassel, B., Wolff, J.M., Frank, R. and Rathjen, F.G. (1992) J. Cell Biol. 118, 149-161.
12. Grumet, M., Mauro, V., Burgoon, M.P., Edelman, G.M. and Cunningham, B. (1991) J. Cell Biol. 113, 1399-1412.
13. Burgoon, M.P., Grumet, M., Mauro, V., Edelman, G.M. and Cunningham, B.A. (1991) J. Cell Biol. 112, 1017-1029.

Fritz G. Rathjen:
Zentrum für Molekulare Neurobiologie,
Hamburg,
Germany

Neuroglian

Neuroglian[1-3] is a homophilic, Ca^{2+}-independent cell adhesion molecule[3] in insects that is a member of the immunoglobulin (Ig) gene superfamily[1,3]; it shares a common ancestor with vertebrate L1[1,4]. The Drosophila protein comes in at least two different forms generated by alternative splicing[2]; the long form is specific for neurons whereas the short form is widely expressed outside of the developing nervous system[2]. The grasshopper gene appears to generate only one form of the protein[3].

Drosophila neuroglian is a transmembrane glycoprotein. The extracellular region of the protein consists of six Ig C2-type domains and five **fibronectin** type III domains[1] (see Figure on page 135). Neuroglian resembles the mouse cell adhesion molecule **L1**[4] in its number and arrangement of extracellular domains; moreover, neuroglian and L1 display 28% amino acid identity extending throughout most of the molecule[1,4].

The *neuroglian* gene in *Drosophila* generates at least two different protein products by tissue specific alternative splicing: a short (more abundant) and a long (less abundant) form[2]. The two protein forms differ in their cytoplasmic domains. Although identical in their extracellular domains, the two neuroglian protein forms share only the first 68 amino acids of their cytoplasmic domains. The short form of the protein continues for another 17 amino acids; the long form extends for another 62 amino acids These extra 62 amino acid residues have a unique amino acid composition: 77% of all residues are either glycine, serine, alanine or proline. The long form of neuroglian is restricted to the surface of neurons in the CNS and neurons and some support cells in the PNS; in contrast, the short form is expressed on a wide range of other cells and tissues[2] (Figure).

The neuroglian homologue was recently cloned in grasshopper using PCR (Grenningloh and Rehm, unpublished). The deduced grasshopper and *Drosophila* neuroglian proteins have approximately 65% amino acid identity over most of their extracellular domains. Whereas the *Drosophila neuroglian* gene generates two different forms of the protein with different cytoplasmic domains (easily detectable by immunoblotting)[2], the grasshopper gene appears to generate only one protein form as detected by immunoblotting (Grenningloh et al., unpublished). This single form in grasshopper is expressed in many tissues of the embryo, much as is the short form of the *Drosophila* protein.

The genetic analysis of *neuroglian* is at an early stage. Several mutations in the *neuroglian* gene have been identified and initially characterized[1,3], including a hypomorphic allele, *l(1)VA142*, as well as a protein null inversion with a breakpoint in the *neuroglian* gene, *l(1)RA35*, both

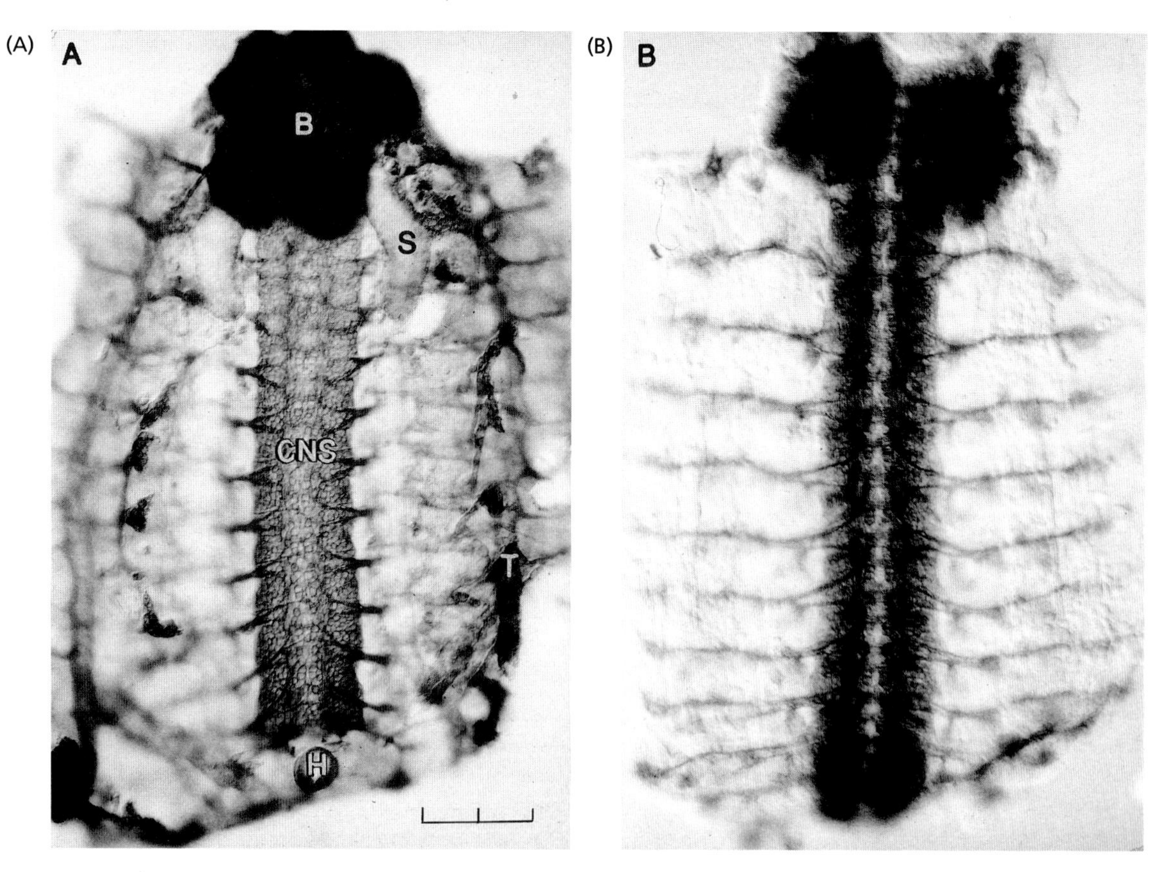

Figure. Expression of neuroglian protein in the *Drosophila* embryo. (A) and (B) show 12 hour *Drosophila* embryos dissected onto glass slides and stained with MAb 1B7 (A) or MAb BP-104 (B). MAb BP-104 specifically recognizes the long form of the neuroglian protein, whereas MAb 1B7 recognizes an epitope common to both forms. Neuroglian protein is detected by MAb 1B7 in the brain (B), central and peripheral nervous system (CNS), salivary gland (S), trachea (T), and the hindgut (H). In contrast, MAb BP-104 (B) stains only the CNS and PNS and none of these other tissues. Bar 50 μm; 38 μm.

of which are lethal during the late embryonic or early larval stage of development. Analysis of homozygous *neuroglian* mutants reveals that at the gross anatomical level, the structure of the CNS and PNS develops in a relatively normal way. There are, however, more subtle phenotypes in the nervous system associated with *neuroglian* mutant embryos. For example, in the PNS the five lateral chordotonal neurons in each abdominal hemisegment normally line up in a tight row maximizing cell-cell contacts. In *neuroglian* mutant embryos this group of five chordotonal neurons are more loosely and randomly organized resulting in a misalignment of their dendritic extensions[3].

■ PURIFICATION

Both *Drosophila* neuroglian protein forms can be copurified by affinity chromatography using the anti-neuroglian monoclonal antibody 1B7[1,2].

■ ACTIVITIES

Transfection of *Drosophila* S2 cells and *in vitro* cell aggregation experiments show that both forms of *Drosophila* neuroglian and grasshopper neuroglian can mediate homophilic, Ca^{2+}-independent cell adhesion[3] (Bieber et al. and Grenningloh et al.; unpublished).

■ ANTIBODIES

Two different monoclonal antibodies against *Drosophila* neuroglian have been reported, one (1B7) recognizing both protein forms and the other (BP-104) recognizing only the larger form of the protein[1,2]. A rat polyclonal antiserum has been generated against affinity purified neuroglian protein.

GENES AND MUTATIONS

The complete cDNA sequence of the shorter protein form of neuroglian has been published[1] (GenBank M28231) as well as the long from specific cDNA sequences[2]. The *Drosophila neuroglian* gene has been localized[1] to the X chromosome at position 7F. Two homozygous lethal mutations have been identified for *Drosophila neuroglian*[1]; one is hypomorphic, *l(1)VA142*, whereas the other is a protein null mutation, *l(1)RA35*.

REFERENCES

1. Bieber, A.J., Snow, P.M., Hortsch, M., Patel, N.H., Jacobs, J.R., Traquina, Z.R., Schilling, J. and Goodman, C.S. (1989) Cell 59, 447-460.
2. Hortsch, M., Bieber, A.J., Patel, N.H. and Goodman, C.S. (1990) Neuron 4, 697-709.
3. Grenningloh, G., Bieber, A.J., Rehm, E.J., Snow, P.M., Traquina, Z.R., Hortsch, M., Patel, N.H. and Goodman, C.S. (1990) Cold Spring Habor Symp. Quant. Biol. 55, 327-340.
4. Moos, M., Tacke, R., Scherer, H., Teplow, D., Früh, K. and Schachner, M. (1988) Nature 334, 701-703.

Michael Hortsch and Corey S. Goodman:*
Howard Hughes Medical Institute,
Department of Molecular and Cell Biology,
University of California, Berkeley, CA, USA
**Present address:*
Department of Anatomy and Cell Biology,
University of Michigan, Ann Arbor,
Michigan, USA

Neurotactin

Neurotactin[1-4] is a Drosophila type II integral membrane glycoprotein that mediates heterophilic cell adhesion. The protein is localized at points of cell-cell contacts and its expression is restricted to phases of cell proliferation and differentiation. Neurotactin belongs to a group of proteins that share a serine esterase protein domain, without retaining the enzyme active center.

Neurotactin is a *Drosophila* cell surface glycoprotein with an apparent molecular mass of 135 kDa. Sequence analysis[2,4] indicates that it contains a single hydrophobic transmembrane domain and no signal peptide. The N-terminal domain of the protein is 324 amino acids long and is localized in the cytoplasm. It is highly hydrophilic (40% charged amino acids), contains two PEST sequences, characteristic of proteins that are rapidly degraded[5], and several putative phosphorylation sites at serine and threonine, in accordance with the observation that neurotactin is phosphorylated in intact cells[3]. The C-terminal, extracellular domain is 500 amino acids long and contains six potential N-linked glycosylation sites. It is structurally related to serine esterases, although the active center serine, that is necessary for enzymatic activity, is not conserved. The structural relation between neurotactin and serine esterases is underscored at two levels: a strong sequence similarity exhibited by a stretch of 200 amino acids and conservation at roughly equivalent positions of six cysteine residues that are engaged in intracellular disulphide bridges in the acetylcholinesterase protein[6]. The reason for the incorporation of the serine esterase protein domain into other functionally unrelated proteins (neurotactin, glutactin[7], thyroglobulin[8]) is currently unknown. It has been speculated that the domain might be involved in protein interactions to form homo- or heteromultimers[9]. The extracellular domain of neurotactin also contains three copies of the tripeptide leucine-arginine-glutamate, a motif that forms the primary sequence of the adhesive site of vertebrate s-**laminin**[10].

The expression of the neurotactin molecule at points of cell-cell contact supports the hypothesis that it might act as a cell recognition and adhesion molecule. In fact, cell transfection experiments[3] indicate that cell expressing neurotactin may bind to a heterologous ligand expressed by primary *Drosophila* embryonic cells. Neurotactin is only detected during cell proliferation and differentiation in the embryonic and larval-pupal stages of the *Drosophila* life cycle, and it is found mainly in neural tissues (Figure 1). Nonneuronal expression is seen in certain mesodermal derivatives and in some epidermal cells of imaginal discs (Figure 2).

PURIFICATION

Neurotactin is routinely purified from embryo lysates by antibody immunoprecipitation[1].

ACTIVITIES

Neurotactin mediates heterophilic cell adhesion[3]. In a cellular binding assay, *Drosophila* S2 cells transfected with neurotactin cDNA are able to bind to embryonic cells in a specific manner. This binding is inhibited by anti-neurotactin antibodies.

ANTIBODIES

Monoclonal antibodies have been described[1-4]. Several others have been produced against the extracellular and the cytoplasmic domains of neurotactin (Piovant, unpublished). Polyclonal antibodies have been raised against the

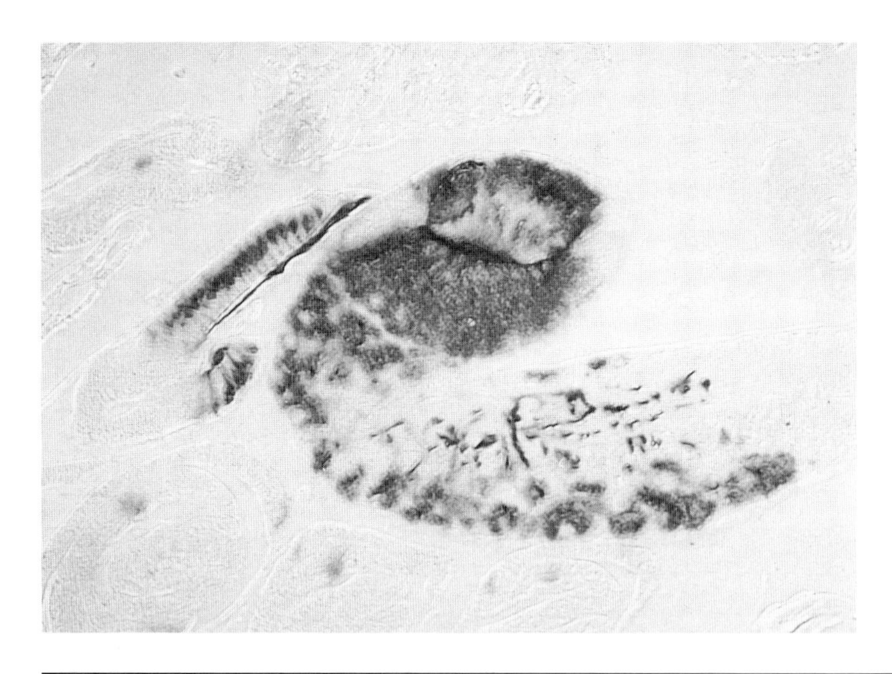

Figure 1. Immunoperoxidase staining of a section showing neurotactin expression in the developing imaginal central nervous system within the larval brain of *Drosophila*.

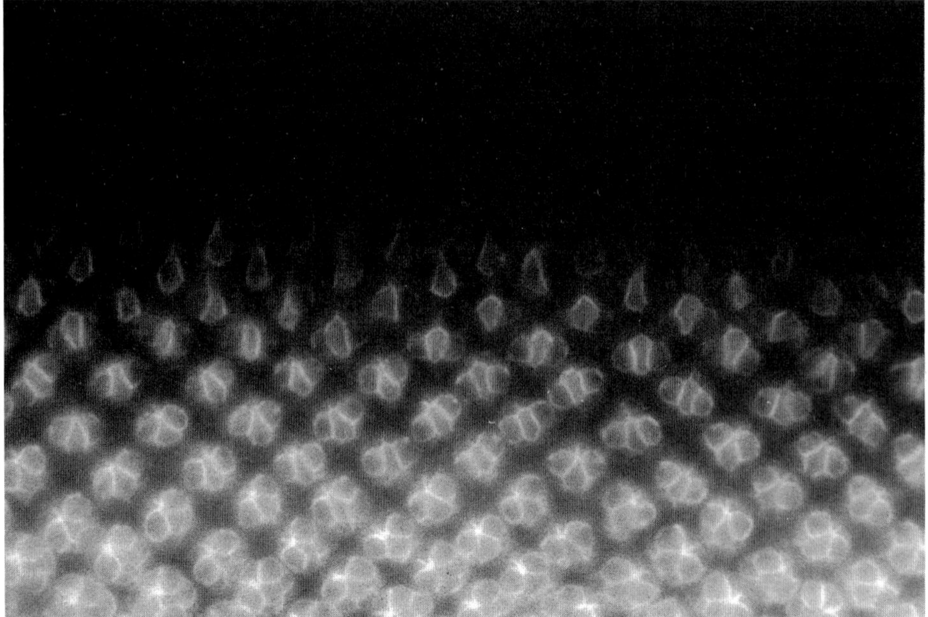

Figure 2. Immunofluorescence staining of neurotactin in the clusters of photoreceptor cells within the eye imaginal disc in the *Drosophila* third instar larva.

whole molecule and parts of the extracellular domain (Piovant and Jiménez, unpublished).

GENES

The sequence of two neurotactin cDNAs is available (EMBL data library X5387 and X54999)[2,4]. The gene maps at the 73C1-2 double band on the third chromosome of *Drosophila*.

REFERENCES

1. Piovant, M. and Léna, P. (1988) Development 103, 145-156.
2. De la Escalera, S., Bockamp, E.-O., Moya, F., Piovant, M. and Jiménez, F. (1990) EMBO J. 9, 3593-3601.
3. Barthalay, Y., Hipeau-Jacquotte, R., de la Escalera, S., Jiménez, F. and Piovant, M. (1990) EMBO J. 9, 3603-3609.
4. Hortsch, M., Patel, N.H., Bieber, A.J., Traquina, Z.R. and Goodman, C. (1990) Development, 110, 1327-1340.

5. Rogers, S., Wells, R. and Rechsteiner, M. (1986) Science 234, 364-368.
6. MacPhee-Quigley, K., Vedvick, T., Taylor, P. and Taylor, S.S. (1986) J. Biol. Chem. 261, 13565-13570.
7. Olson, P.F., Fessler, L.I., Nelson, R.E., Sterne, R.E., Campbell, A.G. and Fessler, J.H. (1990) EMBO J. 9, 1219-1227.
8. Mercken, L., Simons, M.J., Swillens, S., Massaer, M. and Vassart, G. (1985) Nature 316, 647-651.
9. Sikorav, J.-L., Krejci, E. and Massoulie, J. (1987) EMBO J. 7, 1865-1873.
10. Hunter, D.D., Porter, B.E., Bulock, J.W., Adams, S.P., Merlie, J.P. and Sanes, J.R. (1989) Cell 59, 905-913.

Fernando Jiménez:
Centro de Biología Molecular,
CSIC-UAM,
Madrid, Spain

Michel Piovant:
Laboratoire de Génétique et Biologie Cellulaire,
CNRS, Marseille,
France

PECAM-1 (CD31)

PECAM-1 is a newly described cell adhesion molecule and is a member of the immunoglobulin (Ig) gene superfamily. It is distributed on the surface of human platelets, in certain white cell subclasses and at the intercellular junctions of endothelial cells. It is likely to mediate a number of adhesive interactions with the vessel wall.

PECAM-1 (Platelet/Endothelial Cell Adhesion Molecule 1) is a 130 kDa plasma membrane glycoprotein found on the surface of human platelets[1], at endothelial cell intercellular junctions[2] and on white cells of the myeloid lineage[3]. Composed of 711 amino acids in its mature form, PECAM-1 is highly glycosylated, with nearly 40% of its molecular mass made up of carbohydrate residues, largely sialic acid. PECAM-1 appears to be differentially glycosylated, as its mobility in SDS-PAGE varies slightly depending on the cell type from which it is derived. PECAM-1 is a member of the Ig superfamily, and is most closely related to the cell adhesion molecule (CAM) subfamily of Ig-like proteins[1]. It is comprised of a large extracellular domain containing six Ig homology units, a single transmembrane domain, and a relatively long cytoplasmic tail that becomes phosphorylated in response to cellular activation. Interaction with the underlying cytoskeleton is suspected, but has not yet been rigorously documented.

Biochemical and immunochemical studies[1] have shown that PECAM-1 is identical to the previously described CD31 leukocyte differentiation antigen that is expressed in certain tumour cells of the myeloid lineage, including the myelomonocytic cell lines U937, KG-1, and RC-2A[3,4]. PECAM-1 does not appear to be expressed on erythroid cells, resting T- or B-lymphocytes, or other cells outside the vasculature. Although PECAM-1 is only diffusely distributed on the surface of subconfluent endothelial cells in culture, it concentrates at the cell-cell borders when they contact each other, suggesting that PECAM-1 may play a role in contact activated inhibition of cell growth. Human umbilical vein endothelial cells express up to one million molecules per cell at confluency. Antibodies to the bovine form of PECAM-1, EndoCAM[5], block formation of normal cellular junctions in cultured bovine aortic endothelial cells, further supporting the notion of a role for this molecule in establishing and maintaining the integrity of the vessel wall.

In addition to its presence at the intercellular junctions of cells that grow in monolayers, normal circulating monocytes and neutrophils also express PECAM-1 at a density of approximately 50,000-100,000 molecules/cell. Stockinger et al.[6] have shown that granulocyte activation with f-Met-Leu-Phe down-regulates PECAM-1 expression on granulocytes, but not monocytes. The physiological significance of this observation is not clear, but down-regulation of PECAM-1 may be important in decreasing the adhesiveness of neutrophil/endothelial interactions during transmigration of these cells into the extravascular spaces during the inflammatory process.

Only 5,000-10,000 copies of PECAM-1 are present per circulating platelet, and none of the monoclonal and polyclonal anti-PECAM-1 antibodies examined to date interfere with the platelet aggregation process. An earlier report[7] suggests that the anti-PECAM-1 monoclonal antibody HEC-75 may actually induce platelet aggregation, but this needs to be confirmed using a larger panel of antibodies. Although its role in platelet function remains largely undefined, it is possible that platelet PECAM-1 may function only in the precursor cell, the megakaryocyte, either by participating in platelet formation, or by somehow anchoring the megakaryocyte within the bone marrow stroma.

There appear to be multiple mechanisms by which PECAM-1 can mediate cell-cell interactions. In cells that grow in monolayers (endothelial cells, PECAM-1 transfected fibroblast lines, et.), localization of this molecule to intercellular junctions appears to be dependent upon homophilic binding of PECAM-1 to another PECAM-1 molecule on a adjacent cell[8] (Figure). In contrast, PECAM-1 transfected murine L-cells grown in suspension aggregate with nontransfected L-cells, suggesting that heterophilic mechanisms are responsible for at least some PECAM-1 mediated cellular interactions. The molecular nature of the ligand(s) for PECAM-1 in suspension cells is not yet known.

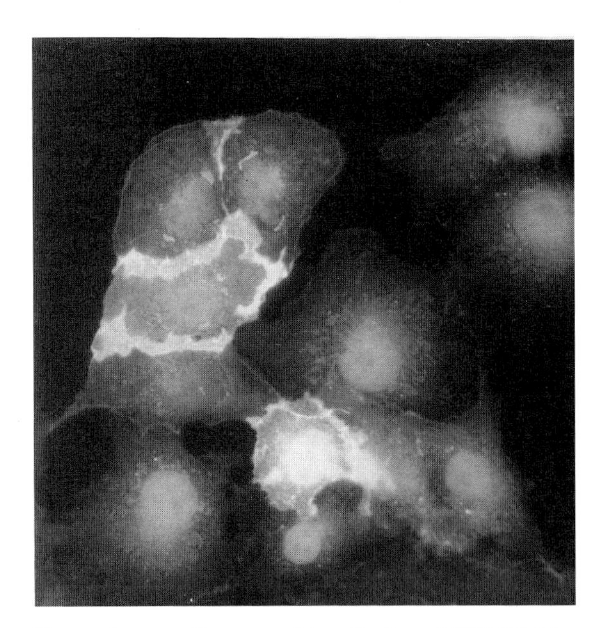

Figure. Immunofluorescent localization of PECAM-1 to the intercellular junctions of adjacently transfected COS-7 cells. PECAM-1 transfected COS cells were fixed, permeabilized, and stained with the anti-PECAM-1 monoclonal antibody, PECAM-1.3, followed by a secondary FITC-conjugated anti-mouse antibody. Adjacently transfected cells show strong border localization of PECAM-1, but no localization occurs between transfected and non-transfected cells in contact with each other, suggesting a homophilic mechanism of attachment via PECAM-1 (photograph kindly supplied by Amy Goldberger, Ph.D., The Blood Center of Southeastern Wisconsin).

PURIFICATION

PECAM-1 is routinely purified from outdated human platelet concentrates using monoclonal antibody affinity chromatography. Due to its high carbohydrate content, PECAM-1 is also reactive with a number of commercially available lectins, including wheat germ agglutinin (WGA), *Ulex europaeus* agglutinin (UAE I), *Ricinus communis* agglutinin RCA_{120}), and *Phaseolus vulgaris* erythroagglutinin (PHA-E). Triton X-100, octylglucoside, or other non-ionic detergents must be present at all times to retain solubility.

ANTIBODIES

Both polyclonal and monoclonal antibodies reactive with different extracellular domains of human PECAM-1 are available. Crossreactivity with nonprimate species is limited. A large number of anti-CD31 monoclonal antibodies have also been described as part of the Fourth International Workshop on Human Leukocyte Differentiation Antigens[9]. An anti-PECAM-1 monoclonal antibody is commercially available from British Biotechnology, LTD.

GENES

Human PECAM-1 was originally cloned from a human umbilical vein endothelial cell cDNA library[1] (GenBank Accession Number M-28526), and more recently from phorbol-myristate-acetate induced HL-60 cells[6,10].

REFERENCES

1. Newman, P.J., Berndt, M.C., Gorski, J., White II G.C., Lyman, S., Paddock, C. and Muller, W.A. (1990) Science 247, 1219-1222.
2. Muller, W.A., Ratti, C.M., McDonnell, S.L. and Cohn, Z.A. (1989) J. Exp. Med. 170, 399-414.
3. Goyert, S.M., Ferrero, E.M., Serematis, S.V., Winchester, R.J., Silver, J. and Mattison, A.C. (1986) J. Immunol. 137,3909-3914.
4. Lyons, A.B., Cooper, S.J., Cole, S.R. and Ashman, L.K. (1988) Pathology 20, 137-146.
5. Albelda, S.M., Oliver, P.D., Romer, L.H. and Buck, C.A. (1990) J. Cell Biol. 110, 1227-1237.
6. Stockinger, H., Gadd, S.J., Eher, R., Majdic, O., Schreiber, W., Kasinrerk, W., Strass, B., Schnabl, E. and Knapp, W. (1990) J. Immunol. 145, 3889-3897.
7. van Mourik, J.A., Leeksma, O.C., Reinders, J.H., de Groot, P.G. and Zandbergen- Spaargaren, J. (1985) J. Biol. Chem. 260, 11300-11306.
8. Albelda, S.M., Muller, W.A., Buck, C.A. and Newman, P.J. (1991) J. Cell Biol. 114, 1059-1068.
9. Knapp, W. (ed.) (1989) In: Leukocyte Typing IV (Oxford University Press).
10. Simmons, D.L., Walker, C., Power, C. and Pigott, R. (1990) J. Exp. Med. 171, 2147-2152.

Peter J. Newman:
Blood Research Institute,
The Blood Center of Southeastern Wisconsin,
Milwaukee, Wisconsin, USA

Steven M. Albelda,
The Wistar Institute,
Philadelphia, Pennsylvania, USA

PH-20 Protein

PH-20 is a sperm specific surface protein which has a required function in sperm adhesion to the zona pellucida of the egg. It is present on both the plasma membrane and secretory granule (acrosomal) membrane and is anchored in both membranes by covalent linkage to glycosylphosphatidyl inositol.

In guinea pigs, PH-20 protein is synthesized as a single chain with a molecular mass of 66 kDa on SDS-PAGE[1,2]. Based on the cDNA sequence, N-terminal sequencing and N-glycanase digestion, mature PH-20 is 468 amino acids long, contains six N-linked glycosylation sites of which five or six are probably used, and has 12 cysteines, eight of which are clustered near the C-terminus. The cDNA sequence indicates PH-20 is a novel protein with no significant homology to other known proteins. Biochemical analysis of the protein, as well as genomic Southern blots, Northern blots and sequencing of various cDNAs, have thus far revealed only a single protein structure, mRNA and gene, indicating that PH-20 on the plasma and acrosomal membranes may be structurally identical[3,4].

PH-20 can be detected by immunofluorescence in round spermatids, first on the acrosomal membrane and subsequently on the plasma membrane[1]. PH-20 undergoes posttranslational modifications later in the life of the sperm, during final differentiation in the epididymis and during fertilization after sperm exocytosis (acrosome reaction). The initial 66 kDa form is reduced to a size of 64 kDa in sperm that have completed differentiation in the epididymis (cauda epididymal sperm)[2]. The second processing step is an endoproteolytic cleavage that generates two fragments held together by disulphide bond(s), so that when the protein is run on reducing SDS-PAGE it separates into two bands at 41-48 kDa and 27 kDa[3]. Although this second processing step may begin while sperm are traversing the epididymis, the bulk of this endoproteolytic cleavage occurs at the time of the acrosome reaction[2,3].

The localization of PH-20 to specific sperm surface domains, its level of surface expression and its ability to diffuse in the membrane are developmentally regulated. On testicular sperm, PH-20 on the plasma membrane is found over the whole cell[1,2]. On cauda epididymal sperm, before acrosome reaction, it is localized to the posterior head domain[5]. After acrosome reaction, which results in insertion of the inner acrosomal membrane into the plasma membrane, plasma membrane PH-20 migrates out of the posterior head into the inner acrosomal membrane. There the plasma membrane PH-20 population joins the other population of PH-20 that is initially associated with the inner acrosomal membrane and therefore is revealed by the acrosome reaction. The revealing of the second population of PH-20 increases the level of PH-20 at the surface 2.5-fold, a change that may regulate the avidity of sperm-zona adhesion[5,6].

PH-20 on the plasma membrane has highly restricted diffusion early in sperm development and free diffusion at the end of development when it functions in sperm-zona adhesion. The measured diffusion coefficients vary over a 250-fold range: 1.9×10^{-11} cm^2/sec (testicular sperm), 1.8×10^{-10} cm^2/sec (cauda epididymal sperm before acrosome reaction) and 4.9×10^{-9} cm^2/sec (cauda epididymal sperm after acrosome reaction)[7,8].

PURIFICATION

PH-20 is purified routinely from octylglucoside extracts of acrosome-intact or acrosome-reacted cauda epididymal sperm by affinity chromatography using an anti-PH-20 monoclonal antibody (mAb) coupled to Sepharose[3]. PH-20 has also been purified from mixtures of spermatogenic cells and testicular sperm using the same protocol[1].

ACTIVITIES

The precise activity of PH-20 in sperm-zona adhesion remains to be established[4]. Acrosome-reacted sperm adhesion to the zona pellucida is blocked by mAbs to certain epitopes of PH-20[9] and by specific removal of PH-20 from the sperm surface with phosphatidyl inositol specific phospholipase C (PI-PLC). Immunization of female guinea pigs with purified PH-20 results in infertility in all immunized animals[10]. PH-20 purified from octylglucoside extracts of sperm, retains the ability to bind the three available anti-PH-20 mAbs[3] but thus far has not been demonstrated to bind to the zona pellucida.

ANTIBODIES

Three anti-PH-20 mAbs have been described[9,11] and two rabbit polyclonal antisera[4] and many guinea pig polyclonal antisera have been generated[10].

GENES

Guinea pig PH-20 cDNA has been cloned and sequenced (GenBank X56332). Cross-species Southern blots show that the gene is conserved among mammals[4].

REFERENCES

1. Phelps, B.M. and Myles, D.G. (1987) Devel. Biol. 123, 63-72.
2. Phelps, B.M., Koppel, D.E., Primakoff, P. and Myles, D.G. (1990) J. Cell Biol. 111, 1839-1847.
3. Primakoff, P., Cowan, A.E., Hyatt, H., Tredick-Kline, J. and Myles, D.G. (1988) Biol. Reprod. 38, 921-934.
4. Lathrop, W.F., Carmichael, E.P., Myles, D.G. and Primakoff, P. (1990) J. Cell Biol. 111, 2939-2949.

5. Myles, D.G. and Primakoff, P. (1984) J. Cell Biol. 99, 1634-1641.
6. Cowan, A.E., Primakoff, P. and Myles, D.G. (1986) J. Cell Biol. 103, 1289-1297.
7. Phelps, B.M., Primakoff, P., Koppel, D.E., Low, M.G. and Myles, D.G. (1988) Science 240, 1780-1782.
8. Cowan, A.E., Myles, D.G. and Koppel, D.E. (1987) J. Cell Biol. 104, 917-923.
9. Primakoff, P., Hyatt, H. and Myles, D.G. (1985) J. Cell Biol. 101, 2239-2244.
10. Primakoff, P., Lathrop, W., Woolman, L., Cowan, A. and Myles, D.G. (1988) Nature 335, 543-546.
11. Primakoff, P. and Myles, D.G. (1983) Develop. Biol. 98, 417-428.

■ *Diana G. Myles and Paul Primakoff:*
Department of Physiology,
University of Connecticut,
Health Center,
Farmington, CT 06030, USA

Selectins

The selectins are an emerging family of cell-cell adhesion proteins, all members of which are involved in leukocyte-endothelial interactions. In contrast to the integrins, cadherins, and the super family of immunoglobulins, whose functions primarily rely on protein-protein interactions, the selectins mediate adhesive interactions through recognition of highly specific cell surface carbohydrates.

The family of selectins, or LEC-CAMs[1,2], has three members to date, which have various designations (Table 1). These cell surface-associated glycoproteins are Type I transmembrane proteins, which exhibit a common organization of protein motifs: an N-terminal Ca^{2+}-type (C-type) lectin domain of ~117 residues, an EGF motif of ~34 residues, a series (2-9) of contiguous complement regulatory domains of ~60 residues, a transmembrane domain, and a short cytosolic tail (Figure). C-type lectin domains are found in a variety of animal proteins, almost all of which are known to manifest Ca^{2+}-dependent lectin activity[3]. Complement regulatory domains are frequently, but not always, associated with binding of complement factors C3b or C4b[4]. The acronym LEC derives from the tandem arrangement of the (L)ectin, (E)GF and (C)omplement domains in these three proteins[1]. Comparison of the lectin and EGF domains of the three selectins reveals homologies of 60-70%, which are substantially higher than the similarities with comparable domains in other proteins[5-8]. Genes for the three proteins constitute a gene cluster (<340 kB) on chromosome 1 in both human and mouse[9]. A striking parallel in the lectin domains of the selectins, as compared with other C-type lectin domains, is a very high concentration of lysine residues. This feature is consistent with the general finding that the carbohydrate ligands for all of the selectins require sialic acid for function[10-12]. A more detailed description of the particular characteristics of each selectin follows.

Table 1
Nomenclature for LEC-CAMs

Family Name	Other Designations
L-selectin	*gp90^{MEL}, *gp100^{MEL}, *gp110^{MEL}, LECAM-1, TQ-1, Leu-8, LAM-1, DREG
E-selectin	ELAM-1
P-selectin	GMP-140, PADGEM, CD62

* identified in mouse; all other designations refer to human molecules

■ L-SELECTIN

This protein was first defined as a lymphocyte homing receptor in the mouse with the monoclonal antibody, MEL-14[13]. This antibody reacts with a diffuse 90-95 kDa band (i.e., gp90^{MEL}) on the lymphocyte surface and blocks lymphocyte attachment to the specialized endothelium of high endothelial venules (HEV) of lymph

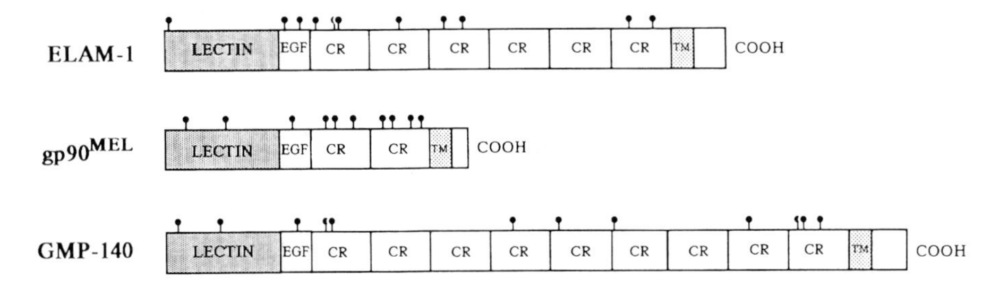

Figure. The selectin family. The organization of protein motifs is shown. Potential sites for N-glycosylation are denoted by small circles on stems.

node. Normally, the binding of blood-borne lymphocytes to HEV initiates the migration of the lymphocytes into the parenchyma of lymph nodes and thus brings potentially responsive lymphocytes into contact with sequestered and processed antigens. gp90^{MEL} contains 10 potential N-linked glycosylation sites[5,6]. Approximately 40% of the mass of the molecule resides in carbohydrate chains, none of which are O-linked[14]. The molecule is reported to be covalently modified by ubiquitin, but this claim has not been rigorously substantiated[14]. In the mouse, L-selectin is present on the majority of peripheral lymphocytes, and, in addition, is found on eosinophils, neutrophils (gp100^{MEL}) and monocytes (gp110^{MEL})[15]. On the neutrophil surface, the molecule is rapidly down-modulated by shedding in response to various chemotactic factors and phorbol esters. Antibody blocking experiments suggest a role for the molecule in the early stages of neutrophil binding to vascular endothelium at sites of acute inflammation[16]. A single gene encodes the same major protein product in both neutrophils and lymphocytes, with differences in molecular weight apparently due to differences in posttranslational modifications[17]. A human homologue to mouse L-selectin is defined on lymphocytes (80-85 kDa) and neutrophils by a variety of monoclonal antibodies (LAMs, TQ1, Leu-8, DREGs)[7,18,19]. Sharing the same domain structure, the mouse and human molecules are 77% identical at the protein level and 79% identical at the nucleotide level[7,8].

PURIFICATION

gp90^{MEL} can be purified to homogeneity from detergent lysates of mouse spleen by immunoaffinity chromatography on a MEL-14 mAb column[5]. The yield is approximately 0.1 μg per spleen.

ACTIVITIES

Purified gp90^{MEL} has been shown to interact with endothelial cells of lymph high endothelial venules and block lymphocyte attachment, thus providing direct evidence for a homing function[20,21]. The interaction is Ca^{2+}-dependent and is inhibited by several anionic carbohydrates, consistent with the presence of a C-type lectin domain. The endothelial ligand requires sialic acid for function. An ELISA assay is available which measures the interaction between gp90^{MEL} and a model carbohydrate ligand[22]. A recombinant chimeric molecule, consisting of the extracellular domains of gp90^{MEL} and the Fc-region of human IgG, retains the ability to bind to HEV and to react in the ELISA assay[21]. The chimera has been used as the basis of a histochemical stain for ligands, and for the purification of the HEV-ligands[23].

ANTIBODIES

A rabbit polyclonal antibody raised against gp90^{MEL} recognizes L-selectin from several species including mouse, rat, and human (Rosen, Singer, and Yednock, unpublished). Several of the monoclonal antibodies, mentioned above, are available. MEL-14 maps to the lectin domain of mouse L-selectin[24], and TQI and Leu-8 map to the lectin domain of the human homologue[25]. Leu-8 (Becton-Dickinson) and TQ1 (Coulter) are commercially available. The anti-Ly-22 monoclonal antibody is an alloantibody which reacts with gp90^{MEL} from mouse strains possessing the Ly-22 marker[26]. The antibody blocks lymphocyte adhesion to endothelial cells without apparently affecting lectin function, suggesting that another domain of the molecule (i.e. EGF) may also be essential for adhesive function.

GENES

Full length cDNAs for mouse[5,6] and human[7,8] are published. The complete sequence of the gene encoding the human LEC-CAM1 has also been reported[17] (GenBank M32406-32414).

E-SELECTIN

E-selectin or ELAM-1 (endothelial leukocyte adhesion molecule-1) is biosynthetically induced on cultured endothelial cells after several hours of stimulation with inflammatory mediators such as IL-1, TNF, and bacterial lipopolysaccharide and functions as a receptor for neutrophils and the promyelocytic cell line HL-60[27,28]. In response to IL-1 or TNF, the protein is rapidly induced (~1 hr) and turned over (~3 hrs). The ELAM-1 mRNA is also transiently induced but then follows a slower decline (~1 day). The molecular mass of ELAM-1 on SDS-PAGE is 115 kDa, of which 37 kDa are removed by N-glycanase digestion. 11 potential N-glycosylation sites in the primary sequence of the extracellular domain are sufficient to account for this loss of mass[28]. The kinetics of ELAM-1 induction and its *in vivo* distribution, suggest a possible role for ELAM-1 in the adhesion of neutrophils to the vascular endothelium at sites of inflammation. Additionally, the ability of ELAM-1 to bind to certain tumor cells raises the possibility of its involvement in metastasis[29,30]. Carbohydrate ligands for ELAM-1[10] have been identified[10] as sialyl Lewis X: NeuAcα2,-3Galβ1,4[Fucα1,3]GlcNAcβ1,3Galβ- and the closely-related CD65/VIM-2 structure: NeuAcα2,3Galβ1,4GlcNAcβ1,3Gal-β1,4[Fucα1,3]GlcNAcβ1,3Gal-. Fucose and sialic acid are both essential for the function of the ELAM-1 ligand while a specific protein backbone is apparently not essential for recognition[10].

PURIFICATION

The molecule has been immunoprecipitated from metabolically labelled cultures of activated endothelium but a purification of chemical quantities has not been reported[27]. A chimera of ELAM-1 and the CH2 and CH3 domains of human IgG (ELAM-Rg) has been expressed[25].

■ ACTIVITIES

ELAM-Rg when coated onto plastic dishes supports the adhesion of granulocytes, HL-60 cells, THP-1 cells, and certain carcinoma cells[25].

■ ANTIBODIES

The adhesion blocking monoclonal antibody H18/7 maps to the lectin domain of ELAM-1[25]. The nonadhesion blocking H4/18 requires the first three complement regulatory repeats for binding[25]. Adhesion-blocking monoclonal antibodies are commercially available from British Biotechnology, Inc.

■ GENES

cDNAs for ELAM-1 are published[28,30]. Full length cDNAs are commercially available from British Bio-technology.

■ P-SELECTIN

GMP-140 (granular membrane protein-140), also known as PADGEM (platelet activation dependent granule-external membrane protein) and CD62, is found in secretory granules of platelets and endothelial cells. Activation of platelets by **thrombin** or of endothelial cells by thrombin or histamine causes a rapid redistribution of GMP-140 to the plasma membrane[31,32]. Expression on the endothelial surface is transient, peaking in three minutes and declining to basal levels within 20 min. Expression on activated platelets is more stable. 30% of the mass of GMP-140 is carbohydrate, probably all occurring as N-linked chains[33]. The primary sequence contains 12 consensus sites for addition of N-linked chains[31]. Two alternative forms of GMP-140 are predicted from cDNAs: one lacking the seventh complement regulatory domain and a soluble form without the transmembrane domain. The different forms are predicted to derive from alternatively spliced mRNAs encoded by a single gene. GMP-140, when externalized on activated platelets, mediates platelet binding to neutrophils, HL-60 cells, monocytes, and U937 cells, but not to lymphocytes or erythrocytes[32]. This activity may reflect a mechanism for the removal of activated platelets from the blood. GMP-140 also mediates neutrophil binding to phorbol ester- or histamine-stimulated endothelial cells, which may provide a mechanism for the rapid targeting of neutrophils or monocytes to sites of inflammation[2]. The soluble form of GMP-140 prevents the CD18-dependent binding of stimulated neutrophils to cultured endothelial cells[34]. This activity may reflect an anti-adhesion mechanism in the body to limit inflammatory reactions. Based on antibody inhibition studies (anti-CD15) and competition by a soluble inhibitor (LNF III), the Lewis X structure (see above) has been suggested as the carbohydrate recognition determinant for GMP-140[35]. The finding that the ligand for GMP-140 is sialidase sensitive argues for a more complex structure involving sialic acid[11,12].

■ PURIFICATION

GMP-140 can be purified by immunoaffinity chromatography of detergent lysates of human platelets[32].

■ ACTIVITIES

GMP-140 coated on plastic wells promotes attachment of neutrophils and HL60 cells[2]. The binding requires calcium, as expected from the C-type lectin domain. PADGEM incorporated into phospholipid vesicles will bind to neutrophils and U937 cells[32]. COS cells transfected with a GMP-140 cDNA can bind to HL-60 cells[2].

■ ANTIBODIES

A large number of adhesion blocking and nonblocking monoclonal antibodies that are reactive with GMP-140 have been described[2,32]. Adhesion blocking polyclonal antibodies have been produced[32,34].

■ GENES

cDNAs for the multiple forms of GMP-140 are published[31].

■ REFERENCES

1. Stoolman, L. M. (1989) Cell 56, 907-910.
2. Geng, J. G., Bevilacqua, M.P., Moore, K.L., McIntyre, T.M., Prescott, S.M., Kim, J.M., Bliss, G.A., Zimmerman, G.A. and McEver, R.P. (1990) Nature 343, 757-760.
3. Drickamer, K. (1988) J. Biol. Chem 263, 9557-9560.
4. Reid, K. and Day, A. J. (1989) Immunol. Today 10, 177-180.
5. Lasky, L. A., Singer, M.S., Yednock, T.A., Dowbenko, D., Fennie, C., Rodriguez, H., Nguyen, T., Stachel, S. and Rosen, S.D. (1989) Cell 56, 1045-1055.
6. Siegelman, M. H., Van de Rijn, M. and Weissman, I.L.. (1989) Science 243, 1165-1172.
7. Tedder, T. F., Isaacs, C.M., Ernst, T.J., Demetri, G.D., Adler, D.A. and Disteche, C.M. (1989) J. Exp. Med. 170, 123-133.
8. Bowen, B. R., Nguyen, T. and Lasky, L.A. (1989) J. Cell Biol. 109, 421-427.
9. Watson, M. L., Kingsmore, S.F., Johnston, G.I., Siegelman, M.H., Le, B.M., Lemons, R.S., Bora, N.S., Howard, T.A., Weissman, I.L., McEver, R.P. and Seldin, M.F. (1990) J. Exp. Med. 172, 263-272.
10. Brandley, B. K., Swiedler, S.J. and Robbins, P.W. (1990) Cell 63, 861-863.
11. Corral, L., Singer, M.S., Macher, B.A. and Rosen, S.D. (1990) Biochem. Biophys. Res. Commun. 172, 1349-1356.
12. Moore, K.L., Varki, A. and McEver, R.P. (1991) J. Cell Biol. 112, 491-499.
13. Gallatin, M., St. John, T., Siegelman, M., Reichert, R., Butcher, E. and Weissman, I.. (1986) Cell 44, 673-680.
14. Siegelman, M., Bond, M.W., Gallatin, W.M., St. John, T., Smith, H.T., Fried, V.A. and Weissman, I.L.. (1986) Science 231, 823-829.
15. Berg, E. L., Goldstein, L.A., Jutila, M.A., Nakache, M., Picker, L.J., Streeter, P.R., Wu, N.W., Zhou, D. and Butcher, E.C. (1989) Immunol. Rev. 108, 1-18.

16.Jutila, M. A., Rott, L., Berg, E.L. and Butcher, E.C. (1989) J. Immunol. 143, 3318-3324.
17.Ord, D. C., Ernst, T.J., Zhou, L.J., Rambaldi, A., Spertini, O., Griffin, J. and Tedder, T.F. (1990) J. Biol. Chem. 265, 7760-7767.
18.Tedder, T. F., Penta, A.C., Levine, H.B. and Freedman, A.S. (1990) J. Immunol. 144, 532-540.
19.Kishimoto, T. K., Jutila, M.A. and Butcher, E.C. (1990) Proc. Natl. Acad. Sci. (USA) 87, 2244-2248.
20.Geoffroy, J. S. and Rosen, S. D. (1989) J. Cell Biol. 109, 2463-2469.
21.Watson, S. R., Imai, Y., Fennie, C., Geoffroy, J.S., Rosen, S.D. and Lasky, L.A. (1990) J. Cell Biol. 110, 2221-2229.
22.Imai, Y., True, D.D., Singer, M.S. and Rosen, S.D. (1990) J. Cell Biol.111, 1225-1232.
23.Imai, Y., Singer, M.S., Fennie, C., Lasky, L.A. and Rosen, S.D. (1991) J. Cell Biol. 113, 1213-1221.
24.Bowen, B., Fennie, C. and Lasky, L.A. (1990) J. Cell Biol. 110, 147-153.
25.Walz, G., Aruffo, A., Kolanus, W., Bevilacqua, M. and Seed, B. (1990) Science 250, 1132-1135.
26.Siegelman, M. H., Cheng, I.C., Weissman, I.L. and Wakeland, E.K. (1990) Cell 61, 611-622.
27.Bevilacqua, M. P., Pober, J.S., Mendrick, D.L., Cotran, R.S. and Gimbrone Jr., M.A. (1987) Proc. Natl. Acad. Sci. (USA) 84, 9238-9242.
28.Bevilacqua, M. P., Spengeling, S., Gimbrone Jr., M.A. and Seed, B. (1989) Science 243, 1160-1165.
29.Rice, G. E. and Bevilacqua, M. P. (1989) Science 246, 1303-1306.
30.Hession, C., Osborn, L., Goff, D., Chi, R.G., Vassallo, C., Pasek, M., Pittack, C., Tizard, R., Goelz, S., McCarthy, K., Hopple, S. and Lobb, R. (1990) Proc. Natl. Acad. Sci. (USA) 87, 1673-1677.
31.Johnston, G. I., Cook, R.G. and McEver, R.P. (1989) Cell 56, 1033-1044.
32.Larsen, E., Celi, A., Gilbert, G.E., Furie, B.C., Erban, J.K., Bonfanti, R., Wagner, D.D. and Furie, B. (1989) Cell 59, 305-312.
33.Johnston, G. I., Kurosky, A. and McEver, R.P. (1989) J. Biol Chem. 264, 1816-1823.
34.Gamble, J. R., Skinner, M.P., Berndt, M.C. and Vadas, M.A.. (1990) Science 249, 414-417.
35.Larsen, E., Palabrica, T., Sajer, S., Gilbert, G.E., Wagner, D.D., Furie, B.C. and Furie, B. (1990) Cell 467-474.

Steven D. Rosen:
Department of Anatomy and Program in Immunology,
University of California,
San Francisco, CA 94143-0452, USA

TAG-1

TAG-1 is a 135 kDa glycoprotein expressed transiently on the surfaces of subsets of developing axons in the initial stages of axonal growth[1,2]. It has six immunoglobulin (Ig) like (C2 type) domains and four fibronectin type III repeats and is closely related to other glycoproteins expressed during axonogenesis[3]. Purified TAG-1 promotes neurite extension when presented as a substrate in vitro and expression of the molecule in nonadherent cells causes cell aggregation[3,4]. TAG-1 is anchored to neuronal surface membranes via a glycosyl-phosphatidylinositol (GPI) linkage and is also released from neurons, suggesting that TAG-1 may function as a substrate adhesion molecule when released into the extracellular environment[3].

The primary structure of TAG-1, deduced by cDNA cloning of the TAG-1 gene[3], indicates that TAG-1 is a member of the Ig superfamily and, in particular, the subfamily of axonal glycoproteins that have both Ig like domains and **fibronectin** type III domains (FNIII) (Figure). Other members of this family include **NCAM**[5], the **L1** like proteins (L1 (NILE)[6], NgCAM (G4)[7], NrCAM[8]), **F11** like proteins (F11[9], F3[10], contactin[11]) and, in *Drosophila*, **fasciclin II**[12] and **neuroglian**[13]. In common with some other members of this family[14,15], TAG-1 can cause cell-cell adhesion by crosslinking cell membranes through homophilic binding[4]. Purified TAG-1 promotes axon extension when presented as a substrate *in vitro*[3]. It is not known whether this latter function of TAG-1 is mediated by a homophilic or heterophilic interaction. How these functions relate to the subdomains of TAG-1 is not known, although the presence of an Arg-Gly-Asp (RGD) sequence in the second FNIII domain suggests that TAG-1 may bind **integrins**[3], a process that can be involved in axon extension[16].

In addition to the cell surface GPI linked form of the protein, a substantial fraction of TAG-1 is released from neurons in culture[3,17]. It is likely that the two forms of TAG-1 arise through differential posttranslational processing of the same precursor protein[3,17]. Light microscopic[1] and ultrastructural[18] studies of surface bound TAG-1 indicate that the protein is present on neuronal cell bodies, axons and growth cones, although this distribution appears to vary according to the developmental stage of the neuron. Where TAG-1 is present in axon fascicles its localisation is punctate and appears at points of membrane contact with a regular periodicity[18].

TAG-1 is expressed on a subset of axons within the developing nervous system[2]. Expression of TAG-1 on these axons is transient and changes; TAG-1 expression coincides with major changes in axonal trajectory suggesting that TAG-1 may be involved in axon growth and guidance[1,2]. *In situ* hybridisation analysis of TAG-1 mRNA, and studies of TAG-1 protein expression *in vitro*, suggest that regulation of TAG-1 expression can occur both transcriptionally and posttranslationally[3,17].

PURIFICATION

TAG-1 has been purified from both embryonic brain lysates[1] and conditioned medium from cell lines expressing transfected TAG-1 genes[19] by immunoaffinity chromatography using monoclonal anti-TAG-1 antibodies[1].

ACTIVITIES

Although the function of TAG-1 *in vivo* is unknown, *in vitro* studies have shown that TAG-1 can promote neurite outgrowth and cause cell-cell adhesion.

ANTIBODIES

Both monoclonal and polyclonal antibodies to rat TAG-1 have been described[1,2], some of which crossreact to murine TAG-1. The anti-rat TAG-1 monoclonal 1C12 will be available from the "Developmental Hybridoma Bank" (Johns Hopkins University, Baltimore, MD. Tel. (301) 955 8485).

GENES

The complete cDNA sequence of rat TAG-1[3] is available from GenBank (M31725). TAG-1 has been mapped to mouse chromosome 1[20].

REFERENCES

1. Dodd, J., Morton, S.B., Karagogeos, D., Yamamoto, M. and Jessell, T.M. (1988) Neuron 1, 105-116.
2. Yamamoto, M., Boyer, A., Crandell, J., Edwards, M. and Tanaka, H. (1986) J. Neurosci. 6, 3576-3594.
3. Furley, A.J., Morton, S.B., Manalo, D., Karagogeos, D., Dodd, J. and Jessell, T.M. (1990) Cell 61, 157-170.
4. Hynes, M., Furley, A.J. and Jessell, T.M. unpublished observations.
5. Cunningham, B.A., Hemperly, J.J., Murray, B.A., Prediger, E.A., Brackenbury, R. and Edelman, G.M. (1987) Science 236, 799-806.
6. Moos, M., Tacke, R., Scherer, H., Teplow, D., Fruh, K. and Schachner, M. (1988) Nature 334, 701-703.
7. Burgoon, M.P., Grumet, M., Mauro, V., Edelman, G.M. and Cunningham, B.A. (1991) J. Cell Biol. 112, 1017-1029.
8. Grumet, M., Mauro, V., Burgoon, M., Edelman, G.M. and Cunningham, B.A. (1991) J. Cell Biol. 113, 1399-1412
9. Brummendorf, T., Wolff, J.M., Frank, R. and Rathjen, F.G.. (1989) Neuron 2, 1351-1361.
10. Gennarini, G.G., Rougon, F., Vitiello, P., Corsi, B.C. Di and Goridis, C. (1989) J. Neurosci. Res. 22, 1-12.
11. Ranscht, B. and Dours, M. (1988) J. Cell Biol. 107, 1561-1573.
12. Harrelson, A.L. and Goodman, C.S. (1988) Science 242, 700-708.
13. Bieber, A.J., Snow, P.M., Hortsch, M., Patel, N.H., Jacobs, J.R., Traquina, Z.R., Schilling, J. and Goodman, C.S. (1989) Cell 59, 447-460.
14. Rutishauser, U., Hoffman, S. and Edelman, G.M. (1982) Proc. Natl. Acad. Sci. (USA) 79, 685-689.
15. Keilhauer, G., Faissner, A. and Schachner, M. (1985) Nature 316, 728-730.
16. Bixby, J.L., Lilien, J. and Reichardt, L.F. (1988) J. Cell Biol. 107, 353-361.
17. Karagogeos, D., Morton, S.B., Casano, F., Dodd, J. and Jessell, T.M. (1991) Development 112, 51-67.
18. Yamamoto, M., Hassinger, L. and Crandell, J.E. (1990) J. Neurocytol. 19, 619-627.
19. Hynes, M., Felsenfeld, D.F., Furley, A.J. and Jessell, T.M. unpublished observations.
20. Jenkins, N. and Copeland, N. personal communication.

Andrew J. W. Furley and Thomas M. Jessell:
Center for Neurobiology and Behavior,
HHMI at Columbia University,
New York, NY, USA

VCAM-1

VCAM-1 (vascular cell adhesion molecule-1) is an immunoglobulin (Ig) superfamily adhesion protein present on the membranes of endothelial cells that have been stimulated by proinflammatory substances. It binds the integrin protein VLA4 ($\alpha_4\beta_1$), present on lymphocytic and monocytic cells. It is probably involved in recruiting these cells from the bloodstream to sites of infection and/or inflammation in the tissues, and may also be important for lymphohemopoeisis.

VCAM-1 (also called INCAM-110) was originally cloned by a functional assay from a cDNA library of IL-1 stimulated human umbilical vein endothelial cells[1]. It binds a variety of lymphoid and monocytic cell lines[1-3], as well as melanoma cell lines[4], via the **integrin** VLA4[5]. Two forms of VCAM-1 mRNA are present in stimulated HUVECs; the originally reported sequence predicts six Ig homologous domains of the H or C2 type[6], while the more abundant 7-domain form[7-9] is identical except for an additional domain in the middle, probably encoded by an exon that is occasionally skipped during splicing to generate the minor 6-domain form (Figure). The protein has a classic N-terminal signal sequence that is clipped to generate the mature protein and a predicted hydrophobic transmembrane region followed by a short (19 amino acid) cytoplasmic tail. The 6-domain form is ~95 kDa while the 7-domain form is ~110 kDa. The 6-domain form has six predicted N-linked glycosylation sites, while the 7-domain form has seven. Both 6- and 7-domain forms of the protein bind VLA4-bearing cells when expressed on the surface of cos7 cells[9].

Cultured endothelial cells can be stimulated to produce

```
                                                     *                                                                           *
6-DOM  7-DOM                                 G  h f C        P  hWp                                      p L f        DsG  o C    N
  1      1             FKIETTPESRYLAQIGDSVSLTCSTTGCESPFFSWRTQIDSPLNGK------------VTNEGITSTLTMNPVSFGNEHSYLCTATCESRKLEKGIQVEIYS
  2      2            FPKDPEIHLSGPL-EAGKPITVKCSVA-DVYPFDRLEIDLLKGDHLMKSQEFLEDADRKSLETKSLEVTFTPVIEDIGKVLVCRAKLHIDEMDSVPTVRQAVKEL
  3      3        QVYISPKNTVISVNPSTKL-QEGGSVTMTCSSEGLPAPEIFWSKKLDNGNLQH------------LSGNATLTL-IAMRMEDSG-IYVCEGVNLIGKNRKEVELIVQ
         4              EKPFTVEISPGPRIAAQIGDSVMLTCSVMGCESPSFSWRTQIDSPLSGK------------VRSEGINSTLTLSPVSFENEHSYLCTVTCGHKKLEKGIQVELYS
  4      5           (A)FPRDPEIEMSGGL-VNGSSVTVSCKVP-SVYPLDRLEIELLKGETILENIEFLEDTDMKSLENKSLEMTFIPTIEDTGKALVCQAKLHIDDMEFEPKQRQSTQTL
  5      6        YVNVAPRDTTVLVSPSSIL-EEGSSVNMTCLSQGFPAPKILWSRQLPNGELQP------------LSENATLTL-ISTKMEDSG-VYLCEGINQAGRSRKEVELIIQ
  6      7           VTPKDIKLTAFPSESV-KEGDTVIISCTCGNV--PET-WIILKKKAETGDTVL------------SIDGAYTIRKAQLKDAG-VYECESKNKVGSQLRSLTLDVQGREN

                  NKDYFSPELLVLYFASSLIIPAIGMIIYFARKANMKGSYSLVEAQKSKV*
                                  tm

                                                                          f - aliphatic - L,I,V
                                                                          h - hydrophobic - L,I,V,M,Y,F
                                                                          o - aromatic - Y,F,W
                                                                          p - polar - K,R,H,D,E,Q,N,T,S
                                                                          s - small - A,G,S,T,V,N,D
```

Figure. Domain structure of VCAM-1. Domains present in the 6- and 7-domain forms of VCAM-1 are indicated. Conserved residues of Ig homology units of the H or C2 type ar indicated at the top of the figure and underlined in each domain (adapted from references 1, 6, and 9). The alanine residue in parentheses (A) is not present in the 7-domain form. The transmembrane region (tm) is underlined. Note the extensive homologies of domain 1 with 4, 2 with 5, and 3 with 6.

VCAM-1 by the inflammatory cytokines IL-1 and TNF[1], bacterial endotoxin (LPS)[2,4], and the T-cell stimulatory cytokine IL-4[10,11]. In response to TNF or IL-1, VCAM-1 mRNA can be detected at 1-2 hrs after addition of cytokine, is present at maximal levels by 2.5 hrs, and is sustained at substantial levels for at least 72 hrs in the continued presence of cytokine[1,9]. Binding activity for the T-cell line Jurkat[1] and for anti-VCAM-1 mAb 4B9[2] was found to reach maximal levels 4-6 hrs after stimulation, and to be sustained at substantial levels for at least 48 hrs.

In vivo, VCAM-1 has been implicated in a number of physiological and pathological processes, including lymphocyte and monocyte extravasation, production and maturation of B lymphocytes, and perhaps metastasis of certain solid tumours. In support of its role in recruitment of mononuclear cells to sites of inflammation, wound healing, and/or infection, VCAM-1 is found on endothelial cells in a variety of inflamed tissues, lymphoid dendritic cells, synovial lining cells, some tissue macrophages, and reactive mesothelial cells[12]. It is required for adhesion to endothelial cells of CD18⁻ lymphocytes from LAD (leukocyte adhesion deficiency) patients, who show nearly normal lymphocyte extravasation but deficient neutrophil extravasation due to lack of **LFA** (CD11/CD18) dependent adhesion of neutrophils to endothelium[13]. VCAM-1 is present on endothelium of atherosclerotic plaques in a rabbit model system[14]. With regard to its role in B-cell function, VCAM-1 has been implicated in adhesion of human B cells to lymphoid germinal centres in the spleen[15]. In murine bone marrow, a VCAM-like antigen is present on stromal cells; mAbs to this antigen interfere with B lymphocyte formation in long-term bone marrow culture[16]. Finally, several melanoma cell lines express VLA4 and bind to VCAM-1, suggesting that this adhesion pathway could play a role in metastasis of the solid tumours from which these lines are derived[4].

PURIFICATION

Purification of VCAM-1 has not been described.

ACTIVITIES

VCAM-1 binds VLA4-bearing cells, such as Ramos, Jurkat, HL60, U937, and peripheral blood lymphocytes and monocytes. This binding can be measured by a simple assay[1,17].

ANTIBODIES

Monoclonal antibodies to VCAM-1 of human[2,4,11,18], macaque[10], rabbit[14], and mouse[16] have been described; most of those described in the literature can block adhesion to VLA4-bearing cells.

GENES

cDNA sequences have been published for the 6-domain form[1] (GenBank M30257) and 7-domain form[7-9] (GenBank M60335).

REFERENCES

1. Osborn, L., Hession, C., Tizard, R., Vassallo, C., Luhowskyj, S., Chi-Rosso, G. and Lobb, R. (1989) Cell 59, 1203-1211.
2. Carlos, T.M., Schwartz, B.R., Kovach, N.L., Yee, E., Russo, M., Osborn, L., Chi-Rosso, G., Newman, B., Lobb, R. and Harlan, J.M. (1990) Blood 76, 965-970.
3. Rice, G.E., Munro, J.M. and Bevilacqua, M.P. (1990) J. Exp. Med. 171, 1369-1374.
4. Rice, G.E. and Bevilacqua M.P. (1989) Science 246, 1303-1306.
5. Elices, M.J., Osborn, L., Takada, Y., Crouse, C., Luhowskyj, S., Hemler, M.E. and Lobb, R.R. (1990) Cell 60, 577-584.
6. Hunkapillar, T. and Hood, L. (1989) Advances in Immunol. 44, 1-63.
7. Polte, T., Newman, W. and Venkat Gopal, T. (1990) Nucl. Ac. Res. 18, 5901.
8. Cybulsky, M.I., Fries, J.W.U., Williams, A.J., Sultan, P., Davis, V.M., Gimbrone Jr., M.A. and Collins, T. (1990) Am. J. Path. 138, 815-820.
9. Hession, C., Tizard, R., Vassallo, C., Schiffer, S.G., Goff, D., Moy, P., Chi-Rosso, G., Luhowskyj, S., Lobb, R. and Osborn, L. (1991) J. Biol. Chem. 266, 6682-6685.
10. Masinovsky, B., Urdal, D. and Gallatin, W.M. (1990) J. Immunol. 145, 2886-2895.

11.Thornhill, M.H., Wellicome, S.M., Mahiouz, D.L., Lanchbury, J.S.S., Kyan-aung, U. and Haskard, D.O. (1991) J. Immunol. 145, 592-598.
12.Rice, G.E., Munro, J.M., Corless, C. and Bevilacqua, M.P. (1990) Am. J. Path. 138, 385-393.
13.Schwartz, B.R., Wayner, E.A., Carlos, T.M., Ochs, H.D. and Harlan, J.M. (1990) J. Clin. Invest. 85, 2019-2022.
14.Cybulsky, M.I. and Gimbrone, M.A. (1991) Science 251, 788-791.
15.Freedman, A.S., Munro, J.M., Rice, G.E., Bevilacqua, M.P., Morimoto, C., McIntyre, B.W., Rhynhard, K., Pober, J.S. and Nadler, L.M. (1990) Science 249, 1030-1033.
16.Miyake, K., Medina, K., Ishihara, K., Kimoto, M., Auerbach, R. and Kincade, P. (1991) J. Cell Biol. 114, 557-565.
17.Bevilacqua, M.P., Pober, J.S., Mendrick, D.L, Cotran, R.S. and Gimbrone Jr., M.A. (1987) Proc. Nat. Acad. Sci. 84, 9238-9242.
18.Graber, N., Venkat Gopal, T., Wilson, D., Beall, L.D., Polte, T. and Newman, W. (1990) J. Immunol. 145, 819-830.

■ *Laurelee Osborn:*
Biogen Inc.,
Cambridge, MA, USA

Index